Liquid Crystalline Order
in Polymers

Contributors

Donald G. Baird
J. J. Beres
Alexandre Blumstein
Y. Bouligand
C. R. Desper
Bernard Gallot
Edward C. Hsu

J. Preston
E. I. Rjumtsev
Edward T. Samulski
N. S. Schneider
I. N. Shtennikova
V. N. Tsvetkov
J. H. Wendorff

LIQUID CRYSTALLINE ORDER IN POLYMERS

Edited by

Alexandre Blumstein

Department of Chemistry
Polymer Program
University of Lowell
Lowell, Massachusetts

ACADEMIC PRESS   New York   San Francisco   London   1978

A Subsidiary of Harcourt Brace Jovanovich, Publishers

ACADEMIC PRESS, INC.
111 Fifth Avenue, New York, New York 10003

United Kingdom Edition published by
ACADEMIC PRESS, INC. (LONDON) LTD.
24/28 Oval Road, London NW1 7DX

Library of Congress Cataloging in Publication Data

Main entry under title:

Liquid crystalline order in polymers.

      Includes bibliographies.
      1.   Polymers and polymerization.   2.   Liquid crystals.
I.   Blumstein, Alexandre.
QD381.L57   547'.84   77-6589
ISBN 0-12-108650-X

Contents

7 Rheology of Polymers with Liquid Crystalline Order

Donald G. Baird

8 Liquid Crystalline Order in Biological Materials

Y. Bouligand

9 Mesomorphic Structure in Polyphosphazenes

N. S. Schneider, C. R. Desper, and J. J. Beres

List of Contributors

Numbers in parentheses indicate the pages on which the authors' contributions begin.

Donald G. Baird * (237), Monsanto Textiles Company, Pensacola, Florida

J. J. Beres (299), Department of Polymer Science and Engineering, University of Massachusetts, Amherst, Massachusetts

Alexandre Blumstein (105), Department of Chemistry, Polymer Program, University of Lowell, Lowell, Massachusetts

Y. Bouligand (261), E.P.H.E., Histophysique et Cytophysique, C.N.R.S., Centre de Cytologie Expérimentale, Ivry-sur-Seine, France

C. R. Desper (299), Polymer and Chemistry Division, Army Materials and Mechanics Research Center, Watertown, Massachusetts

Bernard Gallot (191), Center for Molecular Biophysics, Orléans, France

Edward C. Hsu (105), Department of Chemistry, Polymer Program, University of Lowell, Lowell, Massachusetts

J. Preston (141), Monsanto Textiles Company, Monsanto Triangle Park Development Center, Inc., Research Triangle Park, North Carolina

E. I. Rjumtsev (43), Institute of Macromolecular Compounds, Academy of Sciences of USSR, Leningrad, USSR

Edward T. Samulski (167), Department of Chemistry and Institute of Materials Science, University of Connecticut, Storrs, Connecticut

N. S. Schneider (299), Polymer and Chemistry Division, Army Materials and Mechanics Research Center, Watertown, Massachusetts

I. N. Shtennikova (43), Institute of Macromolecular Compounds, Academy of Sciences of USSR, Leningrad, USSR

V. N. Tsvetkov (43), Institute of Macromolecular Compounds, Academy of Sciences of USSR, Leningrad, USSR

J. H. Wendorff (1), Deutsches Kunststoff Institut, Darmstadt, West Germany

* Present address: Virginia Polytechnic Institute and State University, Blacksburg, Virginia, 24061.

Foreword

For many years we have known that certain solutions of rigid chain polymers could show nematic (or cholesteric) ordering, and the field appeared to many of us as devoid of mysteries. But suddenly, inventions of physical chemists (both in the U.S. and the U.S.S.R.) have brought the subject to the forefront: "magic fibers" can be obtained from certain nematic phases containing rodlike macromolecules as explained by J. Preston in Chapter 4. The impact of these findings reminds me of the days when the group at RCA discovered the "dynamic scattering" of nematics in electric fields and immediately produced a great increase in research on liquid crystals all over the world. Data on polymeric mesophases have been accumulating very quickly; this book gives us the first detailed survey of what is known in the field. In particular, it has become increasingly clear that many *flexible* polymers also show remarkable organization:

(1) Chains with mesogenic side groups display nematic and smectic ordering which is described by Blumstein and Hsu in Chapter 3. In the nematic phases the backbone often has a disturbing effect on the order; but fortunately, nematic alignment can still be found in a number of cases in both homopolymers and copolymers.

(2) A very interesting field for future research is that of chains in which the *backbone* is made up of mesogenic moieties connected by suitable flexible links. Chapter 1 by J. H. Wendorff contains some examples of such structures, and I hope to hear more about them in the future! Here we would expect weaker disturbances of the nematic order. On the other hand, in reticulated structures, we should find a remarkable coupling between nematic and mechanical properties—e.g., an isotropic–nematic phase transition induced by stress, with resulting nonlinear mechanical responses.

(3) It may well be that certain of the "magic fibers" are not completely rigid when in dilute solution, but become simultaneously rigid and nematic at higher concentrations. This effect, which I associate with the heroic motto "United we stand, divided we fall," is currently under theoretical study. On the experimental side, the careful measurements on dilute solutions performed by the Strasbourg and Soviet schools and described here by V. Tsvetkov (Chapter 2) are fundamental. Of course, they should be supplemented later by studies on chain elasticity in the concentrated nematic phase.

At this point we meet one aspect of a general problem: how to connect the local (single chain) viewpoint with the global mesomorphic properties. One attractive example is discussed here by E. Samulski (Chapter 5): the helical pitch in cholesteric PBLG solutions can to some extent be predicted from a simple discussion based on single chain polarizabilities. Another question of similar nature is related to the Frank constants of the nematic phase. In a rigid rod system both splay and bend are "hard." In a system of flexible coils that become nematic through local interactions, I expect the bend to become "soft," the splay remaining "hard": this may provide a simple distinction. There should be many other observations of this kind.

Certain *geometric concepts* will be essential in the analysis of polymeric liquid crystals; some of these are very clearly illustrated in the discussion by B. Gallot on block copolymers (Chapter 6). One crucial parameter is A, the "area per polar head" of lipid–water systems or its long chain counterpart which I characterize by a chain current J ($J = 1/A$). In block copolymers AB we often expect that there will be no chain ends inside one region (A or B). Then J is a conserved current and this implies many important properties for layers, rods, etc. Inside a layer we can also measure the quadrupolar alignment Q discussed here by Tsvetkov (Chapter 2). What is the relation between J and Q in dense polymeric systems?

Another geometrical problem is the classification of the phases observed: for instance, the remarkable phase shown by some phosphazenes (Chapter 9) and muscle structures (Chapter 8) are probably "canonical" phases (with strict periodicity in a plane normal to the optical axis) rather than nematic. This distinction is clearly stated by the authors, and is not purely academic. (A nematic can be twisted by cniral solutes and it also has a strong intrinsic turbidity, but these two features are not expected in a canonical phase.)

A third kind of geometrical problem is related to the *entanglements* of long chains when they are at least partly flexible. Entanglements are not readily detected through structural studies but they play an essential part in the dynamics of chains. The rheological properties of these polymeric phases are thus expected to be very complex but they will carry important information; on the other hand they dominate the field of fiber applications. They are reviewed here in clear terms by D. G. Baird (Chapter 7).

This volume gives us a broad coverage of the properties of these new materials. It presents a balanced combination of innovative chemistry and theoretical approach: Professor Blumstein, who is one of the pioneers of the field, has now, as an editor, made this information accessible to a large scientific community, and should be heartily congratulated.

P. G. de Gennes

Preface

The field of liquid crystalline order in polymers has come in recent years to the forefront of activity. The impetus was provided by the development of ultrahigh strength/high modulus fibers spun from nematic solutions of macromolecules with rigid backbones.

The topic of liquid crystalline order in systems containing rigid synthetic macromolecular chains, as interesting and as important as it may be, is but a limited area of a much vaster field that includes various aspects of liquid crystalline order in natural and synthetic macromolecules. In spite of the interest, most of the literature concerning liquid crystalline order in polymers is scattered throughout a great number of scientific journals and symposia and the patent literature. Thus we felt that the time had come to review the field and to concentrate the information available in a book entirely devoted to this subject. The book should also provide a balance of topics which seem to have been strongly shifted in favor of topics of immediate industrial concern.

Each chapter of the book provides a self-contained and up-to-date "state of the art" review of one important area of the field. Chapter 1 (Wendorff), although devoted to scattering in polymer systems with liquid crystalline order, provides the reader with an introduction to the field of liquid crystals and a general overview of the topics to be treated in some of the following chapters. Chapter 2 (Tsvetkov, Rjumtsev, and Shtennikova) treats the origin of liquid crystalline order in macromolecules by describing the in-depth study of conformation of such macromolecules in their unassociated state. Such studies not only are at the root of understanding the intramolecular order of macromolecules with mesogenic segments, but are also important in the understanding of their intermolecular ordering in bulk or in concentrated solution. The chapters which follow describe successively the liquid crystalline order in polymers with mesogenic side groups (Blumstein and Hsu) and rigid backbones (Preston), in polypeptides (Samulski), and in block copolymers (Gallot). The rheology of such systems is described in Chapter 7 (Baird).

The diversity of the field of liquid crystalline order in macromolecules is well illustrated in the three chapters that encompass the formation of lyotropic liquid crystallinity in block copolymers (a topic of considerable technical importance), liquid crystalline order in biological materials

(Bouligand), and mesomorphic order in the realm of polymers with in-organic backbones (Schneider, Desper, and Beres). Professor Bouligand's chapter vividly shows the extraordinary variety of liquid crystalline organizations, be they at the molecular level of biomembranes, at the supermolecular level of cell organelles, or at the macroscopic level of various tissues. It demonstrates that liquid crystalline order in biological systems plays an important part in the functions and properties of living matter.

It is my hope that this book will provide the polymer scientist, the materials scientist, and the biologist with a valuable source of information. I also hope that the graduate student will be stimulated by this book to study a new and fascinating interdisciplinary field which, judging from the great diversity of natural and synthetic materials displaying liquid crystalline order, holds high hope for the future.

I would like to take this opportunity to express my thanks to the authors for their efforts and to the publisher and the staff for efficient cooperation.

1

Scattering in Liquid Crystalline Polymer Systems

J. H. Wendorff

Deutsches Kunststoff Institut
Darmstadt, West Germany

I. Introduction

Polymers that display structures that are intermediate between the three-dimensionally ordered crystalline state and the disordered isotropic

1

fluid state have been given considerable interest in the last few years for the following reasons:

(i) These polymers can be considered as model systems for isotropic amorphous polymers, where semiordered regions have been proposed to exist, and for drawn amorphous polymers, where by the drawing process a high degree of orientational order can be introduced.

(ii) Polymers that form partially ordered melts or partially ordered solutions can be processed in such a way that fibers with a very high degree of orientational order and chain extension are obtained. This specific structure leads to superior mechanical properties of the material.

(iii) Polymers with ordered melts are of technological importance for the same reasons that low molecular weight systems with anisotropic melts are used widely today in technical products such as electrooptical displays. Additional advantages are expected for polymeric systems, since parameters such as side group structure formations or copolymer formations will lead to a large variety of properties of semiordered melts.

(iv) Finally, if the polymerization is carried out in semiordered liquid phases of the monomer or in semiordered solutions, it is possible to study the polymerization process in ordered systems. This will allow a deeper insight into the reaction mechanism. Also, by this procedure, macroscopically ordered polymers may be obtained if the monomer phase is macroscopically ordered, for instance, due to an external field or due to forces exhibited by specific surfaces. In this case interesting optical properties are expected for the polymeric system.

For these reasons a variety of different studies has been made of polymers with potentially ordered melts or solutions in the last few years. The polymers studied were polymers in which the ordered structure was displayed by the side groups. In this case the structure of the main chain is only of secondary importance. It is intended to give in this chapter a survey of structures that are encountered in low molecular weight liquid crystalline systems, of methods that can be used to analyze these structures, and of studies of structural analysis of polymers with semiordered melts. A considerable part of the chapter will be devoted to a description of the molecular structure displayed by these phases and of specific structural studies of these phases. The author felt that because of recent advances in the structural analysis of polymeric systems and liquid crystalline systems (such as neutron small angle scattering on partially deuterated samples), and in view of the small number of detailed structural studies of polymers with semiordered melts, a general survey of the problem would be useful.

II. Characteristic Properties of Liquid Crystalline Phases

A. *Physical Properties*

Liquid crystalline phases have been known to exist in low molecular weight systems since the end of the nineteenth century. Reinitzer (1888) observed a peculiar melting behavior for a number of esters of cholesterol. He found that the crystals of the substances melted sharply to form not an isotropic melt but an opaque melt. The conclusion drawn from this observation was that some type of order still existed in the molten state. Reinitzer furthermore observed that the opacity vanished at a higher temperature called the clearing temperature. Two years later, Lehmann (1890) reported that ammonium oleate and *p*-azoxyphenetole exhibited turbid states between the truly crystalline and the truly isotropic fluid state. Subsequent studies by these authors indicated that these opaque systems were thermodynamically stable states (Gray, 1962). They also concluded that the structure of these phases had to be intermediate between the crystalline and the fluid state.

To describe this strange behavior of the fluid state, Lehmann introduced the term "Flüssige Kristalle" (liquid crystals). Later Friedel (1922) proposed that it would be more reasonable to refer to these phases as mesophases or mesomorphic phases. It was soon recognized that mesophases may arise as a result of temperature variations, thus the term thermotropic mesophases is used. Mesophases may, however, also result from changes of the concentrations of solutions. With increasing concentration, a transition from a nonordered isotropic solution to an ordered anisotropic solution can take place. For such systems the term "lyotropic liquid crystals" is used. Lyotropic systems show also thermotropic behavior: on heating they become isotropic. A mesomorphic phase characterized by a reversible phase transition is referred to as "enantiotropic." "Monotropic" liquid crystalline phases are stable only in the supercooled state. Monotropic phases are therefore characterized by a transition temperature that is lower than the transition temperature of the neighboring phase.

Since the early work on liquid crystalline phases, a large amount of work has been performed in order to elucidate the peculiar structure and the specific properties of the mesophases. The studies were primarily concerned with the detection of new mesomorphic substances and with measurements of their optical properties. The supermolecular structure was in general studied by means of the polarizing microscope (Gray, 1962). Conclusions with respect to the molecular structure of these phases were drawn from the supermolecular structure and from the miscibility behavior of different components (Sackmann and Demus, 1973). A surprisingly low number of structural studies were published, which used scattering techniques such as X-ray, electron, or light scattering. Nevertheless, the basic structural features of the liquid crystalline modifications of the low molecular weight systems are understood today.

B. Structural Properties

1. General Description of Order

A general description of liquid crystalline structures has to include the physical structure on a molecular scale, which is characterized by the positional as well as the orientational order of neighboring molecules or parts of neighboring molecules and the supermolecular arrangement of assemblies of molecules or parts of molecules. The supermolecular structure, which in polymer science is often called the morphology and which in the case of liquid crystalline phases is exclusively called the texture, has also to be characterized with respect to the positional and orientational order of the assemblies mentioned. The molecular structure and the texture of a mesophase determine the physical and technological properties of this phase.

As already stated, the molecular structure of the liquid crystalline phases is known to be intermediate between the three-dimensional crystalline order and the disorder of the isotropic fluid state. The crystalline order is characterized by the long range positional order. The properties of the crystalline material, such as the thermal or the mechanical, are strongly dependent on the existence of the long range order (Kittel, 1959). For nonspherical particles, in addition to the positional order the orientational order has to be characterized. For an ideal crystalline material, a long range orientational order also exists.

In most cases the long range positional order and the long range orientational order cease to exist at temperatures higher than the melting point. In the melt the positional order is of a short range nature and is due mainly to correlations between neighboring particles. The short range order is usually described by a pair correlation function that can be considered as a one-dimensional representation of the three-dimensional order of the fluid state. The pair correlation function is directly related to properties of the fluid state, such as pressure or compressibility (Egelstaff, 1967). The pair correlation function $g(r)$ may be cast into the form

$$\rho g(r) = N^{-1} \left\langle \sum_{i \neq j} \delta(r + r_i - r_j) \right\rangle \tag{1}$$

where $g(r)$ is the pair correlation function, ρ the density of the fluid, and

r_i and r_j the positions of the particles i and j, respectively. The pair correlation function $g(r)$ thus describes the probability of finding a particle of the fluid at the position r if a particle is positioned at the origin. In principle, scattering experiments can be performed in order to determine $g(r)$.

In the case of nonspherical particles, the orientational order has to be studied. This order can be described by an orientation correlation function $f_\theta(r)$ (Stein and Wilson, 1962)

$$f_\theta(r) = \langle \tfrac{1}{2}(3\cos^2\theta_{ij} - 1)\rangle \qquad (2)$$

where θ_{ij} is the angle between the axes of fluid particles i and j separated by the distance r.

It has been known for some time that not all molecular systems completely lose their order at the melting point. There are systems that make the transition from the crystalline state to the disordered fluid state in more than one step. In some systems the orientational order is lost at a phase transition, whereas the positional order is still maintained above the transition temperature. The positional order is lost at one or more of the higher temperature transitions. These are the so-called plastic crystals, neopentane for example. Molecules that tend to form plastic crystals usually have a spherical shape. There are other systems that lose their positional order at a phase transition while still maintaining their orientational order, which is lost in one or more steps at higher transition temperatures. Above the clearing point, the state of an isotropic melt is obtained. These systems are mesomorphic and the molecules that compose them have a disklike or rodlike shape. It seems that a certain stiffness of the molecules is a necessary requirement for the tendency to form liquid crystalline phases (Gray, 1962). In some liquid crystalline systems the positional order is not lost at one transition but is lost stepwise at separate transitions. Different liquid crystalline phases can thus be defined on the basis of their orientational and positional orders (De Gennes, 1975).

2.   Liquid Crystalline Molecular Structures

a. The Nematic Phase.   In the nematic phase the centers of gravity of the particles are arranged at random, consequently no positional long range order exists. Within volume elements of the macroscopic sample the axes of all particles are oriented in a specific direction. Near the smectic–nematic transition temperature, there may be an additional ordering (positional order). The formation of cybotactic nematic aggregates has been reported by de Vries (1970).

b. The Cholesteric Phase.   The cholesteric phase is often thought of as a modification of a nematic phase, since its molecular structure is assumed to be similar to the latter. No positional order but only an orientational order exists in the cholesteric phase. In contrast, however, to the nematic phase the cholesteric phase is characterized by the fact that the direction of the long axes of the molecules change continuously within the sample. This leads to a twist about an axis perpendicular to the long axes of the molecules. If the pitch of the helical structure agrees with the wavelength of the visible light, selective reflection of monochromatic light can be observed. This effect leads to the irridescent colors often observed in cholesteric phases.

c. The Smectic Phase.   In the smectic phases the centers of gravity of the elongated molecules are arranged in equidistant planes and smectic layers are formed. The planes are allowed to move perpendicularly to the layer normal and within the layers different arrangements of the molecules are possible. The long axes of the molecules may be parallel to the layer normal or tilted with respect to it. A two-dimensional short range order or a two-dimensional long range order may exist within the smectic layers. The smectic modifications are labeled according to the arrangement of the particles within the layers, the symbols A–H being used. Investigations of miscibility between different liquid crystalline modifications allow the distinction between various smectic phases and between smectic, cholesteric, and nematic phases (Sackmann and Demus, 1973).

3.   Liquid Crystalline Textures

The textures of liquid crystalline phases determine the optical properties of these materials to a great extent. The wide range of applications of these systems depends on the ease with which textural changes and therefore changes in optical properties can be brought about by mechanical, thermal, electric, and magnetic forces. The macroscopic orientations of the molecules in the sample determine the textures. In the case of the so-called homeotropic texture, the particles are arranged with their long axes parallel to the film normal throughout the macroscopic sample, whereas in the so-called homogeneous texture, the long axes are oriented parallel to the film surface. The different textures encountered in liquid crystalline phases will be described in detail.

The observed textures are directly related to the molecular structure of the material, as described by Frank (1958). It was even possible to derive the molecular structure of the liquid crystalline modifications from the observations of the textures. Excellent reviews on textures and on the orientational order that have led to these textures are given by Gray (1962) and Sackmann and Demus (1973).

Specific textures and textural changes may be induced by a variety of different methods. These methods include the use of electric or magnetic fields, shearing of the material, or changing the temperature of the sample.

For the latter case, it was observed that textural changes even within one mesophase closely resemble phase transitions in nonmesomorphic systems, such as crystallization from the supercooled melt. The transitions are reversible and large intervals of supercooling can be obtained. It is possible to describe the transformation behavior on the basis of Avrami's theory (Price and Wendorff, 1972; Jabarin and Stein, 1973).

III. Structural Analysis by Scattering Methods

A. Basic Features of Structural Analysis

The most direct method of analyzing a structure is the direct observation of it by visual inspection under the optical, the electron, or the scanning electron microscope. However the contrast between structural features is often small. Furthermore, in the case of molecular structures, the resolution of the methods mentioned is not sufficient. In these cases one is forced to use less direct methods of structural analysis, namely scattering techniques. The intensity and the intensity distribution of the radiation scattered by a material is determined by the structure of this material. Therefore, from the scattering pattern conclusions can be drawn with respect to the structure.

The existing scattering methods, such as X-ray, electron, light, or neutron scattering, are sensitive to various structural parameters that characterize the material. In general, one method is insufficient for structural analysis. In each case one has to determine which structural parameters are needed to describe the nature of a specific molecular or supermolecular arrangement and then choose the appropriate scattering technique. In the following sections the structural parameters characterizing liquid crystalline phases and liquid crystalline polymer phases are described and the experimental methods by which these parameters may be obtained are discussed. The structural parameters pertinent to liquid crystalline systems are—as already mentioned—the positional short range and long range orders, the orientational short range and long range orders on a molecular and supermolecular scale. Also for polymeric systems the conformation of the chain has to be defined.

B. Determination of the Molecular Structure

1. The Positional Long Range Order

The positional long range order, characteristic of crystalline samples, may be studied by different methods such as electron, elastic neutron, or

X-ray scattering. The basic requirement for structural analysis by scattering methods is that the wavelength of the radiation used for the experiments must not be larger than the dimension of the structure to be analyzed. The shape and the intensity of the diffraction pattern yields information about the lattice as well as about the distribution of the molecules within the unit cell. The intensity $I(\mathbf{s})$ which is scattered by a periodic structure can be represented by the equation

$$I(\mathbf{s}) = |F(\mathbf{s})|^2 |G(\mathbf{s})|^2 \tag{3}$$

where $s = |\mathbf{s}| = (4\pi \sin \vartheta)/\lambda$; $\mathbf{s}$ is the scattering vector, which is determined by the direction of the incident and the scattered beam; 2ϑ the angle between the incident and the scattered beam; and λ the wavelength of the radiation. The lattice factor $G(\mathbf{s})$ depends on the geometry of the lattice whereas the structure factor $F(\mathbf{s})$ depends on the distribution of positions of the molecules within the unit cell relative to the lattice points. The specific shape of the structure factor depends on the kind of radiation that is used in the experiments. The factor $G(\mathbf{s})$ determines the position of the reflections in the scattering diagram whereas the factor $F(\mathbf{s})$ determines the intensity of the reflections. One gets information about the two terms in expression (3) if one studies the position as well as the intensity of the reflections.

In general the ordered structure of a material does not have an infinite extension; polycrystalline materials outnumber single crystals. Therefore the crystal size has to be considered as an important structural parameter. Furthermore, the crystalline state is not ideal, lattice defects occur in the material and will influence its properties. Lattice defects may arise from the oscillations of the particles in the crystalline lattice under the influence of thermal energy. These defects, called defects of the first kind (Guinier, 1963), do not destroy the long range positional order of the lattice. There are, however, other kinds of defects, called defects of the second kind, that destroy the long range positional order to a great extent. It was shown by Hosemann and Bagchi (1962) that a useful concept for describing these defects is the concept of paracrystalline distortions. According to this concept one can determine the size L of the crystal in a particular direction and the type and degree of lattice distortions in this direction by analyzing the integral width $\delta\beta$ of the reflections, which are related to the particular direction, as a function of the order of the reflection h. Defects of the first kind do not result in the dependence of the width of the reflection on the order h, whereas the width increases with increasing order h if defects of the second kind are present. According to the concept of paracrystalline distortions, one can relate the integral width of the reflection to the size L of the crystal and to the relative fluctuation of the average lattice dimension l by the equation

$$\delta\beta^2 = L^{-2} + (\pi g h)^4 / l^2 \tag{4}$$

where $\delta\beta$ is the integral width of the reflection, g the relative fluctuation of the average lattice dimension $\delta l/l$, and h the order of the reflection. In the case of the melt, g is of the order of 0.2.

Additional information about specific lattice distortions, which may be of interest for liquid crystalline phases, can be obtained by studying oriented samples. Oriented samples of a polymeric material are usually obtained by drawing or rolling the material. In the case of liquid crystalline polymers, it is often possible to obtain highly oriented samples by applying a magnetic or electric field to the polymerizing system. The long axes of the monomer units will in general be oriented parallel to the field direction. It is also possible to study samples where a homogeneous or homeotropic texture has been induced by the methods just described. One obtains single crystals with respect to the orientational order. This state facilitates the structural analysis, especially the analysis of the defects.

For discussing the influence of the defects on the scattering pattern, we shall consider the case of an uniaxially oriented specimen (fiber) that will be represented by a two-dimensional lattice with the lattice vectors 1 (horizontal direction) and 2 (vertical direction), the Miller indices are called (h, k). The effect of a limited crystal size on the scattering pattern will be omitted. In analogy to Bonart (1957) and Hosemann and Bagchi (1962), the lattice distortions and the resulting scattering diagrams are represented in Fig. 1. In the case of an undistorted lattice (Fig. 1a), the width of all

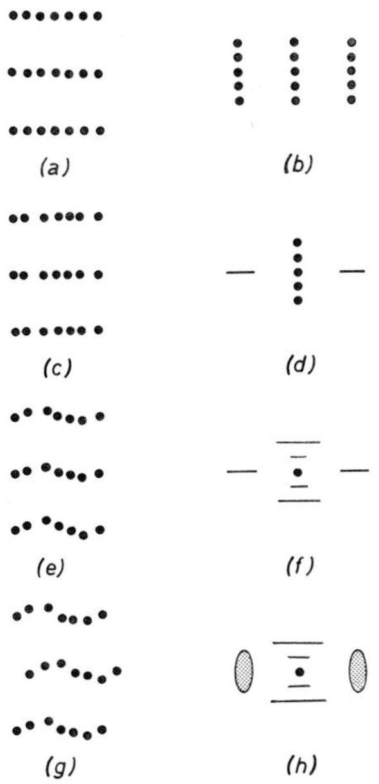

Fig. 1. Schematic illustration of the influence of different lattice defects on the structure (a, c, e, g) and on the corresponding scattering diagrams (b, d, f, h). The defects are described in the text.

reflections in the horizontal as well as in the vertical direction are independent of the order of the reflection (Fig. 1b).

If paracrystalline distortions exist in direction 1 (Fig. 1c), the horizontal width of the reflections with the indices $h \neq 0$ will increase with increasing h (Fig. 1d). If in addition to these distortions consecutive lattice points in direction 1 are shifted in the vertical direction in such a way that a buckled layer structure results (Fig. 1e), the horizontal width of the reflections characterized by $k \neq 0$ will increase with increasing order of the reflection, whereas the vertical width will remain constant (Fig. 1f).

If stacking faults also occur, that is, if consecutive lattice planes in direction 2 are shifted in a horizontal direction (Fig. 1g) [this should easily be possible in smectic phases], an increase of the vertical width of the reflections with $h \neq 0$ will be observed (Fig. 1h). A disorientation of the particles relative to the fiber axis will result in arclike reflections. Thus from the shape of the scattering pattern of oriented samples, information on specific lattice distortions may be derived.

2. The Positional Short Range Order

The spatial arrangement of the particles in noncrystalline materials, that is, in samples that display only a short range order, can be represented by the pair correlation function $g(r)$ defined earlier. The correlation function $g(r)$ is related to the Fourier transform of the scattered intensity $I(s)$:

$$g(r) \propto \int sI(s)\sin(sr)\,ds \qquad (5)$$

where s is the scattering vector defined previously. If the correlation function is known exactly, one can calculate macroscopic properties of the material, using the function. However it is not possible to gain information about the three-dimensional structure of the fluid system from the correlation function directly, since it is a one-dimensional function. A basic method used to gain more insight into the basic structure is to construct a structural model in such a way that the correlation function calculated from this model agrees with the experimentally determined correlation function. This method is extensively used for normal liquids and should be especially valuable for the case of the nematic and cholesteric structure, where no long range order exists. The method has, however, some disadvantages. Experimentally, the scattering curve is obtained only over a limited range of scattering vectors s, whereas in order to carry out the Fourier transform, the scattering function should be known over an indefinite range of the scattering vector. Furthermore corrections for the incoherent scattering and for the instrumental noise background have to be made. These corrections may lead to spurious peaks in the pair correlation function, which are easily misinterpreted on the basis of a structural model. Careful measurements will nevertheless yield useful information about the nature of the short range order. The method has been used to analyze low molecular weight liquid crystal systems by Chistyakov (1967). Vainshtein and Chistyakov (1964) pointed out that in dealing with systems of rods it may be more useful to study the cylindrical rather than the radial distribution function.

3.   The Orientational Order on a Molecular Scale

The orientational order in fluid systems can be determined by fitting a structural model of the fluid state in such a way as to get a good agreement between the theoretical and the experimental pair correlation function. This method requires, however, that the scattering curve be very accurate, which is normally not the case. One therefore prefers more direct methods of determining the orientational order. A widely used method is the light scattering method. The absolute intensity and the shape of the depolarized scattering curve (H_v scattering) are related to the orientation correlation function of the sample. If, for instance, the correlation function extends over a distance which is short in comparison with the wavelength of the light, one can determine the correlation volume V_c, which contains basic information about the extension of the correlation function:

$$V_c = 4\pi \int_0^\infty f_\theta(r) r^2 \, dr \tag{6}$$

where $f_\theta(r)$ is the orientation correlation function. In the simple case of a fluid characterized by aggregates of parallel particles with the optical anisotropy δ_0 and in which the aggregates have a random orientation, the absolute intensity of the depolarized scattering is given by

$$I = (16\pi^4/15\lambda_0^4)((n^2 + 2)/3)^2 N_{cm^3} \delta_0^2 z \tag{7}$$

where n is the index of refraction, λ_0 the wavelength of the light, z the number of anisotropic units in the correlation volume V_c (Stein and Wilson, 1962; Van Aartsen, 1971), and N_{cm^3} the number of aggregates per cm^3.

Light scattering experiments of this kind have been performed with considerable success in order to analyze the orientational order in low molecular weight liquid crystalline systems (Stinson and Litster, 1970), in normal liquids (Dettenmaier et al., 1977), and also in amorphous polymers (Fischer and Dettenmaier, 1973; Fischer et al., 1974, 1976). Information about the orientational order can also be obtained from experiments on the electric or magnetic birefringence. According to the simple Born–Langevin theory (Schütz, 1936), the experimentally observed birefringence arises from a correlated partial orientation of optically and magnetically or electrically anisotropic particles under the influence of the applied field. The overall birefringence depends on the orientation correlations in the material. This method has been used to study the orientational order in liquid crystalline systems and in amorphous polymers (Maret *et al.*, 1974; Meeten, 1974; Fischer *et al.*, 1974).

4. The Chain Conformation

The chain conformation in polymer solutions can be easily studied by means of light scattering or small angle X-ray scattering, due to the fact that a difference exists between the polarizability and the electron density of the polymer chain and the solvent. In the bulk material no such difference exists, therefore light scattering and X-ray diffraction methods can no longer be used for determining the chain conformation.

Neutron small angle scattering has turned out to be the most important method for the studies of conformation of chains in solid polymers. The basic equation for the structure amplitude $S(\mathbf{s})$ of a molecule in solution is

$$S(\mathbf{s}) = |a_r - a_s|^2 \left\langle \left| \sum_{j,m} \exp(i\mathbf{s}\mathbf{r}_{jm}) \right| \right\rangle \tag{8}$$

where $\mathbf{r}_{jm}$ determines the relative position of the particles j and m, respectively, and where a_r and a_s are the scattering amplitudes of the repeat unit and the solvent molecule, respectively. The difference $(a_r - a_s)$ is zero if the repeat unit and the solvent are of the same nature. The important point is that in the case of neutron small angle scattering, it is possible to determine the chain conformation in the bulk material if a mixture of deuterated molecules and hydrogenated molecules is prepared. One studies either a dilute solution of the hydrogenated species in the deuterated matrix or a dilute solution of the deuterated species in the hydrogenated matrix. The scattering curve can be analyzed in the same way as light scattering curves or small angle X-ray curves are analyzed for the case of polymer solutions. Debye has shown that for randomly coiled molecules, each of which can be represented by a Gaussian distribution of chain elements, the following equation holds:

$$I(s)/I(0) = (2/v^2)(v - 1 + e^{-v}) \tag{9}$$

where $v = s^2 \langle R^2 \rangle$, $\langle R^2 \rangle$ is the mean square radius of gyration, and $I(s)$ and $I(0)$ are the scattered intensities at s and $s = 0$, respectively.

From the scattering curve, the radius of gyration can be obtained; it also contains information about the distribution of the chain elements in small volumes of the sample (Debye, 1947). Neutron small angle scattering has been used with considerable success for the case of amorphous polymers as well as partially crystalline polymers (Kirste *et al.*, 1972; Fischer *et al.*, 1974; Cotton *et al.*, 1974).

C. Determination of the Supermolecular Structure

The textures of liquid crystalline phases are most conveniently studied by means of the polarizing microscope. Usually thin films between glass slides are studied; in some cases, thin slices are also cut from the bulk material and studied. The macroscopic orientational order present in the samples leads to characteristic patterns, which have been extensively described in the literature. In addition to this method, light scattering methods have been used to study the orientation correlations on a macroscopical scale (Stein *et al.*, 1968; Rhodes *et al.*, 1970). The observed scattering results primarily from correlations in orientation of anisotropic elements. A quantitative analysis of the shape of the scattering curve and the polarization of the scattered light yields information about the dimensions of the correlated regions, their shape, and their arrangement.

The orientational order on a supermolecular scale can be tested by small angle or wide angle X-ray diffraction techniques. Each grain in a polycrystalline or semiordered sample normally has a crystallographic orientation different from that of its neighbors. Considered as a whole, the orientation of all grains may be randomly distributed in relation to some selected frame of reference. The scattering pattern is then characterized by homogeneous rings around the origin. However, if the orientation of the grains tends to cluster to a greater or lesser extent about some particular orientation or orientations, the pattern will no longer be characterized by homogeneous rings but by arcs or even diffraction spots. The degree and kind of the orientation may be derived from the shape and position of the arcs or spots on the scattering diagram. Thus the orientational order characteristic for a liquid crystalline texture can be studied by means of X-ray diffraction methods.

D. Structural Analysis in Liquid Crystalline Systems

1.   Molecular Structure

a. The Nematic Phase.   The molecular structure of thermotropic crystalline phases has been studied almost exclusively by X-ray diffraction. For this reason only the X-ray scattering diagrams, characteristic of different mesophases, will be discussed here for the case of oriented and unoriented samples.

In the case of the nematic phase, a weak diffuse inner ring is observed, which is related to the length of the molecules, and a strong diffuse ring at larger angles, which is a measure of the distance between neighboring molecules. The large width of the reflections indicates that only a short range positional order exists in the nematic phase. Splitting of the rings is observed in oriented samples. De Vries (1970) reports the existence of cybotactic clusters in the nematic phase close to the nematic–smectic transition with a local smectic organization of the molecules. This order leads to the increase of the intensity of the inner ring.

   b. The Cholesteric Phase. The diffraction pattern of the cholesteric phase is very similar to that of the nematic phase. There again indications of the existence of cybotactic groups may be obtained. The inner reflection increases in intensity if the cholesteric–smectic transition is approached.

   c. The Smectic Phase. In the smectic A modification, which is an optically uniaxial modification, it was found that the long axes of the molecules are parallel to the smectic layer normal. Only a short range order exists within the layers, therefore, they can be considered as two-dimensional fluids. The scattering pattern of the unoriented samples is characterized by one or more sharp inner reflections, which are due to the smectic layers, and by one diffuse outer ring, which is due to the short range order within the layers. In oriented samples the inner rings degenerate to reflections lying on the meridian. The outer reflection degenerates to a broad arclike reflection lying on the equator.

   The structure of the smectic C modification is similar to that of the A modification; within the layers, only a short range order exists. However, the C modification is an optically biaxial modification. The long axes of the molecules are tilted with respect to the layer normal. If optically active molecules are added to the phase, one observes that the direction of the tilt precesses around the layer normal and a helical structure is induced. The X-ray diffraction pattern of the unoriented sample is identical to the pattern of the smectic A modification. In the case of oriented samples, the tilt leads to a splitting of the outer ring into doublets that are positioned around the equator.

   In the smectic B modification, the layers are no longer two-dimensional fluids but seem to be two-dimensionally ordered solids. In most cases a hexagonal close-packed structure is observed; however, other structures may also occur. The long axis of the particles within the layers may be parallel to or tilted with respect to the layer normal. Sharp inner and outer reflections are characteristic of this modification. The orientation of the sample leads to a splitting of the reflection rings in the same way as discussed in the case of the smectic A modification. At present the difference between the smectic B modification and the truly crystalline state is not well understood.

   In addition to the smectic modifications mentioned, a variety of other smectic modifications have been reported to exist. They are called D, E, F, G, etc., modifications. They are less well characterized with respect to their molecular structure than the A, B, and C modifications (De Gennes, 1975).

2.   The Supermolecular Structure

a. The Nematic Phase.   The following part of the chapter is devoted to the description of the optical character of the textures observed for thin films between glass slides or for thin slices of a material under the polarizing microscope. The orientational order, which determines the textures, is also discussed.

In thin samples of a nematic material, one observes dark flexible filaments under the optical microscope. These are caused by lines of singularities in the molecular alignment (Friedel, 1922). The term "black filaments" is used for this texture. A characteristic texture of the nematic phase is the "Schlieren texture," which is caused by a nonhomogeneous orientation of the particles of the material. One observes dark brushes that start from point defects.

In a homeotropic texture, the field of view under the polarizing microscope is black in ideal cases. The optical axes and, consequently, the long axes of the molecules are oriented perpendicular to the plane of the thin films. The optical axes of the molecules are oriented parallel to the plane of the film if the samples exhibit the homogeneous texture. Under the microscope one observes large homogeneous birefringent regions.

The nematic marbled texture consists substantially of a great number of nearly homogeneous regions with different orientation of the optical axes.

b. The Cholesteric Phase.   The most characteristic texture of the cholesteric phase is the "planar" texture, which is also called the "Grandjean" texture. It is characterized by the existence of a cholesteric single crystal where the direction of the helical axis is perpendicular to the plane of the film. Bragg reflection of light occurs on planes with equal direction of the long axes of the molecules, resulting in the observation of iridescent colors. The pitch of the helical structure, which determines the optical properties of the phase, may be influenced by temperature, additives, or external forces.

Just below the clearing point one may observe a texture, in which the helical axis is parallel to the plane of the cholesteric film. One can directly observe the pitch of the helix, provided that it is large enough to be resolved. This texture has been referred to as "fingerprint" texture (de Gennes, 1975). In thicker samples the "focal conic" texture is usually obtained. Characteristic of this texture is the occurrence of an arrangement of fine dark lines. The lines form ellipses and hyperbolas or parts of ellipses and hyperbolas. The specific pattern is caused by the existence of a lamellar structure that can be deformed in such a way that the distance between the lamellar planes stays constant. In the case of the cholesteric phase the lamellar structure is due to the helical structure; it is thus a supermolecular structure.

c. *The Smectic Phase.* The smectic A and C modifications also exhibit a focal conic texture. The lamellar structure is due to the smectic layers, thus it is a molecular structure. Focal conic textures may differ in their appearance. One distinguishes the fan-shaped, the broken fan-shaped, and the polygon textures. No focal conic textures are expected for the smectic B modification since the layers cannot be deformed. The smectic C modification can exhibit a Schlieren texture since the amount of the tilt of the long axes of the molecules is fixed at fixed temperatures, whereas the direction of the tilt may still vary.

The smectic B modification and also the other modifications may show a "mosaic" texture, where homogeneous regions with nonregular boundaries are observed under the polarizing microscope. The optical axes of all particles within one region are parallel; different regions have different orientations.

Homeotropic and homogeneous textures are also observed for smectic phases. The optical pattern agrees with that described earlier.

IV. Liquid Crystalline Order of the Main Chain in Polymers

A. Polymers with Flexible and Semiflexible Chains

1. Polymers with Flexible Chains

The anisotropic geometrical shape and the stiffness of a particle are fundamental requirements for forming liquid crystalline phases. Since in the case of chain molecules an anisotropic geometrical form always exists, chain flexibility must be considered to measure the tendency of a particular polymer to form mesophases. Chain flexibility is a property that cannot be described in a straightforward way. It is determined by the ease with which neighboring repeat units along the chain can orient themselves independent of each other. A decrease of this property will lead to changes of the conformation of the chain and therefore of the dimension of the chain. Since the dimension of the chain can be measured by scattering methods, it is possible to gain information about the chain stiffness using these techniques. Polymers such as polyethylene, polystyrene, poly(methyl methacrylate), polycarbonate, or poly(ethylene terephthalate) can be considered as flexible-chain polymers. The molecular structure of the melt or of the glassy state of these polymers has been discussed for many years. There are basically two different opinions about the molecular structure that have been expressed in various structural models for the amorphous phase of these polymers. In principle, one can distinguish between two different classes of models, the coil model and the bundle model.

In a coil model (Flory, 1953, 1969), it is assumed that the amorphous state of flexible polymers is homogeneous. The configurational state of the chain is supposed to be the same as in a dilute solution, namely, a random coil. In a bundle model, it is assumed that the amorphous phase is hetero-geneous and anisotropic on a molecular scale. Domains with liquid crystal-line order are assumed to exist. The model is based on the experimental observations that the amorphous and the crystalline density of a material are very similar, that under the electron microscope a nodular structure in amorphous polymers is observed and that dark field microscopy indicates existence of some orientational order within these nodules (Pechold and Blasenbrey, 1970; Robertson, 1965; Yeh and Geil, 1967; Yeh, 1972).

However, studies of short range order by electron and X-ray diffraction, studies of orientational order by light scattering and magnetic birefrin-gence, studies of the morphology by small angle X-ray scattering and light scattering, as well as studies of the chain conformation in the amorphous phase by small angle neutron scattering showed that the concept of the coil model is in agreement with all experimental data. This is not the case with the bundle model (Symposium, 1974). Accordingly, the chain conformation is apparently identical with that of a chain in a θ solvent. The orientational order is determined only by the correlation between successive repeat units of a chain. This order can be accounted for on the basis of the theory of rotational isomerism (Tonelli et al., 1970). Furthermore, the amorphous phase is homogeneous and similar to a normal fluid. Thus these results indicate that flexible-chain polymers do not display regions with liquid crystalline order in the noncrystalline state.

There are some indications, however, that special treatments of amor-phous samples, such as drawing of the sample, may lead to a structure that differs from that of the normal amorphous phase. If a polymer such as poly(ethylene terephthalate) is drawn under special conditions, it is possible to introduce a series of different intermediate states, which are often referred to as paracrystalline states, prior to the formation of the truly crystalline state. These intermediate states resemble liquid crystalline states. Bonart (1966) reported the development of the crystalline structure by drawing. The structure is formed through a nematic hexagonal arrangement, which, at a higher draw ratio, is transformed into a smectic packing. The different crystallographic planes appear at different draw ratios.

2. Polymers with Semiflexible Chains

Miller (1960) first reported the detection of a form of isotactic polypropy-lene, that is neither amorphous nor crystalline. This state is characterized by an aggregation of molecules or segments, in which portions of the individual chains maintain the helical structure found in the crystalline state, but in which there is little or no lateral order sufficient to be called crystalline. Miller referred to this state either as the paracrystalline state or as an unstable smectic state of the polymer. The structure was characterized by X-ray scattering, volumetric measurements, dynamic mechanical studies, as well as NMR studies. In the case of the X-ray scattering studies, one observes a second scattering maximum that is not present in the diagram of the truly amorphous material of atactic polypropylene.

Several polymers characterized by a one-dimensional or a two-dimensional order are known to exist. Statton (1959) concluded from a study of several of these polymers that a two-dimensional order could be attributed to a regular lateral packing of linear molecules with no longitudinal order whereas a one-dimensional order could be attributed to a longitudinal periodicity with no lateral order.

Studies of the structure of poly(vinyl carbazole) (Kimura *et al.*, 1970) revealed that an ordered arrangement of the rigid rodlike chains exist in the solid polymer, which does not seem to be truly crystalline. According to this work, the chains are packed approximately in a pseudohexagonal array. A lateral spacing of 10.7 Å, which is experimentally observed, is interpreted as the ($10\overline{1}0$) diffraction of this lattice. This indicates that the nearest chain-to-chain distance is 12.6 Å. Since no layer lines are observed, it is concluded that the chains of the ordered arrangement cannot have a longitudinal order but only a lateral order (chain-to-chain order) of the pseudohexagonal lattice.

A structure very similar to that of poly(vinyl carbazole) was reported by Bohn *et al.* (1961) for the case of atactic poly(acrylonitrile). The formation of a specific structure with a lateral order only was attributed to dipolar intramolecular repulsions, which led to a chain stiffening. The chain may be thought of as a rigid rod with a diameter of approximately 6 Å. The lateral order and the longitudinal disorder of the system was studied by X-ray diffraction techniques on oriented samples. Sharp equatorial reflections but no layer lines were observed; instead, a very diffuse scattering off the equator was observed. It was found that no amorphous halo was present in the case of poly(acrylonitrile). The small angle scattering diagram was characterized by a diffuse scattering that had been attributed to microholes in the material. It was suggested that the polymer is a "100% laterally ordered" material. The ordered polymer was reported to exhibit a glass transition temperature in the same range of temperatures as the amorphous polymer.

An advanced lateral order and a missing longitudinal order is reported also for poly(vinyl trifluoroacetate) and poly(α,p-dimethylstyrene) (Bohn *et al.*, 1961; Baker and Pape, 1952).

The structure of poly(diethylsiloxane) was analyzed by Beatty *et al.* (1975). These authors observed that the structure is intermediate between a crystalline and a truly amorphous structure. Based on NMR data, adiabatic heat capacity studies, dielectric studies, light scattering experiments, as well as optical observations and X-ray scattering data, they described the existence of a preordered state in the boundary between the crystalline and the amorphous phase. The term viscous crystalline phase is used to stress the lack of rigidity and the existence of deviations of properties of such phases from the properties characteristic for liquid crystalline phases. For example, the material studied showed a two-phase structure (preordered phase–amorphous phase) in a certain temperature range. A large supercooling of the amorphous phase is possible since the transformation into the preordered state is slow compared to the very fast transition normally found in liquid crystalline systems. A true crystalline phase may develop from the amorphous phase or the preordered phase. Based on the results of X-ray scattering, it was postulated that the polymer chain maintains a unique interchain distance, which in this particular case is 8.78 Å, but in which a regular spacing in the direction parallel to the chain direction is not maintained. Slight shifts from a regular arrangement along the long axes of the molecules would result in such a structure with a lateral order but without longitudinal order.

Several poly(phosphazenes) were studied by Allen *et al.* (1970) and later by Singler *et al.* (1974, 1975). Allen *et al.* reported that most of the samples studied exhibited two first-order transitions as observed by thermal studies. The two transitions were separated by a temperature range of the order of 150 to 200°C. Optical studies revealed that the ordered structure is not lost at the first transition but persists throughout the temperature range up to the second transition temperature. This appears to be the true melting temperature.

The exact nature of these transitions is still open to questions. In the case of poly[bis(trifluoroethoxy)phosphazene], the wide angle X-ray diffraction pattern undergoes a dramatic change above 80°C. Above this temperature only diffuse rings are observed that are presumed to represent a side group packing order. The transition seems to resemble that of a transition into a mesomorphic phase. In the case of poly(2,2,2-trifluoroethoxyphosphazene), the X-ray pattern observed above the first transition temperature shows only one reflection that corresponds to a lateral dimension of 11 Å, whereas the room temperature phase is characterized by a three-dimensional order.

Singler *et al.* (1974, 1975) reported the results of a study on several poly-(acryloxyphosphazenes). Again for some polymers two melting peaks were observed. The gross spherulitic texture and the extinction pattern observed under the polarizing microscope did not disappear above the first transition temperature. The material softens at this transition temperature and can easily be molded.

Recently, Desper and Schneider (1976) published a paper about investigations on mesomorphic structures at elevated temperatures in meta and para forms of poly[bis(chlorophenoxy)phosphazene]. A pseudohexagonal type of structure is indicated in the temperature range between the first and the second transition temperatures. The absence of all but the first few (h,k,l) reflections is attributed to the presence of a certain degree of disorder in the 001 projection of the structure. The two isomeric forms of the polymer exhibit a pseudohexagonal structure, as derived from the X-ray data, which is stable over a wide temperature range at elevated temperatures. The structure is characterized by an ordered packing of parallel chains in two dimensions. This leads to sharp equatorial reflections. Diffuse meridional reflections indicate that disorder prevails in the third direction. The authors point out that a close resemblance of this structure with the structure of liquid crystalline phases of low molecular weight systems and with plastic crystals exists. A more detailed description of these interesting polymers is in Chapter 9.

Roviello and Sirigu (1975, 1976) studied aliphatic polyesters of 4,4'-dihydroxy-α, α'-dimethylbenzalazine, which showed thermotropic mesophases. The number of the carbon atoms in the aliphatic chain was varied among 8, 10, and 12. The polymers turned out to be soluble in different solvents, the exact molecular weight could not be determined. X-ray diffraction studies showed that the polymer was partially crystalline. By thermal analysis and optical studies it could be shown that the polymers exhibit a first-order transition, which can be attributed to the melting of the crystalline structure, and a second transition, which can be identified as the clearing point of a liquid crystalline melt. The anisotropic fluid appeared under the polarizing microscope as nematic or smectic. By slow cooling it was possible to preserve some features of the texture down to room temperature at which the polymer is in the solid state. The X-ray diffraction pattern of the solid phase depends on the thermal history of the sample. Different diagrams were obtained for the original sample and for a sample cooled down from the anisotropic melt. Rapid cooling from the isotropic melt resulted in the formation of the original structure. If the material in the liquid crystalline state was sheared during the cooling process down to room temperature, an ordering of the microcrystals of the solid polymer was observed.

Roviello and Sirigu (1975, 1976) also synthesized aliphatic homopolyesters containing the $-C_6H_4-C(CH_3){=}CH-C_6H_4-$ group and aliphatic copolyesters with aliphatic chains of different length containing the $-C_6H_4-C(CH_3){=}CH-C_6H_4-$ group or the $-C_6H_4-C(CH_3){=}N-N{=}C(CH_3)-C_6H_4-$ aromatic group. The polymers showed thermotropic mesophases with a melting and a clearing point. They could be quenched from the liquid crystalline state with retention of the liquid crystalline structure. Almost any trace of the crystalline phase is reported to have vanished in these quenched polymers. The X-ray diffraction pattern consists essentially of three maxima. They are attributed to a close packing of the polymer chains and a cybotactic nematic or smectic arrangement of the aromatic groups.

Huang *et al.* (1976) studied the liquid crystalline properties of poly(phenyl-ethylisocyanides) and investigated the nature of the supermolecular structure of homopolymers of poly(α-phenylethylisocyanide) (poly(α-PEI)) and poly-(β-phenylethylisocyanide) (poly(β-PEI)) in the solid and liquid crystalline states. The techniques used to analyze the structure were electron and X-ray diffraction as well as optical microscopy. Liquid crystalline phases were reported to exist not only in highly concentrated solutions of the polymer but also in films that were cast from these solutions. Slowly evaporated solutions gave rise to films reflecting iridescent colors with microscopically visible periodicities and high optical rotation. Circular dichroism and optical rotatory dispersion studies indicated that a helical structure for the poly(α-PEI) system existed. Additional studies supported this conclusion. The arrangement of the helices was characterized by X-ray and electron diffraction.

B. Polymers with Rigid Chains

1. General Considerations

Polymers with rigid chains are expected to show a tendency towards the formation of anisotropic solutions and melts for the reasons discussed earlier.

The statistics of rigid rods in dilute solutions have been developed by Onsager (1949), Zimm (1946), and Ishihara (1950, 1951) based on the evaluation of the second virial coefficient. Onsager (1949) also treated other simple shapes of the particles. The application of these treatments is limited since they are valid only for dilute solutions and for systems with simple shapes of the molecules. Flory (1956) also considered particles with the shape of rigid rods; he extended the statistical treatment, however, to higher concentrations of the rigid particles by using a modified lattice model. He derived a general expression for the free energy of mixing as a function of the mole number, the axial ratio of the solute particles and the disorientation parameter. He predicted a separation of the system into an isotropic phase and a somewhat more concentrated anisotropic phase at a critical concentration. The phase separation arises as a consequence of the asymmetry of the particles without any attractive interaction energy being involved.

A comparatively small positive energy of interaction causes the concentration of the anisotropic phase to increase sharply while the concentration of the isotropic phase decreases. In the case of semiflexible polymers, the properties of the solutions are determined mainly by the length of the rigid segments between the joints. The overall length of the chain is of less importance. Thus an increase in the flexibility of the chain leads to a decrease of the tendency to form mesophases.

DiMarzio (1961) treated the statistics of the orientation effects in linear polymer molecules, using methods similar to the one used by Flory (1956) and included in his treatment molecules of arbitrary shapes. He also discussed the spontaneous ordering of rigid systems in solutions and in the bulk molten state. Ordering of the kind characteristic of the mesomorphic state can occur even in the absence of orientation-dependent attractive interactions between the particles.

The requirements for the formation of anisotropic liquid crystalline solutions in polymers with rigid main chains are

    (i)   rodlike confirmation;

    (ii)   molecular weight that exceeds a minimum value, which may be expressed in terms of the axial ratio;

    (iii)   solvent that is capable to dissolve the polymer beyond a critical concentration value.

2. Biological Systems

A great number of biological chain molecules, such as DNA, or viruses, such as tobacco mosaic virus, and synthetic polypeptides are known to possess rodlike conformations with rod dimensions of the order of several hundreds up to several thousands of angstroms in the direction of the long axes and of several decades up to several hundreds of angstroms in directions perpendicular to the long axes. In concentrated solutions these systems are able to form mesophases. Since the formation of mesophases and transitions between phases can be induced by concentration variations, the mesophases are called lyotropic mesophases. In addition to concentration variations, temperature changes can also induce phase transitions. A survey of the literature published in this area is given by Mishra (1975).

In this chapter some basic features related to structural studies are discussed. Solutions of polypeptides, especially solutions of poly(γ-benzyl glutamate) were studied by Robinson (1956, 1961, 1966) and Robinson et al. (1957, 1958), Hermans (1962) has demonstrated the existence of a critical concentration and a phase separation in solutions of poly(γ-benzyl-L-glutamate) in selected solvents. The rigid rodlike character of the chain is due to the helix formation in these solvents. It is observed that on increasing the concentration, a separation into two liquid phases occurs. The phase of higher concentration is birefringent indicating the existence of an orientational order. The birefringent phase often separates in the form of liquid spherulites, which will eventually coalesce to form a continuous phase. The structure of the liquid crystalline phase is believed to resemble a nematic structure; an axis of torsion of uniform pitch at right angles to the long axis of the molecules, however, is superimposed on it. Thus a structure similar to a cholesteric structure exists in these solutions. This leads to a very high optical rotatory power. Different textures were observed for solutions that resembled the textures observed in thermotropic liquid crystals. The structures of liquid crystalline solutions were characterized by X-ray diffraction (Robinson, 1961). It was observed that the reflections became less and less diffuse with increasing concentration of the solutions. A macroscopic orientation of particles resulted when the solutions were introduced into capillaries.

Iizuka (1973, 1975) studied structural properties of liquid crystalline solutions of poly(γ-benzyl glutamate) as well as dried films under the influence of electric or magnetic fields. He used X-ray diffraction techniques and light scattering techniques (Iizuka *et al.*, 1974). He was concerned with the orientation of atomic groups due to the applied field and with the orientation of clusters of the rodlike molecules in these systems. The properties of lyotropic liquid crystals of poly(γ-benzyl glutamate) are described in Chapter 5.

In addition to poly(γ-benzyl glutamate), solutions of other polypeptides as well as solutions of different biological particles such as tobacco mosaic virus (Bawden and Pirie, 1937), which can exhibit liquid crystalline phases, were studied. In the case of tobacco mosaic virus solutions. Bernal and Fankuchen (1941) obtained X-ray scattering diagrams consistent with a two-dimensional hexagonal array formed by the rodlike particles of the virus. The structures and textures obtained in the case of liquid crystalline solutions of biological systems do not differ from those observed in low molecular weight thermotropic liquid crystalline systems.

3. Synthetic High Modulus Organic Systems

During the last few years a great deal of attention has been focused on very high modulus organic fibers (Symposium, 1973; Symposium, 1976; Black and Preston, 1973; Elias, 1975). The significant feature of these fibers is that within the fibers, highly extended polymer chains are supposed to bear the load. Any deviation from the fully extended state, which could be due for example to a helical conformation, would result in a much lower modulus than otherwise possible.

None of the materials discussed here can be melt-spun, since they decompose at higher temperatures than their melting range. They have to be spun from solutions; hydrogen bonding solvents, or strong acids are used in order to prepare spinning solutions. The basic reason why highly extended chains can be achieved in these materials is that the concentrated solutions used in the spinning process show mesomorphic behavior. It is a well-known property of mesophases to orient by shearing. Spinning and drawing techniques applied to these anisotropic solutions are very important in getting the desired high modulus fibers. Not much has been published in the area of spinning techniques for obvious reasons. It is quite certain, however, that the spinning process is done in such a way that a high chain orientation is obtained and maintained all the way up to the final product. The ease of orientation of the systems considered here is thus directly related to the mesomorphic character of the solutions.

Systems that have been studied in detail are aromatic polyamides (for example, poly(p-phenyleneterephthalamide) and poly(p-benzamide), aromatic polyhydrazides, and polyamide hydrazides. These polymers are constituted of segments linked together in such a way that the chains are extended. In the case of the aromatic polyamides the amide links are presumed to be predominantly in the trans configuration. Due to the stiffness of the chain, anisotropic solutions are formed in wide temperature, concentration, as well as molecular weight ranges. Aromatic polyamides are readily solvated by strong acids such as concentrated sulfuric acid, oleum, hydrofluoric acid as well as in mixtures of these with other powerful polar solvents. Aromatic polyhydrazides can be dissolved in dimethyl sulfoxide–lithium chloride systems, polyamide hydrazides in the usual amide solvents (Carter and Schenk, 1975). The parameters that define the stability range of the anisotropic solutions are solvent, polymer concentration, temperature, molecular weight of the polymer, and alkali-metal concentration (Hermans, 1967). The salt interacts with the polymer, thereby optimizing the process of dissolution.

The anisotropic character of the solutions, which probably have a nematic structure, is recognized visually by a haze at rest and by an opalescent appearance under low shear. Birefringent regions can be detected under the polarizing microscope. In general the formation of anisotropic solutions can also be detected by studying rheological and solution properties as a function of the concentration of the polymer. When solutions of increasing concentration are prepared, the bulk viscosity increases rapidly with increasing concentration up to a critical point, where the liquid crystalline phase is formed. The viscosity decreases rapidly as the proportion of this phase is increased (Kwolek *et al.*, 1976; Papkov *et al.*, 1974).

Only a limited amount of work has been published in the field of structural analysis of liquid crystalline solutions. Studies of the orientational order are rare. Fischer and Dettenmaier (1977) studied the depolarized light scattering of solutions of poly(p-phenyleneterephthalamide) in sulfuric acid as a function of the concentration, whereas Maret *et al.* (1977) studied poly(*p*-benzamide) solutions by magnetic birefringence measurements. No definite conclusion was drawn with respect to the orientational order.

Most of the studies published to date were concerned with the evaluation of the chain conformation in dilute solutions. Schaefgen *et al.* (1976) used light scattering techniques in order to study the viscosity–molecular weight relationship for solutions of stiff polyamides such as poly(1,4-phenylene-terephthalamide) and poly(1,4-benzamide) in dialkylamide solvents with LiCl added. For the viscosity law in the case of poly(1,4-benzamide), it was found that two distinct regions existed. Below $M_w = 12,000$ the viscosity law pointed to the existence of stiff chains in solutions, whereas for molecular weights above this value, indications of a wormlike conformation were obtained. With poly(1,4-phenyleneterephthalamide), only a behavior characteristic of wormlike conformations was observed. Similar studies were reported by Arpin and Strazielle (1975) and Arpin *et al.* (1976).

Papkov *et al.* (1974) also performed experiments very similar to the ones performed by Schaefgen *et al.* (1976) on solutions of poly(1,4-benzamide) in *N,N*-dimethylacetamide. Papkov's value for the experimentally observed molecular weight is much larger than the value reported by Schaefgen *et al.* The large values obtained by Papkov *et al.* are attributed to associations in the amide solvent. Solutions of polyamide hydrazides in DMSO were studied by Burke (1973). He observed highly extended chain conformations in these solutions. A full description of conformational studies performed on dilute solutions of stiff and wormlike chains is given in Chapter 2.

No conformational studies in concentrated solutions by small angle neutron scattering have been published up to now. Panar and Beste (1976) studied the structure of poly(1,4-benzamide) solutions in the high concentration range under the polarizing microscope. These solutions formed nematic phases. The solvent used was composed of fully alkylated amides containing several percent of LiCl salt. It was observed that samples of the pure nematic phase of the polymer of low molecular weight may relax to a transparent state, with a random population of linear nematic regions (striations). In a magnetic field of several thousands of gauss, the lines were aligned in the direction of the field. The sample took the characteristics of a uniaxial birefringent crystal. The addition of an optically active solute resulted in a cholesteric structure similar to low molecular weight twisted nematic phases. The cholesteric helix can be unwound by the application of a magnetic field in analogy with low molecular weight cholesteric phases.

The influence of a magnetic field on the macroscopic orientation of anisotropic solutions of poly(*p*-benzamide) was studied by Platonov *et al.* (1975a,b, 1976). The texture of these solutions were studied by Chancic *et al.* (1975) by means of small angle light scattering. The scattering curves were analyzed with respect to the shape (cylinders) and size of anisotropic aggregates of segments in these solutions.

The structure of fibers spun from the anisotropic solutions as already discussed has been the subject of detailed studies. The crystalline structure, the degree of crystallinity, the crystalline orientation, as well as the defects in the crystalline phases were experimentally analyzed by X-ray diffraction, electron diffraction, as well as birefringence studies. It was found that a fibrillar three-dimensionally ordered structure was in general exhibited by these fibers. The chain orientation is supposed to be high. The degree of crystallinity is large and the chains are believed to be highly extended. The superior mechanical properties of the fibers are attributed to these structural features.

The polymers studied in more detail with respect to their crystalline structure, crystalline orientation, chain orientation, and chain extension are poly(amidehydrazides) (Preston *et al.*, 1973; Holland, 1973), poly(*p*-phenyleneterephthalamide) (Northolt and Van Aartsen, 1973; Hinrichsen *et al.*, 1974; Ballou, 1976), and poly(1,4-benzamide) (Carter and Schenk, 1975; Ballou, 1976), as well as several other aromatic polyamides, such as poly(*m*-phenyleneisophthalamide) (Herlinger *et al.*, 1973; Hinrichsen *et al.*, 1974).

V. Liquid Crystalline Order of the Side Groups in Polymers

A. General Survey

A considerable amount of work has been performed in the area of polymers in which the liquid crystalline order is exhibited primarily by the side groups, whereas the order of the main chain is less well defined. This field is treated in Chapter 3. Studies of the structure of comblike polymers with increasing length of the alkane chain in the side group have revealed that interactions between side groups increase drastically with the length of the alkane chain (Turner Jones, 1964; Plate *et al.*, 1968, 1971; Jordan *et al.*, 1971). One would therefore expect that polymers with stiff and sufficiently long side groups will exhibit liquid crystalline properties.

Such polymers are usually prepared from monomers displaying liquid crystalline behavior. In most cases vinyl monomers are used. The polymerization is carried out in the liquid crystalline state, in the isotropic fluid state, or in isotropic or anisotropic solutions. In some cases monomers were taken, which did not display liquid crystalline properties but which were sufficiently long and stiff. It was expected that polymers prepared in this way would have the liquid crystalline structure of the initial phase "locked in" in the polymer phase. The process of locking in the monomer structure

should be possible even if the polymer itself did not display a thermodynamically stable liquid crystalline phase. After melting or dissolving the original
structure, the superstructure and the texture will not reappear in the solid
material. However one could expect other cases in which the liquid crystalline
order characterizes a thermodynamically stable state. For these systems one
should observe the reappearance of the liquid crystalline structure and
texture under favorable conditions of thermodynamic equilibrium.

Most of the papers published on the subject of polymers with liquid
crystalline side groups were concerned with the influence of the positional
and orientational order (on a molecular and supermolecular scale) on the
reaction kinetics of the systems. Only a limited amount of detailed work was
carried out for the purpose of analyzing the actual structure of the original
monomer phase and of the polymer phase. The methods used for characterizing the structure were mainly optical observations under the polarizing
microscope and X-ray diffraction. The X-ray diffraction diagrams were analyzed to provide information about the type of liquid crystalline order in the
polymer, the arrangement of the side groups and the orientation of the side
groups on a macroscopical scale, if the polymerization was carried out between glass slides or under the influence of an external field.

Studies of polymers with side groups with structures related to mesomorphic behavior have been performed since 1961. The polymerization of
different monomers such as styrene or methyl methacrylate in different liquid
crystalline phases of soaps were studied with the purpose of locking in the
anisotropic structure (Husson *et al.*, 1961; Herz *et al.*, 1961, 1963). It was
observed that in the case of ternary systems of soap, water, and monomer,
the liquid crystalline properties of the original phase were destroyed by the
polymerization process. In the case of the binary system of soap and monomer, the structure was maintained during the polymerization process in the
liquid crystalline state. However the pure polymer did not display liquid
crystalline properties. If on the other hand binary mixtures of vinyl soap
and water or the pure vinyl soap were polymerized in their liquid crystalline
state, a polymer phase could often be obtained, which was characterized
by the same structure as that of the monomer phase. No changes were
observed in the X-ray pattern and in the optical character during the
polymerization.

A schematic representation of liquid crystalline structures that are formed
by side groups is given in Fig. 2 for the isotropic fluid state, in Fig. 3 for the
nematic ordering, and in Fig. 4 for the smectic ordering of the side groups.
In the case of the isotropic ordering, a random orientation of the side groups
exists. In the nematic ordering the long axes of the side groups are oriented
parallel to the preferred direction. In the case of the smectic ordering, a

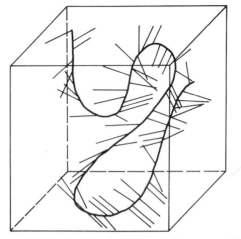

Fig. 2. The structure of the bulk isotropic fluid state of polymers with long side chains.

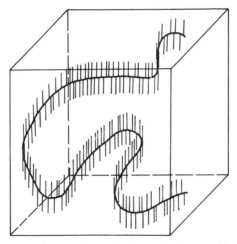

Fig. 3. The structure of the bulk nematic phase of polymers with long side chains.

preferential orientation of the side groups exists, but in contrast to the nematic phase, an ordered arrangement of the centers of gravity of the side groups occurs. This leads to the formation of smectic layers. In the smectic arrangement of the side groups a long range order exists in the direction of the layer normal.

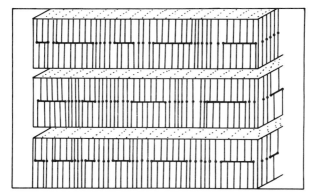

Fig. 4.   The structure of the bulk smectic phase of polymers with long side chains.

B. Polymers with Side Groups Containing the Cholesterol Moiety

Low molecular weight substances containing the cholesterol moiety are known to form liquid crystalline phases readily. Therefore it is not surprising that studies of polymers with side groups containing the cholesterol moiety were performed by several research groups. Most of the published papers state that the polymers obtained do not possess liquid crystalline order but are amorphous (Sadron, 1962; Tanaka *et al.*, 1972; Toth and Tobolski, 1970; de Visser *et al.*, 1970, 1971, 1972, 1975; Hardy *et al.*, 1970). However, recently papers appeared that described a mesomorphic structure for these kinds of polymers (Shibaev *et al.*, 1976; Hsu *et al.*, 1977; Blumstein *et al.*, 1977; Finkelmann *et al.*, 1977a,b).

Sadron (1962) studied the thermal bulk polymerization of cholesteryl methacrylate in the mesomorphic state of the monomer, which showed the characteristics of a cholesteric phase with vivid colors. The soluble polymer exhibited no thermal transitions prior to the decomposition, indicating that no ordered arrangement of the side groups exists. Cholesteryl methacrylate was also polymerized in the presence of air and in vacuum by Tanaka *et al.* (1972) and in the liquid crystalline and isotropic fluid states by Saeki *et al.* (1972). The indications of an ordered arrangement of the side groups were reported. Toth and Tobolski (1970) studied the polymerization behavior of cholesteryl acrylate and cholestanyl acrylate. The soluble polymers showed no thermal transition, indicative of ordered crystalline or liquid crystalline structures. De Visser *et al.* (1970, 1971, 1972, 1975) polymerized cholestanyl acrylate, cholestanyl methacrylate, as well as cholesteryl acrylate in the cholesteric isotropic fluid, and also in the solid state of the monomers. The polymers obtained were analyzed by IR techniques, thermal analysis, and

X-ray diffraction techniques. The polymers were reported to be amorphous since no traces of crystallinity were detected in the Debye–Scherrer X-ray diagrams. No melting point but only a glass transition point were observed for these systems. Irreversible transitions were observed for the polymers in the temperature range from 75 to 125°C. The origin of this transition was unknown.

The system cholesteryl acrylate was studied by Hardy *et al.* (1970). They reported that the polymers obtained were isomorphous with the monomers, if the polymerization was performed below a temperature of 0°C. If the polymerization was carried out, however, above this temperature, it was found that the polymers had a globular shape. The material turned out to be amorphous.

Kamogawa (1972) prepared vinyl polymers containing long cholesteric side groups. It was found that the polymers formed transparent films from solvents such as toluene. Films consisting of a mixture of the polymer with cholesteryl chloride showed mesomorphic behavior at elevated temperatures. Recently Shibaev *et al.* (1976) published a paper in which they described the existence of layers in polymers with the cholesteric moiety attached to the backbone via a flexible bridge. The long side groups are reported to be approximately perpendicularly oriented to the direction of the main chain. Hsu *et al.* (1977) studied the polymerization of cholesteryl *p*-acryloyloxy-benzoate in its cholesteric state, the polymerization of cholesteryl acrylate, and cholesteryl methacrylate above the cholesteric–isotropic transition temperature. In the case of the cholesteryl methacrylate polymer, a smectic structure was found by using small angle and wide angle X-ray difffraction. The main chains were found to be confined to planes and the side groups were oriented perpendicular to this plane. No long range order was reported but only a short range order within the planes. A structure very similar to the one described was also found for the cholesteryl acryloyloxybenzoate polymer whereas the tendency to form smectic structures is less developed in the case of cholesteryl acrylate polymers. The abrupt precipitation of the polymer leads to an amorphous structure. Furthermore copolymers of various compositions of cholesteryl methacrylate with *n*-alkylmethacrylates of different chain lengths were studied (Blumstein *et al.*, 1977). X-ray diagrams characteristic of a smectic structure were observed. The degree of structural order decreased with increasing concentration of the nonmesogenic comonomer.

Finkelmann *et al.* (1977) and Wendorff *et al.* (1977) reported enantiotropic smectic and cholesteric phases for homopolymers and copolymers in which the cholesteryl moiety is attached to the polymer backbone via flexible spacer groups, which consist of $(CH_2)_n$ chains. In the case of homopolymers (n = constant) and in the case of copolymers with $n = 2$ and 6, as well as $n = 6$ and 12, only smectic phases were observed whereas in the case of copolymers with $n = 2$ and 12, a cholesteric phase was observed in a limited concentration range of the comonomers. The molecular structure and the texture, which were analyzed by X-ray scattering and optical studies, were similar to the ones observed for low molecular weight liquid crystals. At room temperature smectic or cholesteric glasses were obtained.

C. Polymers with Various Side Groups

The polymerization of cetyl vinyl esters, which exhibit a smectic phase, were studied by Hardy et al. (1966). The reaction was carried out in the liquid crystalline phase, the isotropic fluid state, and in the solid state of the monomers. It was found that the X-ray diagram of the polymer obtained in the liquid crystalline state corresponded directly to the X-ray diagram of the liquid crystalline monomer phase. Both diagrams were characterized by a sharp and a diffuse reflection. The polymer was birefringent. Within a few hours a transition from the mesomorphic structure to a truly crystalline structure was observed for the polymer system. The close resemblance between the polymer and the monomer structures vanishes if the polymer is melted and subsequently cooled down to room temperature. This indicates that the liquid crystalline arrangement is not an inherent property of the material but is forced onto it by the monomer phase. A variety of different systems with liquid crystalline properties were studied by Amerik et al. (1965, 1967a,b, 1968; Amerik and Krentsel, 1967), Krentsel and Amerik (1971), and Baturin et al. (1972). These authors did not study the structure of the polymers with the exception of the poly(vinyl oleate). In the case of the vinyl oleate (Amerik et al., 1965; Amerik and Krentsel, 1967) the radiation-induced polymerization was carried out in the isotropic fluid state, in the liquid crystalline state, and in the solid state of the monomer phase. The structure of the polymer was found to be greatly dependent on the phase state of the monomer during polymerization. In the liquid phase a transparent polymer was generated which had the properties of an isotropic viscous liquid. If the polymerization was performed in the liquid crystalline state, a crystalline polymer was obtained according to the X-ray pattern. A similar structure was obtained if the polymerization took place in the solid monomer phase. Subsequent studies of this polymer revealed, however, that it was not truly crystalline. It resembled a viscous fluid in not being able to keep its form. NMR data showed that no complete freezing of molecular motions had occurred in the polymer. Therefore this state of the matter was believed to be a mesophase (Amerik and Krentsel, 1967).

Polymers derived from mesophase forming monomers with one or two double bonds attached to it were studied by Liébert and Strzelecki (1973),

Strzelecki and Liébert (1973), and Bouligand *et al.* (1974). The polymers studied in detail were homopolymers from di-(*p*-acryloyloxybenzylidene)-*p*-diaminobenzene (P1), from *p*-acryloyloxybenzylidene-*p*-cyanoaniline (P2), *p*-acryloyloxybenzylidene-*p*-carboxyaniline (P3), copolymers from P1 and P2, and terpolymers from P1, P2, and cholesteryl acrylate. The polymerization was carried out in the nematic, the cholesteric, and the smectic phases of the particular low molecular weight systems. The textures of the resulting polymers were studied under the polarizing microscope either on thin films, if the polymerization was performed between glass slides, or on thin polished sections of the bulk material, if polymerization was carried out in the bulk phase. The homopolymers of P2, which were obtained from polymerization in the liquid crystalline phase, did not exhibit liquid crystalline order. The homopolymers of P1, on the other hand, which were obtained from the polymerization in the nematic or smectic phase of the monomer system, were characterized by a nematic or smectic structure as indicated by the textures of the material. X-ray diffraction data confirmed these results. The copolymers of P1 and P2, polymerized in the nematic phase, were characterized by a nematic structure. The terpolymerization of P1, P2, and cholesteryl acrylate resulted in a polymer with a cholesteric structure as shown by the existence of a cholesteric texture. If polymers of P3 were obtained by polymerization in the temperature range 220–230°C, they were characterized by a smectic texture. If, on the other hand, the polymerization was performed at 350°C, the polymers exhibited a nematic texture. These textures were retained if the samples were cooled to room temperature. The textures of the polymeric systems were found to be very similar to those found for low molecular weight systems.

Cser *et al.* (1976) studied the structures of polymers of different *p*-alkyl-*p*-acryloyloxyazoxybenzenes by small angle and wide angle X-ray diffraction. The polymers were obtained from the polymerization in the bulk monomer phase which displayed a nematic structure. It was found that the polymer could be dissolved in the monomer phase. The polymer was characterized by a nematic structure and at room temperature, the polymer was in the glassy nematic state.

In an early paper Blumstein *et al.* (1969) described the cross-linking of a two-dimensional arrangement of monomer molecules that were absorbed on montmorillonite. The work was performed in order to lock in the quasi-smectic organization of the molecules within the monolayers. A series of papers of Blumstein *et al.* (1971, 1974a,b, 1975a–c, 1976) deals with the polymerization of *p*-methacryloyloxybenzoic acid (MBA) in the isotropic melt, the anisotropic and isotropic fluid states in solutions, the polymerization of *N*-*p*-methacryloyloxybenzylidene-*p*-aminobenzoic acid in the bulk liquid or solid state of the monomers as well as in isotropic solutions

(MBABA), the polymerization of *N*-*p*-butoxybenzylidene-*p*-aminostyrene (BBAS) in its bulk isotropic and liquid crystalline state, and the polymerization of acryloyloxybenzoic acid (ABA) in the bulk material as well as in nematic or smectic solutions. MBA itself does not display a mesomorphic state; it is capable, however, of forming mixed smectic or nematic mesophases with lower melting smectic or nematic *p*-*n*-alkoxybenzoic acids. The phase diagrams of these mixtures were studied in detail by Blumstein *et al.* (1974b). The bulk polymerization of MBA in the isotropic melt led to a polymer that appeared to have small amounts of spherulitic crystalline structure. A much more pronounced crystallinity was detected in poly(ABA) prepared under identical conditions. It was observed that during polymerization, the formation of a Schlieren texture occurred under the polarizing microscope (Blumstein *et al.*, 1976). Out of this texture the growth of spherulites was observed. The polymers of MBA prepared from nematic solutions of *p*-heptyloxybenzoic acid and from smectic solutions of cetyloxybenzoic acid were highly birefringent. From X-ray scattering data and from optical studies, it was inferred that these polymers displayed a smectic structure. If, on the other hand, the polymerization was carried out in the isotropic solution, an amorphous polymer was obtained. A mesomorphic structure, characteristic of the smectic phase, was observed for films cast from polymer solutions in DMF irrespective of the way in which the polymer was prepared. The X-ray analysis indicated that a layered structure with bilayers of molecules strongly tilted in the plane of the layers existed. For polymers of ABA the bulk polymerization in the isotropic fluid state resulted in a polymer with X-ray diagrams characteristic of high crystallinity. Higher reaction rates and shorter heating periods were found to result in less crystalline samples. It was found that crystallinity may also develop in films cast from DMF (dimethylformamide) (Blumstein *et al.*, 1975c).

MBABA exhibits several smectic phases as well as a nematic phase (Blumstein *et al.*, 1975b,c). It was polymerized in its nematic phase, the isotropic fluid state, and in solutions of DMF. It was found that the structure obtained in the polymer depends strongly on the reaction rate and the annealing time. Fast reaction rates and short annealing times led to a polymer with a nematic texture, which existed up to the decomposition temperature (Blumstein *et al.*, 1975b). If the reaction rate was slowed down (Blumstein *et al.*, 1975c) and if the polymer was annealed for some hours, a smectic structure was observed. The structure was analyzed by small angle and wide angle X-ray diffraction. A layered structure with only a short range order within the layers was reported. An abrupt precipitation led to the destruction of the order of the polymer, either of the smectic structure or of the crystalline structure in the case of the polymers of MBA and ABA.

Blumstein *et al.* (1975b) also studied the polymerization of BBAS in the bulk isotropic fluid and nematic states. In both cases a nematic structure was observed for the resulting polymers, as indicated by the X-ray diffraction pattern and by optical studies of the textures displayed by the polymers.

Studies of the polymerization of liquid crystalline monomers and of the structure and liquid crystalline properties of the resulting polymers were performed by Perplies *et al.* (1974a,b, 1975, 1977) and by Wendorff *et al.* (1975). The polymerization of *p*-(4-acryloxybenzylidene-4′-ethoxyaniline) (ABE), which exhibits a nematic phase in the temperature range 78–136°C, was investigated in the nematic phase, the isotropic fluid state, and in isotropic solutions. Furthermore, the polymerization of *p*-(4-methacryloxy-benzylidene-4-ethoxyaniline) (MBE), which exhibits a nematic phase in the temperature range 86–100°C, was studied in the isotropic fluid state and in the nematic phase. It was found that in all cases a smectic polymer was obtained, which was in the glassy state at room temperature. At higher temperatures, a glass transition was observed. Above this temperature a mobile smectic phase occurred. The smectic texture was observed up to high temperatures where a melting process was detected. There were indications that the melting in the case of PMBE consisted of two steps. In the first step, a transition from the smectic to the nematic structure was believed to occur, whereas in the second step the transition to the isotropic fluid state took place. The second temperature can be considered as the clearing point temperature of the polymer. The transitions described were found to be reversible, the crystalline texture reappeared on cooling the polymer.

If the polymerization was carried out in the isotropic fluid state of the monomer phase under the influence of strong magnetic fields (up to 100 kg), one did not get an orientation of the polymer system. If the polymerization under the influence of the magnetic field was carried out in the nematic phase, a high orientation of the polymer was observed. The long axes of the side groups and the layer normals were found to be parallel to the direction of the magnetic field. A detailed structural analysis was carried out on oriented samples by means of small angle and wide angle X-ray diffraction. It was found that the smectic polymer structure is characterized by a well-defined long range order in the direction of the layer normal. The relative fluctuation of the layer distance was about 1.4%. Up to 80 layers were found to be stacked one on top of the other. The individual layers turned out to be flat without any detectable amount of buckling and without observable stacking faults. In the case of PAME the smectic structure was made of noninterpenetrating single smectic layers, whereas in the case of PMBE interpenetrating bilayers were observed. The side groups were perpendicular to the plane of the layer for both systems studied. The short range order within the smectic layers was found to be very similar to that

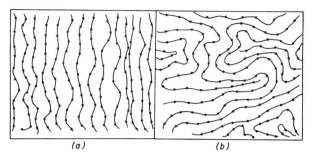

Fig. 5. Schematic representation of the two-dimensional conformation of the main chain in a smectic layer. (a): Bundle model, (b): random chain model.

of the nematic or isotropic phase of the monomers. The smectic ordering of the polymers displayed a scattering diagram analogous to the pattern of the smectic. A modification of low molecular weight compounds. No information could be derived about the actual conformation of the main chain. The X-ray data showed, however, that the main chain had to be rigidly confined to a plane within the smectic layers. On this plane the chain may either form a two-dimensional coil or a two-dimensional bundle structure (Fig. 5). Packing arguments lead to the conclusion that in the case of a two-dimensional arrangement, the bundle structure should be more favorable.

Finkelmann *et al.* (1977) and Wendorff *et al.* (1977) studied polymers in which mesogenic groups with the general structure $-C_6H_4-COO-C_6H_4-R'$ (R' variable) were attached to the polymer backbone via flexible spacer groups consisting of $(CH_2)_n$ chains. The basic idea was to decouple the orientation and the motion of the side chains from those of the backbone by inserting flexible spacer groups of variable length. The precipitated polymers or polymer films cast from solutions were found to be amorphous. On heating above the glass transition temperature liquid crystalline melts were formed, which were either nematic or smectic, depending on the substituent R'. Nematic phases were observed with short substituents R' whereas smectic phases were observed with long substituents R'. The structure and the textures of the polymer liquid crystalline phases were found to be similar to those of low molecular weight liquid crystals. The textures could be frozen in by cooling down to room temperatures; liquid crystalline glasses were obtained.

Structural studies were also performed on samples, which were obtained from nematic monomer phases, where either a homeotropic or a homogeneous texture had been induced. It was observed that in most cases the texture was not destroyed by the onset of the polymerization. Thus highly oriented polymer systems were obtained. Studies of such systems led to additional information about polymer structure and organization. This technique was used by Lorkowski and Reuther (1976), who investigated the polymerization of the nematic monomer 4-acryloyloxybenzoic acid-4-*n*-hexyloxyphenyl ester. The polymerization was carried out in the nematic phase, which was macroscopically unoriented, in the macroscopically oriented nematic phase, as well as in the isotropic fluid state. The orientation of the monomer phase was obtained either by applying a magnetic field or by performing the polymerization between glass slides, treated in such a way as to induce appropriate textures in the monomer phase. It was possible to orient the monomers either parallel or perpendicular to the plane of the slides. The textures of the bulk samples were studied from slices that were cut from the sample. In the case of the polymerization in the isotropic fluid state, no orientation was observed for the polymer phase even if the magnetic field was applied. Schlieren textures were obtained when the polymerization was carried out in the unoriented nematic phase. A strong macroscopic orientation occurred in the polymers that were obtained from the oriented nematic phase.

Apparently the polymerization process was not able to destroy the orientation and the texture of the monomer phase. X-ray diffraction studied revealed that the polymer had a smectic structure. This was indicated by the observation of three sharp reflections at smaller angles and of one broad reflection at larger angles. The side groups were found to be perpendicular to the plane of the smectic layer. It was also observed that arc-shaped reflections occurred if the X-ray beam was perpendicular to the fiber axis of the system, whereas Debye rings were observed if the X-ray beam was parallel to the fiber axis. This result is in agreement with the orientation imposed on the system by the external forces.

Samples that were prepared between glass slides in the unoriented nematic phase showed a marbled texture. If the polymerization was done in the nematic phase at temperatures close to the clearing point, Schlieren textures were observed. Oriented samples, which were obtained from nematic phases with specific textures (homeotropic, homogeneous) imposed on it, behaved like orientational single crystals; they displayed a long range orientational order. In one of these systems the long axes of the side groups were parallel to the film normal, in the other system they were perpendicular to the film normal. No orientational effects were seen when the polymerization between glass slides was performed in the isotropic fluid state of the monomer. This result is in agreement with the fact that low molecular weight systems are not oriented between glass slides if they are in the isotropic fluid state, whereas a strong orientation occurs if they are in the nematic state.

The synthesis and characterization of poly(*p*-biphenyl acrylate) PPBA was described in papers published by Baccaredda *et al.* (1971), Newman *et al.* (1972), and Magagnini *et al.* (1974). It was observed that the atactic polymer was not amorphous but seemed to be characterized by a more or less advanced degree of lateral order. X-ray diffraction studies on the polymer revealed the existence of a single sharp reflection. A similar diagram was observed for the atactic poly(*p*-cyclohexylphenyl acrylate) (PPCPA) whereas the scattering diagram of the atactic poly(*p*-biphenyl methacrylate) (PPBMA) was characterized by an amorphous halo. The intensity of the reflections was increased for the case of PPBA and PPCPA if the samples were annealed. DSC measurements showed that these samples melted over a range of about 5°C, while PPBMA showed no melting peak. The combined results of the thermal and structural studies were interpreted in terms of a high degree of order in these polymers, which is intermediate between that of amorphous and crystalline orders.

VI. Liquid Crystalline Structures Displayed by Block Copolymers

A series of papers have been published by Skoulios and Finaz (1962), Gallot *et al.* (1966), Douy *et al.* (1969, 1970; Douy and Gallot, 1969, 1971), Gervais *et al.* (1972, 1973; Gervais and Gallot, 1973), and Perly *et al.* (1974) that dealt with the existence of liquid crystalline structures in block copolymers in selected solvents. In contrast to low molecular weight liquid crystalline structures and to the structures described previously, one observes a structure which is not formed by individual molecules or segments of molecules but by large supermolecular domains of different compositions and solvent concentration. Due to the size of the domains, the structures were analyzed by means of electron microscopy and small angle X-ray diffraction. The structures were characterized with respect to the size, shape, and arrangement of the domains. It was found that lamellar, hexagonal, and cubic arrangements are possible. The exact nature of the supermolecular liquid crystalline order depends on the copolymer composition and the concentration of the polymer solution. Some of the systems studied are polystyrene–polybutadiene copolymers, polystyrene–poly(ethylene oxide) polymers, polystyrene–polydiene copolymers, and polybutadiene–poly-L-benzyl glutamate copolymers. The liquid crystalline order in block copolymers is treated in detail in Chapter 6.

References

Allen, G., Lewis, G. J., and Todd, S. M. (1970). *Polymer* **11**, 44.
Amerik, Y. B., and Krentsel, B. A. (1967). *J. Polym. Sci., Part C* **16**, 1383.
Amerik, Y. B., Krentsel, B. A., and Konstantinow, I. I. (1965). *Dokl. Akad. Nauk SSSR* **165**, 1097.

Amerik, Y. B., Konstantinow, I. I., and Krentsel, B. A. (1967a). *Vysokomol. Soedin.*, *Ser. A* **9**, 2236.

Amerik, Y. B., Konstantinow, I. I., Krentsel, B. A., and Malachajew, E. W. (1967b). *Vysokomol. Soedin.*, *Ser. A* **9**, 2591.

Amerik, Y. B., Konstantinow, I. I., and Krentsel, B. A. (1968). *J. Polym. Sci. Part C* **23**, 231.

Arpin, M., and Strazielle, C. (1975). *C. R. Acad. Sci.*, *Ser. C* **280**, 1293.

Arpin, M., Debeauvais, F., and Strazielle, C. (1976). *Makromol. Chem.* **177**, 585.

Baccaredda, M., Magagnini, P., Pizzirani, G., and Gusti, P. (1971). *J. Polym. Sci.*, *Part B* **9**, 303.

Baker, W. O., and Pape, N. R. (1952). *High Polym*, **10**, 142.

Ballou, J. W. (1976). *Polym. Prepr. Am. Chem. Soc.*, *Div. Polym. Chem.* **17** (2), 75.

Baturin, A., Amerik, Y. B., and Krentsel, B. A. (1972). *Mol. Cryst. Liq. Cryst.* **16**, 117.

Bawden, F. C., and Pirie, N. W. (1937). *Proc. R. Soc.*, *Ser. B* **123**, 1274.

Beatty, G. L., Pochan, J. M., Froix, M. F., and Hinman, D. D. (1975). *Macromolecules* **8**, 547.

Bernal, J. D., and Fankuchen, I. (1941). *J. Gen. Physiol.* **25**, 111.

Black, B. W., and Preston, J. (1973). "High Modulus Wholly Aromatic Fibers." Dekker, New York.

Blumstein, A., Blumstein, R. B., and Vanderspurt, T. H. (1969). *J. Colloid Interface Sci.* **31**, 236.

Blumstein, A., Kitagawa, N., and Blumstein, R. B. (1971). *Mol. Cryst. Liq. Cryst.* **12**, 215.

Blumstein, A., Blumstein, R. B., Murphy, G. J., Wilson, C., and Billard, J. (1974a). *In* "Liquid Crystals and Ordered Fluids," Proceedings of the ACS Symposium (J. F. Johnson and R. S. Porter, eds.), p. 277. Plenum, New York.

Blumstein, A., Billard, J., and Blumstein, R. B. (1974b). *Mol. Cryst. Liq. Cryst.* **25**, 83.

Blumstein, R. B., Patel, L. R., Clough, S. B., and Blumstein, A. (1975a). *Polym. Prepr.*, *Am. Chem. Soc.*, *Div. Polym. Chem.* **16** (1), 611.

Blumstein, A., Blumstein, R. B., Clough, S. B., and Hsu, E. C. (1975b). *Macromolecules* **8**, 73.

Blumstein, A., Clough, S. B., Patel, L., Lim, L. K., Hsu, E. C., and Blumstein, R. B. (1975c). *Polym. Prepr.*, *Am. Chem. Soc.*, *Div. Polym. Chem.* **16**, 241.

Blumstein, A., Clough, S. B., Patel, L., Blumstein, R. B., and Hsu, E. C. (1976). *Macromolecules* **9**, 243.

Blumstein, A., Osada, Y., Clough, S. B., Hsu, E. C., and Blumstein, R. B. (1977). *Am. Chem. Soc. Div. Polym. Chem.* **18** (2), 14.

Bohn, C. R., Schaefgen, J. R., and Statton, W. O. (1961). *J. Polym. Sci.* **55**, 531.

Bonart, R. (1957). *Z. Kristallogr.*, *Kristallgeom.*, *Kristallphys.*, *Kristallchem.* **109**, 296.

Bonart, R. (1966). *Kolloid Z. Z. Polym.* **213**, 1.

Bouligand, Y., Cladis, P. E., Liébert, L., and Strzelecki, L. (1974). *Mol. Cryst. Liq. Cryst.* **25**, 233.

Burke, J. J. (1973). *J. Macromol. Sci.*, *Chem.* **A7**, 187.

Carter, G. B., and Schenk, V. T. J. (1975). *In* "Structure and Properties of Oriented Polymers" (I. M. Ward, ed.), p. 454. Appl. Sci. Publ., London.

Chancic, O. A., Serkov, A. T., Kalmykova, V. D., and Volochina, A. V. (1975). *Khim. Volokna* **5**, 68.

Chistyakov, I. G. (1967). *Sov. Phys.—Usp.* **9**, 551.

Cotton, J. P., Decker, D., Benoit, H., Farnoux, B., Higgins, J., Jannink, G., Ober, R., Picot, C., and des Cloizeaux, J. (1974). *Macromolecules* **7**, 863.

Cser, F., Nyitrai, K., Seyfried, E., and Hardy, Gy. (1976). *Magy. Kem. Foly.* **82**, 207.

Debye, P. (1947). *J. Phys. Chem.* **51**, 18.

de Gennes, P. G. (1975). "The Physics of Liquid Crystals," Oxford Univ. Press (Clarendon), London and New York.

Desper, C. R., and Schneider, N. S. (1976). *Macromolecules* **9**, 424.

Dettenmaier, M., Fischer, S., and Fischer, E. W. (1977). *Prog. Colloid. Sci.* **62**, 37.

de Visser, A. C., Feyen, J., De Groot, K., and Bantjes, A. (1970). *J. Polym. Sci.*, *Part B* **8**, 805.

de Visser, A. C., De Groot, K., Feyen, J., and Bantjes, A. (1971). *J. Polym. Sci., Part A-1* **9**, 1893.
de Visser, A. C., De Groot, K., Feyen, J., and Bantjes, A. (1972), *J. Polym. Sci., Part B* **10**, 851.
de Visser, A. C., Van Den Berg, J. W., and Bantjes, A. (1975). *Makromol. Chem.* **176**, 495.
de Vries, A. (1970). *Mol. Cryst. Liq. Cryst.* **10**, 31, 219.
Diele, S., Brand, P., and Sackmann, H. (1972). *Mol. Cryst. Liq. Cryst.* **16**, 105.
DiMarzio, E. A. (1961). *J. Chem. Phys.* **35**, 658.
Douy, A., and Gallot, B. (1969). *C. R. Acad. Sci.* **268**, 1218.
Douy, A., and Gallot, B. (1971). *Mol. Cryst. Liq. Cryst.* **14**, 191.
Douy, A., Mayer, R., Rossi, J., and Gallot, B. (1969). *Mol. Cryst. Liq. Cryst.* **7**, 103.
Douy, A., Gervais, M., and Gallot, B. (1970). *C. R. Acad. Sci.* **270**, 1646.
Egelstaff, P. A. (1967). "An Introduction to the Liquid State," Academic Press, New York.
Elias, H. G. (1975). "Neue polymere Werkstoffe 1969–1974," Carl Hansen Verlag, Munich.
Finkelmann, H., Ringsdorf, H., and Wendorff, J. H. (1977). *Makromol. Chem.*, in press.
Fischer, E. W., and Dettenmaier, M. (1973). *Kolloid Z. Z. Polym.* **251**, 922.
Fischer, E. W., and Dettenmaier, M. (1977). *Jahresber. SFB 41, Mainz-Darmstadt, 1976*, p. 144.
Fischer, E. W., Wendorff, J. H., Dettenmaier, M., Lieser, G., and Voigt-Martin, I. (1974). *Polym. Prepr., Am. Chem. Soc., Div. Polym. Chem.* **15** (2), 8.
Fischer, E. W., Wendorff, J. H., Dettenmaier, M., Lieser, G., and Voigt-Martin, I. (1976). *J. Macromol. Sci., Phys.* **B12**, 41.
Flory, P. J. (1953). "Principles of Polymer Chemistry," Cornell Univ. Press, Ithaca, New York.
Flory, P. J. (1956). *Proc. R. Soc., Ser. A* **234**, 73.
Flory, P. J. (1969). "Statistical Mechanics of Chain Molecules," Wiley, New York.
Frank, F. C. (1958). *Discuss. Faraday Soc.* **25**, 19.
Friedel, G. (1922). *Ann Phys. (Paris)* **18**, 273.
Gallot, B., Mayer, R., and Sadron, C. (1966). *C. R. Acad. Sci.* **263**, 42.
Gervais, M., and Gallot, B. (1973). *Makromol. Chem.* **171**, 157.
Gervais, M., Jouan, G., and Gallot, B. (1972). *C. R. Acad. Sci., Ser. C* **275**, 1243.
Gervais, M., Douy, A., and Gallot, B. (1973). *C. R. Acad. Sci., Ser. C* **276**, 391.
Gray, G. W. (1962). "Molecular Structure and Properties of Liquid Crystals," Academic Press, New York.
Guinier, A. (1963). "X-ray Diffraction," Freeman, San Francisco, California.
Hardy, G., Nyitrai, K., Cser, F., Cselik, C., and Nagy, I. (1969). *Eur. Polym. J.* **5**, 133.
Hardy, G., Cser, F., Gallo, A., Nyitrai, K., Bodor, G., and Lengyel, M. (1970). *Acta Chim. Acad. Sci. Hung.* **65** (*3*), 287.
Herlinger, H., Knoell, J., Menzel, H., and Schlaefer, J. (1973). *J. Appl. Polym. Sci.* **21**, 215.
Hermans, J. (1962). *J. Colloid Sci.* **17**, 638.
Hermans, J. (1967). *Adv. Chem. Phys. B* **13** (*10*), 707.
Herz, J., Husson, F., and Luzzati, V. (1961). *C. R. Acad. Sci.* **252**, 3462.
Herz, J., Reiss-Husson, F., Rempp, R., and Luzzati, V. (1963). *J. Polym. Sci., Part C* **4**, 1275.
Hinrichsen, G., Miessen, R., and Reichard, M. (1974). *Angew. Makromol. Chem.* **40/41**, 239.
Holland, V. F. (1973). *J. Macromol. Sci., Chem.* **A7**, 173.
Hosemann, R., and Bagchi, S. N. (1962). "Direct Analysis of Diffraction by Matter," North-Holland Publ., Amsterdam.
Hsu, E. C., Clough, S. B., and Blumstein, A. (1977). *J. Polym. Sci. Lett. Ed.* **15**, 545.
Huang, S. Y., Kiamco, E. A., and Hellmuth, E. W. (1976). *Proc. Prague. Meet. Macromol.* C22-1.
Husson, F., Herz, J., and Luzzati, V. (1961). *C. R. Acad. Sci.* **252**, 3290.
Iizuka, E. (1973). *Polym. J.* **5**, 62.
Iizuka, E. (1975). *Polym. J.* **7**, 650.
Iizuka, E., Keira, T., and Wada, A. (1974). *Mol. Cryst. Liq. Cryst.* **23**, 13.
Ishihara, A. (1950). *J. Chem. Phys.* **18**, 1446.

Ishihara, A. (1951). *J. Chem. Phys.* **19**, 1142.

Jabarin, S. A., and Stein, R. S. (1973). *J. Phys. Chem.* **77**, 399, 409.

Jordan, E. F., Artymyshyn, B., Speca, A., and Wrigley, A. N. (1971). *J. Polym. Sci., Part A-1* **9**, 3349.

Kamogawa, H. (1972). *J. Polym. Sci., Part B* **10**, 7.

Kimura, A., Yoshimoto, S., Akana, Y., Hirata, H., Kustabayashi, S., Mikawa, H., and Kasai, N. (1970). *J. Polym. Sci., Part A-2* **8**, 643.

Kirste, R. G., Kruse, W. A., and Schelten, J. (1972). *Makromol. Chem.* **162**, 299.

Kittel, C. (1959). "Introduction to Solid State Physics," Wiley, New York.

Krentsel, B. A., and Amerik, Y. B. (1971). *Vysokomol. Soedin., Ser. A* **8**, 1358.

Kwolek, S. L., Morgan, P. W., Schaefgen, J. R., and Gulrich, L. W. (1976). *Polym. Prepr. Am. Chem. Soc., Div. Polym. Chem.* **17** (*1*), 53.

Lehmann, O. (1890). *Z. Kristallogr., Kristallgeom., Kristallphys., Kristallchem.* **18**, 464.

Liébert L., and Strzelecki, L. (1973). *Bull. Soc. Chim. Fr.* **2**, 603.

Lorkowski, H. J., and Reuther, F. (1976). *Plaste Kautsch.* **23**, 81.

Magagnini, P. L., Marchetti, A., Materia, F., Pizzirani, G., and Turchi, G. (1974). *Eur. Polym. J.* **10**, 585.

Maret, G., Von Schickfus, M., and Wendorff, J. H. (1974). *Colloq. Int. CNRS* No. 242

Maret, G., Stamm, M., Fischer, E. W., and Wendorff, J. H. (1977). *Jahresber. SFB 41, Mainz-Darmstadt, 1976*, p. 155.

Meeten, G. H. (1974). *Polymer* **15**, 187.

Miller, R. L. (1960). *Polymer* **1**, 135.

Mishra, R. K. (1975). *Mol. Cryst. Liq. Cryst.* **29**, 201.

Newman, B. A., Magagnini, P. E., and Frosini, V. (1972). *In* "Advances in Polymer Science Engineering" (K. D. Pae, D. R. Morrow, and Yu. Chen, eds.), p. 21. Plenum, New York.

Northholt, M. G., and Van Aartsen, J. J. (1973). *J. Polym. Sci., Part B* **11**, 333.

Onsager, L. (1949). *Ann. N.Y. Acad. Sci.* **51**, 627.

Panar, M., and Beste, L. F. (1976). *Polym. Prepr. Am. Chem. Soc., Div. Polym. Chem.* **17** (*2*), 65.

Papkov, S. P., Kulichikhin, V. G., Kalmykova, V. D., and Malkin, A. Y. (1974). *J. Polym. Sci., Polym. Phys. Ed.* **12**, 1753.

Pechhold, W., and Blasenbrey, S. (1970). *Kolloid Z. Z. Polym.* **241**, 955.

Perly, B., Douy, A., and Gallot, B. (1974). *C. R. Acad. Sci., Ser. C* **279**, 1109.

Perplies, E., Ringsdorf, H., and Wendorff, J. H. (1974a). *Makromol. Chem.* **175**, 553.

Perplies, E., Ringsdorf, H., and Wendorff, J. H. (1974b). *Ber. Bunsenges. Phys. Chem.* **78**, 921.

Perlies, E., Ringsdorf, H., and Wendorff, J. H. (1975). *J. Polym. Sci., Polym. Lett. Ed.* **13**, 243.

Perlies, E., Ringsdorf, H., and Wendorff, J. H. (1977). Polymerization of Organized Systems, *Midland Macromol. Monographs* **3**, 149.

Plate, N. A., Shibaev, V. P., Petrukhin, B. S., and Kargin, V. A. (1968). *J. Polym. Sci., C* **23**, 37.

Plate, N. A., Shibaev, V. P., Petrukhin, B. S., Zubov, Y. A., and Kargin, V. A. (1971). *J. Polym. Sci., Part A-1* **9**, 2291.

Platonov, V. A., Litovcenko, G. D., Kulicichin, V. G., Sablygin, M. V., Belousova, T. A., Pozalkin, N. S., Kalmykova, V. D., and Papkov, S. P. (1975a). *Khim. Volokna* **4**, 36.

Platonov, V. A., Litovcenko, G. D., Milkova, L. P., Sablygin, M. V., Kulicichin, V. G., and Papkov, S. P. (1975b). *Khim. Volokna* **5**, 70.

Platonov, V. A., Litovcenko, G. D., Belousova, I. A., Milkova, L. P., Sablygin, M. V., Kulicichin, V. G., and Papkov, S. P. (1976). *Vysokomol. Soyedin., Ser. A* **18**, 221.

Preston, J., Black, W. B., and Hofferbert, W. L. (1973). *J. Macromol. Sci., Chem. A***7**, 45, 67.

Price, F. P., and Wendorff, J. H. (1972). *J. Phys. Chem.* **76**, 276.

Reinitzer, F. (1888). *Monatsh. Chem.* **9**, 421.

Rhodes, M. B., Porter, R. S., Chu, W., and Stein, R. S. (1970). *Mol. Cryst. Liq. Cryst.* **10**, 295.

Robertson, R. E. (1965). *J. Phys. Chem.* **69**, 1575.

Robinson, C. (1956). *Trans. Faraday Soc.* **52**, 571.

Robinson, C. (1961). *Tetrahedron* **13**, 219.

Robinson, C. (1966). *Mol. Cryst. Liq. Cryst.* **1**, 467.

Robinson, C., Ward, J. C., and Beevers, R. B. (1957). *Nature (London)* **180**, 1183.

Robinson, C., Ward, J. C., and Beevers, R. B. (1958). *Discuss. Faraday Soc.* **25**, 29.

Roviello, A., and Sirigu, A. (1975). *J. Polym. Sci., Polym. Lett. Ed.* **13**, 455.

Roviello, A., and Sirigu, A. (1976). *Proc. Prague Meet. Macromol.* C18-1.

Sackmann, H., and Demus, D. (1973). *Mol. Cryst. Liq. Cryst.* **21**, 239.

Sadron, C. (1962). *Pure Appl. Chem.* **4**, 347.

Saeki, H., Iimura, K., and Takeda, M. (1972). *Polym. J.* **3**, 414.

Schaefgen, J. R., Foldi, V. S., Logullo, F. M., Good, V. H., Gulrich, L. W., and Killian, F. L. *Polym. Prepr., Am. Chem. Soc., Div. Polym. Chem.* **17** (*2*) 69.

Schütz, W. (1936). "Magnetooptik," Akad. Verlagsanstalt, Leipzig.

Shibaev, V. P., Freidzon, J. S., and Plate, N. A. (1976). *Dokl. Akad. Nauk SSSR* **227**, 1412.

Singler, R. E., Hagnauer, G. L., Schneider, N. S., Laliberte, B. R., Sacher, R. E., and Matton, R. W. (1974). *J. Polym. Sci., Polym. Chem. Ed.* **12**, 433.

Singler, R. E., Schneider, N. S., and Hagnauer, G. L. (1975). *Polym. Eng. Sci.* **15**, 321.

Skoulios, A., and Finaz, G. (1962). *J. Chim. Phys. Phys.-Chim. Biol.* **59**, 473.

Statton, W. O. (1959). *Ann. N.Y. Acad. Sci.* **83**, 27.

Stein, R. S., and Wilson, P. R. (1962). *J. Appl. Phys.* **33**, 1914.

Stein, R. S., Rhodes, M. B., and Porter, R. S. (1968). *J. Colloid Interface Sci.* **27**, 336.

Stinson, J. W., and Litster, J. P. (1970). *Phys. Rev. Lett.* **25**, 503.

Strzelecki, L., and Liébert, L. (1973). *Bull. Soc. Chim. Fr.* **2**, 597, 605.

Symposium on High Modulus of Wholly Aromatic Fibers (1973). (W. B. Black, ed.). *J. Macromol. Sci. Chem.* *A*7, 3–348.

Symposium on the Structure of Amorphous Polymers (1974). (G. Allen and S. E. B. Petrie, eds.). *Polym. Prepr., Am. Chem. Soc., Div. Polym. Chem.* **15**(2), 1–45.

Symposium Honoring R. W. Morgan (1976). *Polym. Prepr., Am. Chem. Soc., Div. Polym. Chem.* **17**(2), 53–77.

Tanaka, Y., Kabaya, S., Shimura, Y., Okada, A., Kurihara, Y., and Sahakibara, Y. (1972). *J. Polym. Sci., Part B* **10**, 261.

Tonelli, A. E., Abe, Y., and Flory, P. J. (1970). *Macromolecules* **3**, 303.

Toth, W. J., and Tobolski, A. V. (1970). *J. Polym. Sci., Part B* **8**, 289.

Turner Jones, A. (1964). *Makromol. Chem.* **71**, 1.

Vainshtein, B. K., and Chistyakov, I. G. (1964). *Sov. Phys. —Dokl.* **8**, 1044.

Van Aartsen, J. J. (1971). *In* "Polymer Networks" (A. J. Chompf and S. Newman, eds.), p. 307. Plenum, New York.

Wendorff, J. H., Perplies, E., and Ringsdorf, H. (1975). *Prog. Colloid Sci.* **57**, 272.

Wendorff, J. H., Finkelmann, H., and Ringsdorf, H. (1977). *Am. Chem. Soc. Div. Polym. Chem.* **18** (*2*), 5.

Yeh, G. S. Y. (1972). *Pure Appl. Chem.* **31**, 65.

Yeh, G. S. Y., and Geil, P. H. (1967). *J. Makromol. Sci., Phys.* *B*1, 235.

Zimm, B. H. (1946). *J. Chem. Phys.* **14**, 164.

2

Intramolecular Orientational Order and Properties of Polymer Molecules in Solution

V. N. Tsvetkov
E. I. Rjumtsev
I. N. Shtennikova

Institute of Macromolecular Compounds
Academy of Sciences of USSR
Leningrad, USSR

I. Introduction

The great progress made in recent years in the investigations of the structure of crystalline polymers is well known. A favorable factor in these investigations is the possibility of using chain molecules of regular structure and obtaining polymers based on them as single crystals with a distinct three-dimensional order similar to that of common low molecular weight crystalline bodies (Geil, 1963).

Details of the structure of amorphous polymers are not so well known since they do not exhibit a three-dimensional coordination order and, hence, direct methods of investigation of molecular packing, such as X-ray analysis, are not very effective.

At the same time, if we take into account the chain structure of molecules of linear polymers, it might be expected that in the amorphous state, they should exhibit a mutual order in the arrangement of neighboring molecules similar to that existing in amorphous nonpolymeric liquids with linear chain molecules (Stewart, 1929a–c, 1930). Nevertheless, in contrast to crystalline bodies characterized by long range order involving macroscopic volumes of substance, in amorphous liquids and unoriented polymers there is a short range order, i.e., a local order (Frenkel, 1945) involving only the immediate vicinity of the molecule or a part of it. Moreover, since chain molecules are characterized by a specific structure, it is natural to assume that this local order is mainly the orientational and mono-dimensional order with the axis of symmetry parallel to the long axis of the molecule (or of its part). When an amorphous polymer undergoes strong unidirectional stretching, its local order changes to long range uniaxial orientational order. In this case the polymer structure is similar to that of a homogeneously oriented mesomorphic substance (liquid crystal).

In some cases polymers can attain a state similar to that of a liquid crystal without any external orienting force by spontaneous parallel orientation of molecules in concentrated polymer solutions (lyotropic mesomorphism).

The nature of lyotropic liquid crystals may be very complex since these phenomena are observed for widely differing systems including colloid solutions (Ostwald, 1931), soap solutions (Lawrence, 1938), and living tissues. Among biological lyotropic mesophases, concentrated solutions of tobacco mosaic virus, DNA, and polypeptide molecules in helical conformation are well known (Robinson, 1956, 1961).

Recently great attention has been devoted to the lyotropic mesomorphism displayed by concentrated solutions of aromatic polyamides (Kalmykova et al., 1971; Papkov et al., 1973a–c) due to their importance in the production of synthetic fibers (Black, 1973). Polymer liquid crystals of this type are also very interesting because their molecules exhibit the most pronounced specific structure providing a possibility for supermolecular organization with a distinct orientational order. Investigations of conformational properties of molecules of aromatic polyamides have shown that they are characterized by intramolecular orientational order with a high degree of organization (Tsvetkov, 1976). This means that in this case the following general principle known for low molecular weight thermotropic liquid crystals is valid: only those substances can attain the mesomorphic state in which the molecules are of a "rodlike" shape. This is due to the presence of conjugated bonds and aromatic rings included in the chain in the para position (Gray, 1962).

Investigations of conformational properties and intramolecular order of "comblike" polymer molecules lead to the same conclusion; this will be considered in the next section in greater detail.

Thus, the investigation of intramolecular organization of macromolecules is important not only for obtaining information on their structure and conformation but also because this information may eventually be used in the formation of polymer substances with a high degree of orientational order of the mesomorphic type.

Hence, investigations of supermolecular order in polymers should be combined with investigations of order on intramolecular level. The method that was used successfully for the solution of these problems is the investigation of hydrodynamic, optical, and electrooptical properties of polymer molecules in dilute solutions.

II. Orientational–Axial Order in Chain Molecules

The degree of the uniaxial orientational order in a mesomorphic substance is determined by the value (Tsvetkov, 1942)

$$Q = (3\overline{\cos^2 \vartheta} - 1)/2 \qquad (1)$$

where ϑ is the angle of divergence of the long axis of the molecule from the direction of the predominant orientation (from the director) in its thermal rotational vibrations about this direction. The value of $\overline{\cos^2 \vartheta}$ is obtained by averaging over orientations of all molecules.

The existence of the orientational long range order in a nematic liquid crystal leads to the appearance of macroscopic anisotropy of the substance with respect to a number of properties: optical, magnetic, electrical, etc. Hence, measurements of the birefringence of the liquid crystal (Freeeriksz and Tsvetkov, 1934; Tsvetkov *et al.*, 1973a; Rjumtsev *et al.*, 1973; Kovshik *et al.*, 1975), its diamagnetic anisotropy (Tsvetkov, 1939; Tsvetkov and Sosnovski, 1943), or dichroism (Maier and Markau, 1961) can serve as a direct method for the determination of the value of Q.

In a similar manner, the concept of the degree of intramolecular orientational order in a polymer molecule may be introduced.

The molecule of a linear polymer is a chain consisting of consecutively joined elements and the h parameter, the end-to-end vector, is usually introduced to characterize its conformation. An effective universal model for describing the conformational properties of the chain is the "persistent" or "wormlike" chain (Porod, 1949), a line in space with the curvature identical at all points and depending on the persistence length a. This length is determined by the equation

$$\overline{\cos \psi} = \exp(-L/a) \tag{2}$$

Here L is the length of a part of the wormlike chain, such that the angle between the chain direction at the beginning and at the end of the part is ψ (Fig. 1). The value of $\overline{\cos \psi}$ is the average value of all conformations of the molecule corresponding to a given L.

The contour length L (Fig. 1) of the wormlike chain and the mean square end-to-end distance $\overline{h^2}$ for this chain are related to each other by the equation (Porod, 1949)

$$\overline{h^2}/2aL = 1 - (1/x)(1 - e^{-x}) \tag{3}$$

where x is L/a. It follows from Eq. (3) that over the range of high x, the wormlike chain becomes a Gaussian chain for which $\overline{h^2} = 2aL$ and the

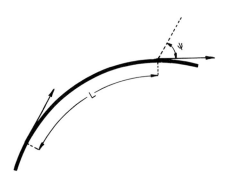

Fig. 1. Persistent curve.

persistence length a is equal to half the length of the Kuhn segment A (Kuhn, 1934, 1936). At low x, it follows from Eq. (3) that h is equal to L, i.e., the conformation of the wormlike chain is linear. The persistence length a (like A which is equal to $2a$) is a measure of the equilibrium rigidity of the polymer chain and characterizes the mean coiling of the chain molecule in a dilute solution under θ conditions (Flory, 1953). For most molecules of well-known synthetic polymers, the values of the Kuhn segment fall in the range of 15–30 Å. These molecules are usually called "flexible-chain molecules." However, polymer chains are known with A values of hundreds and even thousands of angstroms. We shall call them "rigid-chain molecules."

Due to the chain structure of the polymer molecule, its elements are predominantly oriented in the direction of the h vector. Actually, if ϑ is an angle formed by a chain element ΔL and the direction h (Fig. 2), then for any chain conformation corresponding to a given value of h, the average value of $\overline{\cos \vartheta}$ for all chain elements is given by

$$\overline{\cos \vartheta} = h/L \tag{4}$$

This means that for any finite h, the value of $\overline{\cos \vartheta}$ differs from zero and, hence, the direction h is the direction of the preferential orientation of chain elements.

The degree of ordering Q may serve as a measure of the orientational–axial order in the chain molecule. It is determined in the same way as liquid crystals from Eq. (1) where ϑ denotes an angle formed by the chain element and the h axis; however, the averaging of $\overline{\cos^2 \vartheta}$ is made not only over all chain elements but also over all chain conformations at a given value of L.

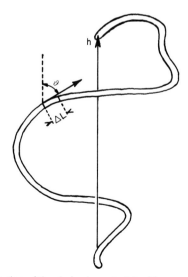

Fig. 2. Orientation of the chain elements ΔL with respect to the h vector.

A statistical method developed by Kuhn (Kuhn and Grün, 1942) makes it possible to express $\overline{\cos^2 \vartheta}$ as a function of h/L:

$$\overline{\cos^2 \vartheta} = \frac{1 - 2(h/L)}{\mathscr{L}^*(h/L)} \tag{5}$$

where $\mathscr{L}^*(h/L)$ is the inverse Langevin function introduced by Kuhn.

If the polymer molecule is modeled by a wormlike chain, then in order to find the value of Q corresponding to this chain, the value of $\overline{\cos^2 \vartheta}$ in Eq. (5) should be averaged over all equilibrium chain conformations (i.e., over all h at a given L). As a result, by using Eq. (3) it is possible to transform Eq. (1) into Eq. (6) (Tsvetkov, 1965a):

$$Q(x) = \frac{\frac{6}{5}[1 - ((1 - e^{-x})/x)]}{\{x - 0.8[1 - ((1 - e^{-x})/x)]\}} \tag{6}$$

At low values of x ($x \ll 1$), we have

$$Q(x)_{x \ll 1} = 1 - \frac{5}{9}x + \frac{85}{324}x^2 - \frac{170}{1458}x^3 + \cdots \tag{6a}$$

At high values of x ($x \gg 1$), we have

$$Q(x)_{x \gg 1} = 6/5x \tag{6b}$$

Equation (6) shows that the degree of the intramolecular orientational–axial order in the chain molecule is a unique function of the x parameter. This function is plotted in Fig. 3 as curve 1. The value of Q is unity (ideal order) at $L \ll a$, i.e., for a rodlike molecule, and decreases with the increase in the chain length L or with the decrease in its equilibrium rigidity (i.e., in the value of a) attaining the asymptotic limit 0 (at, $x \to \infty$).

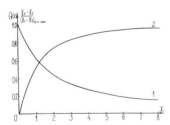

Fig. 3. Degree of orientational–axial order Q (curve 1) and relative optical anisotropy $(\gamma_1 - \gamma_2)/(\gamma_1 - \gamma_2)_{x \to \infty}$ (curve 2) versus parameter $x = L/a$ for a wormlike chain.

The decrease in Q with increasing L reflects the fact that in the chain molecule the correlation between mutual orientations of consecutive units decreases with increasing distance between them, and the lower the value of a the faster is this decrease.

Obviously, when chains of rigid-chain polymers (high a) are of the same length, they are characterized by a high degree of orientational order when compared to flexible-chain molecules (low a). Hence the value of Q for the chain molecule may be calculated from Eq. (6) if one knows the molecular weight M ($L = \lambda M/M_0$, where M_0 is the molecular weight of the monomer unit and λ is its length in the chain direction) and the persistence length a. Since M and a can be determined experimentally by well-known hydrodynamic and optical methods of investigation of dilute polymer solutions (Tsvetkov et al., 1964a; Tsvetkov, 1969; Yamakawa, 1971), it is also possible to determine Q.

III. Orientational Order and Optical Anisotropy of the Chain

Among the characteristics of a molecule, its optical anisotropy is the most direct and sensitive measure of the axial ordering of its structural elements. If in a system consisting of P optically anisotropic elements (each of them being characterized by the difference in the two main polarizabilities $a_1 - a_2 = \Delta a$) these elements become preferably oriented with respect to a certain axis, this system becomes optically anisotropic "as a whole" and the difference between its polarizabilities in the chain direction and in the direction normal to it is given by

$$\gamma_1 - \gamma_2 = P\,\Delta a\,Q = (M/M_0)\,\Delta a\,Q \tag{7}$$

where Q is the degree of the orientational order in the system determined by Eq. (1).

In the case of the polymer molecule, P is the degree of polymerization, Δa the anisotropy of the monomer unit, and $\gamma_1 - \gamma_2$ the anisotropy of the molecule, i.e., the differences between its main polarizabilities in the h direction (the optical axis) and in the direction normal to it. If the molecule is modeled by a wormlike chain, it follows that

$$P\,\Delta a = L\beta = ax\beta = \tfrac{1}{2}(\alpha_1 - \alpha_2)x = \tfrac{1}{2}\Delta a\,Sx \tag{8}$$

where β is the anisotropy per unit chain length, $\alpha_1 - \alpha_2$ the anisotropy of the Kuhn statistical segment, and S the number of monomer units in a segment. In this case, Q is expressed by Eq. (6).

A combination of Eqs. (6)–(8) shows that in the limiting cases, at low and high x, the anisotropy of the wormlike chain is given by

$$(\gamma_1 - \gamma_2)_{x \to 0} = a\beta x = \beta L \tag{9}$$

i.e., it increases proportionally to the chain length, and

$$(\gamma_1 - \gamma_2)_{x \to \infty} = \tfrac{6}{5}a\beta = (\tfrac{3}{5})(\alpha_1 - \alpha_2) \tag{10}$$

i.e., it attains the asymptotic limit equal to $\tfrac{3}{5}$ of the segmental anisotropy.

The dependence of $(\gamma_1 - \gamma_2)/(\gamma_1 - \gamma_2)_{x \to \infty}$ on x for a wormlike chain is shown in Fig. 3 by curve 2.

It follows from Eq. (7) that the specific anisotropy of the macromolecule $(\gamma_1 - \gamma_2)/M$ is equal to the degree of intramolecular order Q multiplied by the specific anisotropy $\Delta a/M_0$ of a chain element (of the monomer unit or Kuhn's segment). Consequently, curve 1 in Fig. 3 represents the variation of the specific anisotropy of the wormlike chain, divided by the constant factor $\Delta a/M_0$, as a function of the parameter x. Since the specific anisotropy of the molecular unit is usually known from its chemical structure, it follows that in order to find Q, it is sufficient to determine $(\gamma_1 - \gamma_2)/M$. Flow birefringence of a dilute solution of macromolecules is a universal, and the most direct, method for determining their optical anisotropy (Tsvetkov, 1964).

IV. Flow Birefringence and Optical Anisotropy of Chain Molecules

In a laminar flow, the chain molecule rotates as a whole under the influence of hydrodynamic forces. Since the average statistical shape of the polymer molecule is not spherical (Kuhn et al., 1951; Solc and Stockmayer, 1971), its rotation in the flow is not uniform and this leads to a preferred orientation of long geometrical axes of molecules at an angle α (the orientation angle) to the flow direction. The direction of the preferred orientation is the axis of optical anisotropy that appears in solution as a result of the orientation of polymer molecules. The sign of flow birefringence (FB) of solution coincides with that of the anisotropy of the chain molecule, since on the average its major geometrical axis coincides with its optical axis (i.e., with the h direction). This rule is also obeyed for low molecular weight liquids for which FB is always positive, since for low molecular weight substances the direction of the greatest geometrical length of the molecule coincides with the direction of its greatest optical polarizability (Silberstein, 1927a–c). However, in polymer solution FB may be either positive or negative, i.e., either the highest or the lowest value of polarizability of the chain molecule may correspond to the h direction. The lowest value is observed when the polarizability of the monomer unit in the direction of the main chain is lower than in the direction normal to it (negatively anisotropic monomer unit, $\Delta a < 0$). This is usually observed for polymers with bulky side groups. Hence, not only the value but also the sign of FB is a sensitive indicator of the microstructure of the polymer chain and of the monomer unit (Tsvetkov et al., 1964a; Tsvetkov, 1964).

From measurements of FB of solutions one obtains experimentally the characteristic values of flow birefringence $[n]$ and the orientation angle $[x/g]$. The former is determined from the equation

$$[n] \equiv \lim_{g \to 0, c \to 0} (\Delta n/gc\eta_0)$$

where Δn is the difference between the two main refractive indices in solution at the concentration C and at the flow gradient g and η_0 the viscosity of the solvent. The characteristic orientation angle is

$$\left[\frac{\chi}{g}\right] \equiv \lim_{g \to 0, c \to 0} \left[\frac{(\pi/4 - \alpha)}{g}\right]$$

where α is the angle determined experimentally between the optical axis of solution and the flow direction.

If hydrodynamic properties of the chain molecule are modeled by a rigid body in which the axis of cylindrical symmetry coincides with the direction of the optical axis of the molecule, then the FB theory (Tsvetkov et al., 1964a; Kuhn et al., 1951) leads to the main relationships (11) and (12)

$$[\chi/g] = (12D_r)^{-1} \tag{11}$$

where D_r is the coefficient of rotatory diffusion of the molecule when its long axis rotates about its short axis, and

$$[n] = 2\pi \mathcal{N}_A (n^2 + 2)^2 (\gamma_1 - \gamma_2) b / 135 n \eta_0 D_r M \tag{12}$$

Here n is the refractive index of the solvent, $b = (p^2 - 1)/(p^2 + 1)$, where p is the ratio of longitudinal dimensions of the molecule to its transverse dimensions, and $\mathcal{N}_A$ is Avogadro's number.

These two equations should be supplemented with a third equation relating the intrinsic viscosity of solution $[\eta]$ to the rotatory diffusion coefficient D_r and the molecular weight M of dissolved macromolecules:

$$[\eta]D_r M \eta_0 = FRT \tag{13}$$

The F coefficient is weakly dependent on the shape of the particles and for two limiting conformations of the chain molecule—the straight rod (Kuhn et al., 1951) and the Gaussian coil (Kirkwood and Riseman, 1948; Riseman and Kirkwood, 1950) ($Q = 1$ and $Q \to 0$)—it is $\frac{2}{15}$ and $\frac{1}{4}$, respectively. On the basis of Eq. (13) and taking into account Eq. (11), the molecular weight of the polymer can be determined from experimental values of $[\eta]$ and $[\chi/g]$.

A comparison of Eqs. (12) and (13) gives

$$[n]/[\eta] = (2\pi/135kTn)(b/F)(n^2 + 2)^2 (\gamma_1 - \gamma_2) \tag{14}$$

Since the b/F coefficient changes only slightly with the chain conformation, Eq. (14) shows that in a homologous series of polymer molecules the dependence of $[n]/[\eta]$ on M virtually reflects the dependence of $\gamma_1 - \gamma_2$ on x shown by curve 2 in Fig. 3.

A comparison of Eqs. (11) and (12) gives

$$[n]/[\chi/g] = Bb(\gamma_1 - \gamma_2)/M \qquad (15)$$

where

$$B = 8\pi \mathcal{N}_A (n^2 + 2)^2 / 45 n \eta_0$$

It follows from Eq. (15) that the ratio $[n]/[x/g]$ is equal to the specific anisotropy of the molecule $(\gamma_1 - \gamma_2)/M$ multiplied by the coefficient Bb and, hence, $[n]/[\chi/g]$ [see Eq. (7)] is proportional to the degree of intramolecular orientational order Q. For a wormlike chain, the dependence of $[n]/[\chi/g]$ on M is obviously determined by curve 1 in Fig. 3.

These equations were used for the interpretation of experimental data from the hydrodynamic and optical properties of molecules of rigid-chain polyamides in solutions.

V. Equilibrium Rigidity of Molecules of Rigid-Chain Polyamides

Investigations of conformational properties of molecules of aliphatic polyamides (nylon 6 and nylon 66) in dilute solutions (Saunders, 1964, 1965; Flory, 1969) have shown that they are typical flexible-chain polymers similar, for example, to polyethylene. This is quite natural since over $\frac{4}{5}$ of the molecule of nylon 6 is formed by the polymethylene chain of high flexibility that (like other carbochain polymers) has considerable freedom of intramolecular rotations about the C—C bonds.

The situation is greatly changed when amide groups in an aliphatic polyamide chain are located very close to each other. This is observed for poly(alkyl isocyanate) (nylon 1, —CONR—) in which the chain consists entirely of amide groups in the head-to-tail sequence and the H at N is replaced by an alkyl radical R. It is known (Pauling, 1960) that high resonance energy in the amide group leads to the quasiconjugation and coplanarity of its structure that in nylon 1 extends to the whole chain, because amide groups are in direct contact with each other. Steric interactions between the radical R and the neighboring carbonyl oxygen may lead to incomplete coplanarity of the chain (Tsvetkov et al., 1975a). Nevertheless, its structure retains regularity with a high degree of orientational order; the amide groups in the chain are sequenced in a regular alternation of cis and trans configurations. A probable structure of the poly(alkyl isocyanate) chain is shown in three projections in Fig. 4. The conjugation energy stabilizing this structure and virtually excluding a possibility of rotation about the C–N bond favors higher equilibrium rigidity of the poly(alkyl isocyanate) chain. Thus, for the poly(butyl isocyanate) chain, the length of the Kuhn segment A is 1000 Å; this corresponds to the number of monomer units in a segment $S = 500$ (Tsvetkov *et al.*, 1968a, 1971a, 1974a). High equilibrium rigidity is shown in a number of specific properties of dilute solutions of these polymers: characteristic dependence of viscosity, translational and rotatory friction, and optical anisotropy on the molecular weight. In particular, the dependence of $[n]/[\eta]$ on M has the shape of curve 2 in Fig. 3 (see also Fig. 14) (Tsvetkov *et al.*, 1971a; Vitovskaia and Tsvetkov, 1976; Tsvetkov *et al.*, 1975b).

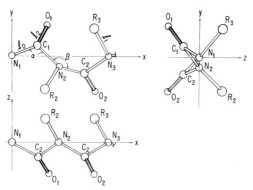

Fig. 4. Probable structure of the poly(alkyl isocyanate) chain. $\alpha = 120°$; $\beta \approx 120°$; $l_0 = 1.4\,\text{Å}$; $l_1 = 1.47\,\text{Å}$; $l_2 = 1.24\,\text{Å}$.

The replacement of an aliphatic side radical by an aromatic (poly(tolyl isocyanate)) or the separation of amide groups along the chain by copolymerization with chloral destroys conjugation in the isocyanate chain and leads to the flexibility corresponding to that of flexible-chain polymers (Tsvetkov *et al.*, 1968a).

In aromatic polyamides (which are of great interest as has already been shown (Black, 1973)), amide groups are separated along the chain by phenyl rings. As a result, C_{al}–C_{ar} and C_{ar}–N bonds appear in the chain, and rotation about them is possible. In principle it might be expected that the equilibrium rigidity of chains of aromatic polyamides should exceed that of polyisocyanate chains.

These expectations were justified by experimental investigations of conformational properties of molecules of poly(*m*-phenylene isophthalamide)

(PMPhIPhA), an aromatic polyamide in which amide groups bonded head-to-head and tail-to-tail are separated by phenyl rings in the meta position (Fig. 5). Sedimentation, diffusion, and viscometry of solutions of fractions of PMPhIPhA in dimethylacetamide (Vitovskaia *et al.*, 1976) have been used to determine the length of Kuhn's segment of this polymer A, which was found to be 50 Å. This corresponds to the number of monomer units (a unit is half of the identity period) in a segment $S = 8.5$, a value characteristic of typical flexible-chain polymers. Similar results have been obtained when FB was investigated for the same PMPhIPhA fractions in sulfuric acid (Tsvetkov *et al.*, 1975c). It was found that over the entire range of M (from 30×10^3 to 300×10^3), the value of $[n]/[\eta]$ is virtually constant ($\approx 30 \times 10^{-10}$ cm sec^2 g^{-1}). This means that the anisotropy of the molecule over this range of M does not change (Eq. (14)) and its conformation is that of a Gaussian coil. If the value of $[n]/[\eta]$ is used quantitatively on the basis of Eqs. (14), (10), and (8) and taking into account that $\Delta a = 41.5 \times 10^{-25}$ cm^3 (see below) (Tsvetkov *et al.*, 1976a), this leads to the value of $S = 8.5$ in accordance with hydrodynamic data for solutions in dimethyl acetamide.

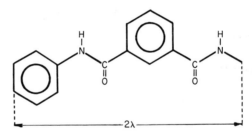

Fig. 5. Chain identity period of poly(*m*-phenylene isophthalamide).

There are some experimental difficulties in the investigation of conformational properties of para-aromatic polyamides in solution because of their poor solubility. This is true for poly(*p*-phenylene terephthalamide) (PPPhTPhA). Its structure (Fig. 6) differs from that of PMPhIPhA in that the phenyl rings in the chain are in the para position. PPPhTPhA is soluble

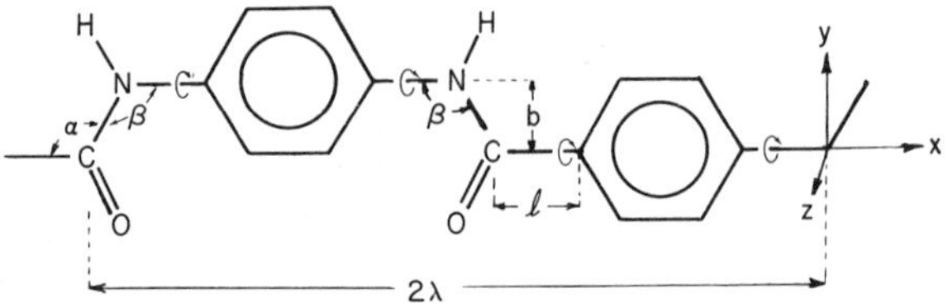

Fig. 6. Chain identity period of poly(*p*-phenylene terephthalamide).

only in concentrated sulfuric acid. Hence, it is virtually impossible to investigate this substance by sedimentation in an ultracentrifuge and the use of the diffusion method is greatly restricted. The method of FB was found to be much more promising (Tsvetkov et al., 1976a). In combination with viscometry (according to Eq. (13)), this method made it possible to determine the molecular weight of a number of samples (with M from 8×10^3 to 60×10^3) and to obtain quantitative information on conformational properties of PPPhTPhA molecules in sulfuric acid. The main results are shown in Fig. 7. White circles show experimental values of $[n]/[\eta]$ versus M. Curve 1 plotted from experimental points corresponds to the theoretical curve 2 in Fig. 3. Its initial slope is $([n]/[\eta])/M_{M\to 0} = 3.4 \times 10^{-12}$ cm sec^2 g^{-2} and its asymptotic limit is $([n]/[\eta])_{M\to\infty} = 4 \times 10^{-8}$ cm sec^2 g^{-1}. According to Eqs. (9) and (14), and taking into account that b is unity and F is 2/15, it can be written:

$$([n]/[\eta])/M_{M\to 0} = (\pi/9kTn)(n^2 + 2)^2(\Delta a/M_0) \tag{16}$$

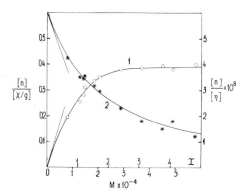

Fig. 7. Values of $[n]/[\eta]$ (curve 1) and $[n]/[\chi/g]$ (curve 2) versus molecular weight M for poly(p-phenylene terephthalamide) in sulfuric acid. Full curves 1 and 2—theoretical dependences: 1, according to Eqs. (6), (7), and (14); 2, according to Eqs. (6), (7), and (15). Points are the experimental data.

Substitution of the experimental value of the initial slope and $M_0 = 119$ into Eq. (16) gives the anisotropy of the molecular unit of PPPhTPhA, $\Delta a = 41.5 \times 10^{-25}$ cm^3. This value is in good agreement with the experimental value since the anisotropy of the benzene ring in the chain is $+30 \times 10^{-25}$ cm^3 (Stuart, 1952) and that of the amide group is close to $+10 \times 10^{-25}$ cm^3 (Tsvetkov et al., 1968a, 1971a, 1974a).

A comparison of Eqs. (10), (14), and (16) shows that when the value of b/F is constant over the entire range of M investigated (this is true for PPPhTPhA), the ratio of the limit of curve 1 to its initial slope is $\frac{3}{5}M_0S$.

Hence, by using the experimental data obtained, we find $S = 165$ and, correspondingly (since $\lambda = 6.5$ Å), $A = \lambda S = 1100$ Å.

Black circles in Fig. 7 represent experimental values of $[n]/[\chi/g]$ versus M for PPPhTPhA. The character of this dependence corresponds to curve 1 in Fig. 3, i.e., it illustrates the change in the orientational order in the PPPhTPhA chain with the increase in its length. Curve 2 represents the theoretical dependence $Q(x)$ plotted from Eq. (6) for a wormlike chain. The scale of x shown in Fig. 7 was chosen to obtain the best pit with the experimental points (taking into account the value of $\Delta a/M_0$ and Eqs. (7) and (15)). A comparison of scales of x and M leads to the value of $S = 200 \pm 20$ that corresponds to the value of $A = 1300 \pm 130$ Å.

Similar investigations have also been carried out for poly(p-benzamide) (PPBA) solutions (Tsvetkov et al., 1975d, 1976b). The molecular structure of this polymer differs from PPPhTPhA in that all amide groups are included regularly in the chain head-to-tail. Although this polymer can dissolve in dimethyl acetamide (with the addition of lithium chloride), experiments have shown (Tsvetkov et al., 1975e) that these solutions are not molecularly disperse and cannot serve for investigations on the molecular level. Consequently, solutions of PPBA in sulfuric acid have been investigated. At relatively low concentration, they are molecularly disperse (Tsvetkov et al., 1975e), and FB was chosen as the method for investigating them. The results can be seen in Fig. 8 showing dependences of $[n]/[\eta]$ on M (curve 1) and of $[n]/[\chi/g]$ on M (curve 2) for PPBA samples. Here the range of M for the samples investigated is more narrow than in the case of PPPhTPhA, therefore curve 1 does not attain the saturation range. However, the character of the corresponding curves in Figs. 7 and 8 is the same. The initial slope of curve 1 or the intercept of curve 2 on the ordinate make it possible to determine the anisotropy of the monomer unit of PPBA the value of which, 41.5×10^{-25} cm^3, coincides with the value of Δa found for PPPhTPhA. The run of curve 2 representing the dependence $Q(x)$ permits the determination of the number of monomer units in a Kuhn segment $S = 320$ that corresponds to the segment length $A = 2100$ Å.

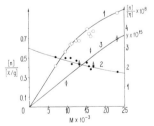

Fig. 8.   Values of $[n]/[\eta]$ (curve 1), $[n]/[\chi/g]$ (curve 2), and $y \equiv K_H[\chi/g]/[n]$ (curve 3) versus molecular weight for poly(p-benzamide) in sulfuric acid.

Another well-known example of polyamide in which the amide groups are separated along the chain by phenyl rings in the para position is poly(p-amide hydrazide) (PPAH, Fig. 9).

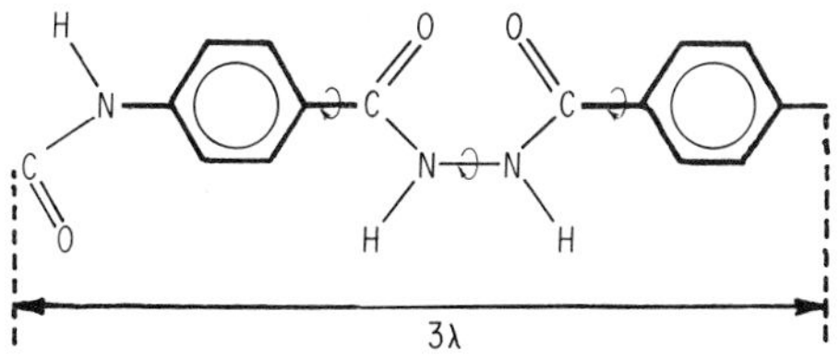

Fig. 9. Chain identity period for poly(p-amide hydrazide).

Investigations of FB for solutions of PPAH samples (with M from 8×10^3 to 40×10^3) in dimethyl sulfoxide allowed the establishment of relationships similar to those in Figs. 7 and 8. The curve of the dependence of $[n]/[\eta]$ on M exhibits the initial slope of 3.1×10^{-12} cm sec^2 g^{-2} and the asymptotic limit of 2.8×10^{-8} cm sec^2 g^{-1}. The ratio of this limit to the initial slope (9×10^3 g) is equal to the molecular weight of a part of the chain with a length of $\frac{3}{5}A$. By using this value and taking into account the fact that the length of that chain part corresponding to the identity period (Fig. 9) is $3\lambda = 15$ Å and its molecular weight is 281, the length of the Kuhn segment for the PPAH molecule, A, is found to be 800 Å.

VI. Structure and Conformation of Molecules of Aromatic Polyamides

The results reported in the previous section show that the way aromatic rings are joined to form molecules of polyamides is of primary importance in the establishment of their conformational properties. Meta-aromatic polyamides are typical flexible-chain polymers, whereas molecules of para-aromatic polyamides are characterized by very high equilibrium rigidity (exceeding a hundredfold the rigidity of flexible-chain polymers) and by the corresponding high degree of intramolecular orientational order Q (see Table 1). At the same time the structure of the amide group is very important for the formation of these properties. This structure is strictly coplanar and in principle can exhibit either trans or cis configuration (e.g., in alkyl iso-cyanates, Fig. 4). If the amide group is in cis configuration, the two bonds adjoining it, about which rotation in the chain is possible, form an angle close to 120° (Fig. 10). When the amide group is in trans configuration, and the valence angles at the nitrogen and carbon atoms are equal (Figs. 5, 6, and 9), these rotating bonds are parallel.

TABLE 1

Degree of Intramolecular Orientational Order Q for Various Polymers[a]

Polymer	$A(\text{Å})$	Range of changes	
		$M \times 10^{-3}$	Q
Poly(γ-benzyl-L-glutamate)	2400	33–330	0.88–0.38
Poly(butyl isocyanate)	1000	15–150	0.76–0.19
Poly(chlorohexyl isocynate)	480	24–240	0.57–0.10
Poly(p-benzamide)	2100	5.5–55	0.86–0.35
Poly(p-phenylene terephthalamide)	1300	5.5–55	0.79–0.22
Poly(p-amide hydrazide)	800	5.5–55	0.69–0.16
Poly(m-phenylene isophthalamide)	50	5.9–59	0.1–0.01
DNA	900	58.3–583	0.72–0.17
Ladder poly(phenyl siloxane)I	200	31–310	0.34–0.04
Ladder poly(phenyl siloxane) II	140	31–310	0.25–0.03
Ladder poly(phenyl siloxane) III	100	31–310	0.19–0.02
Ladder poly(phenyl siloxane) IV	300	31–310	0.44–0.06
Ladder poly(chlorophenyl siloxane)	300	39–390	0.44–0.06
Ladder poly(phenyl isobytylsiloxane) (1:1)	100	28.5–285	0.19–0.02
Ladder poly(phenyl isohexylsiloxane) (1:1)	130	32–320	0.23–0.02
Ladder poly(methyl butenesiloxane) (3:1)	220	29–290	0.36–0.04
Nitrocellulose	230	17.3–173	0.37–0.05
Ethyl cellulose	200	14–140	0.33–0.05
Phenylcarbanilate of cellulose	160	23–230	0.28–0.04
Polystyrene	20	12.5–125	0.04–0.004
Poly(methyl methacrylate)	20	12–120	0.04–0.004
Polyethylene	20	3.3–33	0.04–0.004

[a] The table presents ranges of changes in molecular weight M and degree of order Q corresponding to changes in the chain length from 300 to 3000 Å.

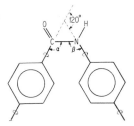

Fig. 10.   Cis configuration of the amide group.

The analysis of experimental data concerning the rigidity of PMPhIPhA ($S = 8.5$) and PPPhTPhA ($S = 200$) molecules suggests that the only stable configuration of the amide group in these polymers is its trans form and that the degree of hindrance to rotation about the C_{ar}–C_{al} and C_{ar}–N bonds in the chain σ^2 is 2.5 (Tsvetkov, 1976). Here $\sigma^2 = S/S_f$ is the ratio of the experimental number of monomer units in the Kuhn segment to their number S_f calculated with the assumption of complete freedom of rotation about the bonds. The value of $\sigma^2 = 2.5$ is very low, it is 2–3 times lower than the values of σ^2 characteristic of typical flexible-chain polymers (Tsvetkov et al., 1964a). This explains high flexibility of the PMPhIPhA chains in which the direction of rotating bonds (when the amide group exhibits trans configuration, Fig. 5) changes by 60° in going from one monomer unit to the next.

In contrast, in molecules of para-aromatic polyamides (when the amide group is in trans configuration and the angles at its C and N atoms are equal (Figs. 6 and 9), the directions of axes of rotation in all units coincide and the chain exhibits a conformation of a "crankshaft" in which all rotation axes are mutually parallel but are not on the same straight line. In this case, irrespective of the degree of hindrance to rotation, the dimensions of the molecule in the direction of the rotation axis (the x axis in Fig. 11) increase proportionally to the degree of polymerization as is observed for all rodlike structures. Transverse dimensions of the molecule are characterized by the projection of its elements on the yz plane (Fig. 11). When random rotations about the x-axis exist in the chain, transverse dimensions of the molecule increase proportionally to the square root of the degree of polymerization (Fig. 11a). When rotations are symmetrically hindered, the φ azimuth changes uniformly along the chain that has the shape of a regular helix (Fig. 11b).

As to the optical properties of the chain shown in Fig. 11, irrespective of the character of internal rotation and of any hindrance to it, they correspond to a system with superior limiting orientational–axial intramolecular order ($Q = 1$). For this order the difference between the two main polarizabilities of the molecule $\gamma_1 - \gamma_2$ is proportional to the degree of polymerization. Experimental data show that for molecules of para-aromatic polyamides, Q is high but smaller than unity and decreases with increasing chain length. Thus, for PPPhTPhA the value of Q changes from 0.71 to 0.22 when M

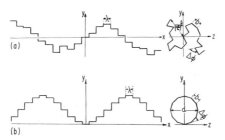

Fig. 11. Chain conformation of a para-aromatic polyamide in two projections: (a) random rotation in the chain; (b) uniform change in rotation azimuth along the chain (symmetrically hindered rotation).

increases from 8,000 to 60,000 (curve 2 in Fig. 8). This corresponds to a high but finite value of $A = 1300$ Å, i.e., to a finite equilibrium chain flexibility.

Finite chain flexibility of para-aromatic polyamide may be caused by two mechanisms: (1) deviation of the amide group from coplanarity due to its deformation during thermal chain motion, and (2) inequality of valence angles α and β at the carbon and nitrogen atoms of the amide group. The first source of flexibility exists in poly(butyl isocyanate) molecules and, as we have seen, leads to the number of monomer units in the Kuhn segment $S_1 = 500$. This exceeds experimental values of S reported for the para-aromatic polyamides investigated. The difference in the values of S and S_1 may be accounted for by the second mechanism. The inequality of angles α and β leads to the appearance of a finite angle $\Delta\vartheta = \beta - \alpha$ between the neighboring rotating bonds in the chain. This leads to the curving of the chain (Fig. 12), and its conformation should be described in terms of the persistent model. According to Eq. (2):

$$\cos \Delta\vartheta = \exp(-\lambda/a_f) = \exp(-2/S_f) = 1 - 2/S_f + \cdots \tag{17}$$

where a_f and S_f are the persistence length and the number of monomer units in a segment, respectively, for a persistent model with complete rotational freedom and λ is the length of the monomer unit. Taking into account the hindrance to rotation, the value of S_2 determined by the finite character of angle $\Delta\vartheta$ is $S_2 = \sigma^2 S_f$. For PPPhTPhA this gives $S_2 = 2.5S_f$. Assuming that the above two mechanisms of flexibility are additive, we find S_2 from the equation

$$1/S_2 = (1/S) - (1/S_1) \tag{18}$$

which for PPPhTPhA ($S = 200$) leads to $S_2 = 330$ and, hence, to $S_f = 330/2.5 = 132$. The substitution of $S_f = 132$ into Eq. (17) gives $\Delta\vartheta = 10°$. This value agrees with the limits of possible values of $\Delta\vartheta = 6° - 12°$ according to data in the literature (Shmueli *et al.*, 1969).

The equilibrium rigidity of the PPBA chain ($S = 320$) markedly exceeds that of the PPPhTPhA molecules. The only difference in the structure of molecules of these two polymers is the inclusion of amide groups into the

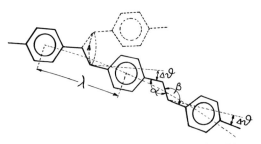

Fig. 12.   Bent chain of para-aromatic polyamide.

chain in the head-to-tail order for PPBA and in the alternating order for PPPhTPhA. This should lead to the existence of the longitudinal component of the dipole moment correlated along the chain in the PPBA molecule and to its absence in PPPhTPhA. It seems possible that the dipole-ordered structure of the PPBA chain favors its greater geometrical and optical order. It is possible that in the PPBA molecule the conjugation in the carbonyl–nitrogen system is transmitted through every phenyl ring increasing the degree of hindrance to rotation about the para-aromatic axes of the chain. Quantitative evaluation of the latter effect from Eqs. (17) and (18) by using angle $\Delta\vartheta = 10°$ and the value of $S_1 = 500$ shows that the value of $S = 320$ found experimentally for PPBA corresponds to the degree of hindrance $\sigma^2 = 6.8$—the value characteristic of typical flexible-chain polymers (Tsvetkov *et al.*, 1964a).

The equilibrium rigidity of PPAH chains is lower than that of the PPPhTPhA molecule. This may be caused both by lower regularity of its chemical structure and by the fact that the rotating bond in its chain (—N—N—) is three times shorter than the two adjacent bonds (Fig. 9).

Although the equilibrium rigidities of molecules of para-aromatic polyamides related to details of their structure exhibit some differences, all these are characterized by the crankshaft conformation responsible for their specific morphological properties. All these molecules exhibit mainly regular alternation of bonds with a relatively free rotation and bonds with a fixed coplanar trans structure. It is these peculiarities of structure that give rise to unique physical properties of para-aromatic polyamides and the properties of fibers with ultrahigh modulus obtained from them.

A similar situation is well known for thermotropic liquid crystals in which normal aliphatic acids even with a long alkyl chain are not nematogenic, whereas in aliphatic diene acids the stable nematic mesophase is readily attained (Maier and Markau, 1961).

Table 1 shows the values of Q, the degree of intramolecular orientational–axial order determined experimentally (curves in Figs. 7 and 8) for several rigid-chain polymers. For thermotropic nematic liquid crystals, experimental values of Q usually range from 0.4 to 0.7 (Tsvetkov, 1942; Tsvetkov and Sosnovski, 1943). For the investigated para-aromatic polyamides, the values of Q are in the same range and even exceed its upper limit. Consequently, it is quite reasonable to call these molecules crystallike molecules.

VII. Equilibrium Rigidity and Intramolecular Order of Some Rigid-Chain Polymers

The conjugation in the chain which is characteristic of molecules of polyamides just considered is not the only mechanism favoring the increasing rigidity and intramolecular order of the chain molecule. Another method of increasing the chain rigidity is cyclization and the corresponding decrease in the possibility of rotation about valence bonds. Cellulose derivatives are a classical example of rigid-chain polymers with chains consisting of glucoside rings. Additional cyclization may be due to intramolecular hydrogen bonds. According to data by several authors (Burchard, 1965; Tsvetkov *et al.*, 1973b, 1974b; Lavrenko *et al.*, 1974; Rjumtsev *et al.*, 1975a,b), the length of segments of various cellulose derivatives amounts to hundreds of angstroms (Table 1), i.e., they correspond to the equilibrium rigidity of rigid-chain polymers. These data are supported by theoretical calculations of conformations of various polysaccharides (Yahindra and Rao, 1970a,b, 1971a,b, 1972; Burchard, 1971).

The introduction of rings into the main chain of synthetic polymers can also increase their rigidity but only if the rings are not separated by a considerable amount of bonds making rotation possible. Thus, the rigidity of the polycarbonate chain containing such bonds virtually does not differ from that of typical flexible-chain polymers (Berry *et al.*, 1967; Tsvetkov *et al.*, 1966; Garmonova *et al.*, 1971).

For attaining a pronounced increase in rigidity, the chain should be almost completely cyclized, and as a result the molecule acquires a "ladder" structure. Ladder polymers of high solubility with a wide range of molecular weights have been synthesized on the basis of double chain poly(silsesquioxanes) with aliphatic and aromatic side groups (Andrianov *et al.*, 1965; Andrianov, 1969, 1971) (Fig. 13). Detailed investigations of their solution properties (Tsvetkov *et al.*, 1969a, 1971b, 1973c, 1975f; Tsvetkov, 1972a) have shown that molecules of these polymers are rigid-chain molecules and the corresponding lengths of segments attain several hundred angstroms (Table 1).

Optical anisotropy (and, correspondingly, FB) of molecules of ladder phenyl siloxanes is high and negative in sign. This indicates that the planes

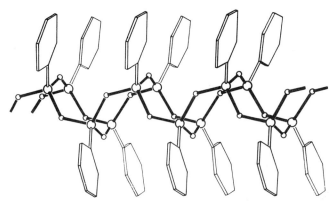

Fig. 13.   Structure of molecules of ladder poly(phenyl siloxane).

of side phenyl rings exhibit orientational order normal to the direction of double chain. In an ideally ordered ladder molecule, intramolecular motion in the main chain is impossible without a distortion in the values of valence angles and lengths of valence bonds. Hence, the fact that experimental values of segment length for ladder poly(siloxanes) are finite (and even much lower than for aromatic polyamides) means that the flexibility of these molecules is due to the deformation of valence angles and bonds of their double-chain network during its thermal vibrations. This mechanism differs widely from that in linear polymers (with single chain) in which the chain is curved due to internal rotations about valence bonds without the deformation of valence angles.

Macromolecules with a ladder structure can also be obtained from purely organic compounds (such as various heterocyclic compounds) (De Winter, 1966; Overberger and Moore, 1970) but investigations of these polymers on a molecular level are virtually absent because of their poor solubility.

Molecules with completely cyclized chains include molecules with a regular secondary structure: DNA (Watson and Crick, 1953), and polypeptides in a helical conformation (Doty, 1956). In this case the cyclization in the chain is caused by hydrogen bonds ensuring the rigidity of the molecular form.

The optical anisotropy of DNA molecules determined from FB is very high and negative in sign ($\Delta a = -190 \times 10^{-25}$ cm^3, $\alpha_1 - \alpha_2 = 5 \times 10^{-21}$ cm^3). This corresponds to a high degree of orientational order of purine and pyrimidine bases that are rigidly bonded to the chain and are normal to its axis (Tsvetkov, 1964; Tsvetkov et al., 1965). The segment length is also large: $A = 900$ Å (Hearst and Stockmayer, 1962; Crothers and Zimm, 1965; Gray et al., 1967).

A still higher equilibrium rigidity is characteristic of helical molecules of the best known polypeptide poly(γ-benzyl-L-glutamate): $A = 2400$ Å (Shtennikova, 1973). Hence, FB in solutions of this polymer and the segmental optical anisotropy of molecules are also very high ($\alpha_1 - \alpha_2 = 22 \times 10^{-22}$ cm^3), although the anisotropy of the monomer unit is low ($\Delta a = 15 \times 10^{-25}$ cm^3). This circumstance and also the positive sign of the anisotropy show that the rigidity and the orientational order of side groups are much lower than those of the main chain of the helical molecule.

All these rigid-chain polymers are characterized by the molecular weight dependence of the optical anisotropy of molecules and correspondingly, of the $[n]/[\eta]$ value of their solutions. The corresponding experimental data are shown in Fig. 14. The general trend of curves in Fig. 14 is similar to that of curve 2 in Fig. 3 and curve 1 in Figs. 7 and 8. Therefore they illustrate the decrease in the degree of intramolecular order with increasing molecular weight. These dependences were used to determine the optical anisotropy, the equilibrium rigidity, and the intramolecular order of these polymers.

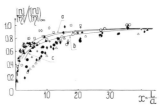

Fig. 14.   Values of $y \equiv ([n]/[\eta])/([n]/\eta])_\infty$ versus $x = L/a$ for some rigid-chain polymers; points are the experimental data : ($\bigcirc$)—nitrocellulose ; ($\times$)—poly(γ-benzyl-L-glutamate) ; ($\bullet$)— ladder poly(phenyl siloxane) I ; ($+$)—ladder poly(phenyl siloxane) II ; ($\triangle$)—ladder poly(phenyl siloxane) III ; ($\blacklozenge$)—ladder poly(phenyl isobutylsiloxane) and poly(phenyl isohexylsiloxane); ($\square$)—ladder methylbutenesiloxane; ($\blacktriangle$)-poly(butyl isocyanate); ($\diamondsuit$)—poly(chlorohexyl iso-cyanate). Curves a,b, and c are plotted according to modern concepts (see Tsvetkov, 1974).

Table 1 permits a comparison to be made between degrees of the orientational order Q for various chain molecules. For this purpose, apart from the segment length, the table presents range of changes in M and Q corresponding to the change in the chain length from 300 to 3000 Å. Even at low M, the values of Q for typical flexible-chain polymers of the vinyl series are lower by one or two orders of magnitude than those characteristic of nematic liquid crystals. This low degree of orientational order can be detected and measured only by using such a sensitive property as optical anisotropy determined from FB.

The values of Q for ladder polymers and cellulose derivatives are much higher; in the region of low molecular weights, they approach the inferior limit of Q values corresponding to the nematic mesophase.

The highest degree of intramolecular order is characteristic of the molecules of poly(γ-benzyl-L-glutamate), DNA, and para-aromatic polyamides. The values of Q for these polymers are the same as for thermotropic mesomorphic substances. This explains the ability of these polymers to form in concentrated solutions a stable lyotropic mesophase (Robinson, 1956, 1961; Kalmykova et al., 1971; Papkov et al., 1973a–c).

Poly(alkyl isocyanates) are also characterized by a high value of Q. Their concentrated solutions presumably do not exhibit lyotropic mesomorphism. This may be attributed to the presence in these polymers of relatively long side groups preventing regular packing of chains into a mesomorphic structure.

VIII. Magnetic Anisotropy and Intramolecular Order of Chain Molecules

In principle, magnetic anisotropy may serve as a measure of the orientational order in a polymer molecule, just as for a nematic liquid crystal (Tsvetkov and Sosnovski, 1943).

Actually, if the difference in magnetic polarizabilities of the monomer unit in the chain direction and normal to it is $\Delta\chi$, then the difference in magnetic polarizabilities of the whole molecule in the h direction and normal to it, $\chi_1 - \chi_2$ (magnetic anisotropy), by analogy with Eq. (7) is given by

$$\chi_1 - \chi_2 = (M/M_0)\Delta\chi\, Q \tag{19}$$

where Q is still determined from Eq. (6).

A classic method for investigating magnetic anisotropy of molecules is birefringence of their solutions in a magnetic field (the Cotton–Mouton effect: MB). According to the theory (Langevin, 1910), the characteristic value of birefringence, the Cotton–Mouton constant $K_H \equiv \Delta n/cH^2$ for molecules with an axial symmetry of magnetic and optical properties, is given by

$$K_H = 2\pi\mathcal{N}_A(n^2 + 2)^2(\gamma_1 - \gamma_2)(\chi_1 - \chi_2)/135kTnM \tag{20}$$

where Δn is the value of birefringence of solution of concentration C in a magnetic field with strength H. Symbols in Eq. (20) correspond to symbols in Eqs. (12) and (14).

It follows from Eq. (20) that for a wormlike chain oriented as a whole in a magnetic field, the K_H value should vary with a change in M proportionally to $(\gamma_1 - \gamma_2)(\chi_1 - \chi_2)/M$, i.e., proportionally to $Q(\chi_1 - \chi_2)$.

Investigations of MB in solutions of chain molecules with magnetically anisotropic phenyl rings in side radicals (polystyrene) (Tsvetkov and Frisman, 1944) or in the main chain (polycarbonate) (Champion *et al.*, 1974) have shown that for these polymers K_H does not depend on M and the value of the effect observed corresponds to values of $\gamma_1 - \gamma_2$, $\chi_1 - \chi_2$, and M in Eq. (20), which are close to the values of Δa, $\Delta\chi$, and M_0 for the monomer unit. It follows from this that polystyrene and polycarbonate molecules in a magnetic field behave as kinetically flexible chains (see below). This means that kinetic units oriented in the field are not molecules as a whole but their parts, the length of which is close to that of the monomer unit. Evidently, the value of K_H measured in solutions of these polymers cannot be used to characterize the orientational order in macromolecules.

Quite different results have been obtained in investigations of MB in PPBA solutions (Tsvetkov *et al.*, 1975e). In these solutions, K_H depends on the molecular weight of the polymer and increases with it. This means that under the action of a magnetic field, orientations of monomer units of the PPBA molecule are correlated along the chain rather than independent of each other. As a result, the molecule rotates in a magnetic field "as a whole," i.e., it behaves as a kinetically rigid chain (see below) and the measured value of K_H depends on the degree of the intramolecular orientational order. This can be proven by comparing the values of MB and FB in solutions of the same PPBA samples.

A comparison of Eqs. (20) and (5) gives

$$K_H[\chi/g]/[n] = \eta_0(\chi_1 - \chi_2)/12kT \tag{21}$$

According to Eq. (21), a combination of magnetooptical and dynamo-optical values $y \equiv K_H[\chi/g]/[n]$ permits the determination of the magnetic anisotropy of the molecule $\chi_1 - \chi_2$. The values of y are shown in Fig. 8 (by slashed circles) versus M for PPBA samples in sulfuric acid. Although the scattering of points is great (owing to experimental difficulties), curve 3, which fits them, is similar to curve 1. This is natural because the dependence of optical and magnetic anisotropy of the wormlike chain on its length is represented by curves 1 and 3, respectively.

The initial slope of the curve of y versus M is (see Eq. (21)) $(\eta_0/12kTM_0)\Delta\chi$. The experimental value of the initial slope of curve 3 is $15 \times 10^{-20} \sec g^{-1}$. Hence, by using the values of $\eta_0 = 0.2$ P and $M_0 = 119$, we find the magnetic anisotropy of the PPBA monomer units $\Delta\chi = 4.5 \times 10^{-29} cm^3$. This co-incides with the diamagnetic anisotropy of the benzene ring with respect to the central axis laying on its plane (according to measurements in crystalline benzene (Lonsdale and Krishnan, 1936)) This means that phenyl rings in the chain are a source of the diamagnetic anisotropy of the PPBA molecules, and the character of curve 3 (similar to curve 1) shows that according to diamagnetic properties (and to optical properties as well) of the PPBA molecules they are crystallike, kinetically rigid molecules with a structure characterized by a high orientational order.

IX. Intramolecular Orientational–Polar Order and Dipole Moment of the Chain Molecule

As a consequence of the chain structure of polymer molecules, in contrast to the nematic mesophase, chain molecules can exhibit the orientational–polar order. This is clear from Fig. 2 (where each ΔL element may be regarded as a vector) and from Eq. (4) in which $\overline{\cos \vartheta}$ is a measure of the orientational order of vectors ΔL with respect to the polar axis $\mathbf{h}$ at a given value of h.

For a wormlike chain characterized by an array of conformations differing in the h value, the value $Q_p \equiv \langle(\overline{\cos \vartheta})^2\rangle$ obtained by averaging of $(\overline{\cos \vartheta})^2$ over all chain conformations may serve as a measure of the orientational–polar order. Hence, taking into account Eqs. (3) and (4), we find

$$Q_p = (2/x)[1 - (1 - e^{-x})/x]$$
$$Q_p = 1 - (x/3) + (x^2/12) - (x^3/60) + (x^4/360) - \cdots \quad \text{at low } x \tag{22}$$
$$Q_p = (2/x)[1 - (1/x)] \quad \text{at high } x$$

Hence, the dependence of Q_p on x resembles the dependence $Q(x)$ represented by curve 1 in Fig. 3 (see also Fig. 17).

The dipole moment of the chain molecule is a physical value directly characterizing the orientational–polar order in it. If the monomer unit of the molecule has the dipole moment μ_0 directed along the chain, it follows that the molecule in a given conformation is characterized as a whole by the dipole moment

$$\mu = P\mu_0 \overline{\cos \vartheta} = (M/M_0)(h/L)\mu_0 \tag{23}$$

the direction of which coincides with the h direction.

Equations (22) and (23) give for the average (overall chain conformations) square of the dipole moment of the molecule $\overline{\mu^2}$:

$$\overline{\mu^2}/M = (\mu_0/M_0)^2 M Q_p = (\mu_0{}^2/M_0)S[1 - (1 - e^{-x})/x] \tag{24}$$

It follows from Eq. (24) that at low x, $(\overline{\mu^2})^{1/2} = \mu_0 P$, i.e., the dipole moment of the wormlike chain increases proportionally to the degree of polymerization. At high x (the Gaussian chain) $\overline{\mu^2}/M = (\mu_0{}^2/M_0)S$, i.e., $\overline{\mu^2}/M$ does not depend on the chain length and is proportional to the number of monomer units in Kuhn's segment. Thus, the general type of the dependence of $\overline{\mu^2}/M$ on M (or on x) resembles the dependence of $\gamma_1 - \gamma_2$ on x shown by curve 2 in Fig. 3.

Equation (24) is also valid when the dipole moment of the monomer unit μ_0 is not directed along the chain but forms with it an angle δ. Nevertheless, in this case the direction of the total dipole moment of the molecule μ may not coincide with the h direction (Kuhn, 1948) (see next section).

Investigation of dielectric polarization (DP) or electric birefringence (EB)—the Kerr effect in polymer solutions—may serve for detection and quantitative determination of dipole moments of polymer molecules. However, an indispensable condition for the success of these experiments is a sufficiently high kinetic rigidity of the polymer chain revealed under the action of an electric field (Tsvetkov, 1972a, 1974).

X. Kinetic Flexibility and Dipole Orientation
of Chain Molecules

The rotational moment to which the total dipole μ of the chain is subjected in an electric field can lead to the orientation of the dipole in the field direction by the rotation of the molecule as a whole (large scale motion). In this case the mechanism of polarization of solution may be called the orientational mechanism. At the same time, since the polymer chain contains polar groups (dipole μ_0), each of them is subjected to the rotational moment in an electric field. If the correlation of rotations of single polar groups is relatively slight, they are oriented in the field virtually independent of each other. As a consequence of these intramolecular rotations (small scale motion), the conformation of the molecule changes, i.e., it undergoes deformation and in this case the mechanism of the polarization of solution may be called deformational. Which of these mechanisms causes the polarization of solutions in every given case depends on the ratio of the time duration necessary for deformation and orientation of the macromolecule. If the orientation time of the macromolecule as a whole τ_0 is shorter than the time necessary for its deformation τ_d, then the orientational mechanism will play a predominant part in polarization and this molecule may be called kinetically rigid. At $\tau_0 > \tau_d$, the solution undergoes deformational polarization and the chain molecule will be kinetically flexible.

Numerous experimental data show (North, 1972) that polar flexible-chain polymers in solutions placed in an electric field undergo deformational polarization, i.e., they behave as kinetically flexible chains. In most cases the kinetic unit oriented in the field independently of other units virtually does not differ from the monomer unit of the chain (Krigbaum and Dawkins, 1974). Consequently, both dielectrical and electrooptical (Le Fevre *et al.*, 1958; Le Fevre and Sundaram, 1963; Aroney *et al.*, 1960) properties of solutions of a flexible-chain polymer are close to those of a solution of the corresponding monomer having equal concentration and do not depend on the degree of polymerization.

Hence, the investigation of dielectrical and electrooptical properties of solutions of flexible-chain polymers cannot yield information concerning the conformational characteristics and orientational order of the structure of their molecules.

Rigid-chain polymers are characterized by quite different electrooptical properties. Experiments show that chain molecules with a relatively high equilibrium rigidity (long Kuhn segments) also exhibit high kinetic rigidity (long deformation time τ_d). Hence, rigid-chain molecules are oriented in an electric field as a whole (the orientational mechanism, characteristic time τ_0), and the equilibrium characteristics and kinetics of the DP and EB processes in solutions of these molecules reflect their dipole, optical, and hydrodynamic properties.

A characteristic feature of solutions of rigid-chain polymers is the dispersion of DP and EB observed at radio frequencies. Figure 15 can serve as an example. It shows the dependence of the dielectric increment $\Delta\varepsilon_\nu$ and the Kerr constant K_ν on the frequency ν of the electric field for fractions of poly(chlorohexyl isocyanate) in tetrachloromethane (Tsvetkov *et al.*, 1971a, 1974a, 1975b). Here $K_\nu \equiv \Delta n_\nu / cE^2$, where Δn_ν is the birefrengence of a dilute solution at a concentration c in a sinusoidal electric field with the frequency

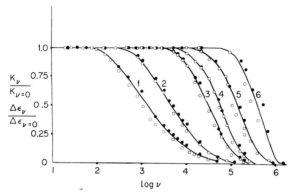

Fig. 15. Values of $K_\nu/K_{\nu=0}$ (O) and $\Delta\varepsilon_\nu/\Delta\varepsilon_{\nu=0}$ (●) versus frequency ν of electric field for fractions of poly(chlorohexyl isocyanate) in tetrachloromethane: $M \times 10^{-4} = 27.4(1)$; 12.5(2); 5.0(3); 3.7(4); 2.3(5); 1.0(6).

ν and the effective strength E. The existence of dispersion of DP and EB in the low frequency range of the field indicates that the dipole polarization proceeds by the orientational mechanism and that its relaxation takes place. Dispersion curves of DP and EB for the same fraction coincide and this shows that mechanisms of both phenomena are identical. The corresponding relaxation times τ calculated according to the Lorentz–Debye equation

$$K_\nu/K_{\nu=0} = \Delta\varepsilon_\nu/\Delta\varepsilon_{\nu=0} = (1 + 4\pi^2\nu^2\tau^2)^{-1}$$

increase drastically with the molecular weight. This shows that the relaxation phenomenon is related to the rotatory motion of polar molecules. This can be proven by calculating the coefficients of rotatory diffusion D_r from the ratio $D_r = (2\tau)^{-1}$ and comparing them with molecular weights M and intrinsic viscosities $[\eta]$ according to Eq. (13). The substitution of experimental values of D_r, $[\eta]$, and M into Eq. (13) permits the calculation of the F coefficients. Their values for fractions of poly(chlorohexyl isocyanate) are shown in Fig. 16 versus M (Tsvetkov et al., 1975b). These values of F are not

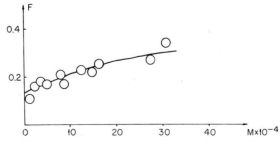

Fig. 16. F coefficient versus molecular weight M for fractions of poly(chlorohexyl isocyanate).

only within the range predicted by the theory but also clearly show the decrease in F with decreasing M in accordance with the parallel change from a random coil shape to a rodlike molecule. Similar results have been obtained for fractions of poly(butyl isocyanate) (Tsvetkov et al., 1971a, 1974a), ladder poly(chlorophenyl siloxane) (Tsvetkov et al., 1971b, 1973c, 1975f), cellulose carbanilate (Tsvetkov et al., 1974b; Lavrenko et al., 1974) and other rigid-chain polymers (Tsvetkov et al., 1973b). This relationship shows that the molecular motion responsible for the Kerr effect and the dielectric polarization in solutions of rigid-chain polymers is similar to the motion responsible for the viscosity of solutions. This motion is the rotation of the kinetically rigid molecule as a whole in an electric field (or a mechanical shear field).

Coefficients of rotatory diffusion D_r with the application of modern theories of rotatory friction of the wormlike chain (Yamakawa, 1971; Hearst, 1963) may be used to determine quantitatively such important structural and conformational characteristics of macromolecules as the length of the monomer unit λ and the number of monomer units S in the Kuhn segment. The data obtained are in agreement with the values of λ and S determined by other methods (light scattering, viscometry, diffusion, and sedimentation). This also confirms the validity of the previous conclusion concerning the character of the molecular motion governing the non-equilibrium electrooptical properties of rigid-chain macromolecules.

XI. Equilibrium Electrooptical Properties of Rigid-Chain Polymers in Solution

Measurements of the dielectric permeability of solutions of rigid-chain polymers under equilibrium conditions (ε_0 at $v \to 0$) and in region of high frequencies (ε_∞ at $v \to \infty$) permit a direct determination of $\overline{\mu^2}/M$ from the well-known Debye equation:

$$\overline{\mu^2}/M = (9kT/4\pi\mathcal{N}_A c)[\{(\varepsilon_0 - 1)/(\varepsilon_0 + 2)\} - \{(\varepsilon_\infty - 1)/(\varepsilon_\infty + 2)\}] \quad (25)$$

Experimental data are shown in Fig. 17 as $\overline{\mu^2}/P$ (where $P = M/M_0$) versus P for poly(butyl isocyanate) (PBIC) (Tsvetkov et al., 1971a, 1974a) (curve 1) and poly(chlorohexyl isocyanate) (PCHIC (Rjumtsev et al., 1975c) (curve 2) in tetrachloromethane. Ratio of the limiting value to the initial slope (according to Eq. (24), it is equal to S) is two times greater for curve 1 than for curve 2 in accordance with the fact that the equilibrium rigidity of the PBIC molecules is twice that of the PCHIC molecules (see Table 1). Curves 3 and 4 show Q_p versus P for PBIC and PCHIC, respectively.

Fig. 17. Value of $\overline{\mu^2}/P$ versus degree of polymerization P for poly(butyl isocyanate) (curve 1) and poly(chlorohexyl isocyanate) (curve 2) fractions in tetrachloromethane. Curves 3 and 4 represent Q_p versus P (from Eq. (24)) for the same polymers, respectively.

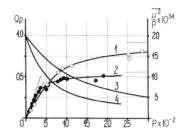

They illustrate a high degree of orientational–polar order in molecules of poly(alkyl isocyanates).

Electric birefringence (EB) is a more complex phenomenon than DP since it depends both on the orientational–polar and on the orientational–axial order in the molecule.

In accordance with the orientational mechanism of this phenomenon, the Kerr constant for solutions of rigid molecules with the axial symmetry of optical polarizability and the dipole–orientational mechanism of EB is given by (Stuart, 1952)

$$K \equiv (\Delta n / cE^2)_{\substack{c \to 0 \\ E \to 0}} = B_0(\gamma_1 - \gamma_2)(\overline{\mu^2}/M)(3\cos^2\theta - 1) \qquad (26)$$

Here θ is the angle formed by the dipole direction in the molecule and the axis of its optical symmetry (the end-to-end distance h). For a system of chain molecules with an array of conformations, $\overline{\cos^2\theta}$ is averaged over all conformations; B_0 is the optical coefficient given by

$$B_0 \equiv \pi \mathcal{N}_A (n^2 + 2)^2 (\varepsilon + 2)^2 / [1215n(kT)^2]$$

Equation (26) clearly shows the difference in the equilibrium properties of dielectric polarization and EB for a solution of the rigid-chain polymer. At a given molecular weight and a given concentration of the solution, the dielectric increment $\Delta \varepsilon$ depends only on the absolute value of the dipole moment of the molecule, whereas the value and the sign of K are also profoundly affected by angle θ. If the dipole of the monomer unit μ_0 of the chain forms an angle δ with the chain, then the roles of its longitudinal ($\mu_{0\parallel} = \mu_0 \cos \delta$) and transverse ($\mu_{0\perp} = \mu_0 \sin \delta$) components in the formation of electrooptical properties of the molecule differ widely. As was shown above, the direction of that part of the total molecular dipole μ, which consists of the sum of longitudinal components of monomer dipoles ($\sum \mu_0 \cos \delta$), coincides with the h vector (Fig. 18a). In contrast, the sum of transverse components ($\sum \mu_0 \sin \delta$) yields the molecular dipole, the direction of which in the Gaussian chain is not correlated with that of h, i.e., $\overline{\cos^2\theta} = \frac{1}{3}$ (Fig. 18b). In this case the correlation between directions μ and h appears only when

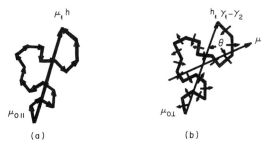

Fig. 18. Direction of dipole moment μ and optical axis γ_1 (or h) in the chain molecule for a component of the dipole moment of the monomer unit parallel (a) and normal (b) to the chain.

L decreases and the chain conformation approaches that of a "curved rod" and may be quantitatively expressed by equations of Langevin chains (Kuhn, 1948). The use of these concepts, Eqs. (3), (6), (7), (24), and (26), and the method reported in reference (Tsvetkov, 1972b) makes it possible to obtain the expression relating the Kerr constant of the solution of the rigid-chain polymer under equilibrium conditions with the parameter $x = L/a$, i.e., with the length and the equilibrium rigidity of the chain

$$K/K_\infty = \frac{[1 - (1 - e^{-x})/x]^2}{1 - (0.8/x)[1 - (1 - e^{-x})/x]} \left\{ 1 - (0.6/x)tg^2\,\delta \right.$$

$$\left. \times \frac{[1 - (1 - e^{-x})/x]^2}{[1 - (1 - e^{-x})/x]\{1 - (0.8/x)[1 - (1 - e^{-x})/x]\}} \right\} \qquad (27)$$

Here K_∞ is the limiting value which is attained in the Gaussian range $(x \to \infty)$

$$K_\infty = B_0\,\tfrac{6}{5}\Delta a(\mu_0{}^2/M_0)S^2\cos^2\delta \qquad (28)$$

At low molecular weights ($x \to 0$), Eqs. (27) and (28) give

$$K_{x\to 0} = B_0 P^2\,\Delta a\mu_0{}^2(3\cos^2\delta - 1)/M_0 \qquad (29)$$

This means that the Kerr constant increases proportionally to the square of the degree of polymerization P^2 as should be expected for a rodlike molecule in which both the dipole moment and the difference in the main polarizabilities $\gamma_1 - \gamma_2$ are proportional to the chain length. In this case the sign of EB may either coincide with that of FB (i.e., with the sign of Δa, see Eqs. (7) and (12)) or be opposite to it depending on the value of angle δ.

In contrast, as Eq. (28) shows, in the Gaussian range EB is caused by the longitudinal component of the dipole moment of the monomer unit $\mu_0\cos\delta$ and the sign of the Kerr constant coincides with that of Δa (since $\cos^2\delta > 0$)

and, hence with the sign of FB. This reveals a general property of Gaussian chains; the coincidence (on the average) of three main directions of the molecule: the direction of its greatest geometrical length, the direction of its orientational–axial order and the direction of its orientational–polar order. This conclusion of the theory is confirmed by all experimental data available on FB and EB in solutions of rigid-chain polymers (Tsvetkov et al., 1971b, 1973b,c, 1974b, 1975b,f; Tsvetkov, 1972a; Lavrenko et al., 1974).

It follows from Eq. (28) that, for a polymer of high molecular weight, K is proportional to S^2. This accounts for a very high value of K for rigid-chain polymers as compared to flexible-chain polymers, since in an electric field the value of S for the latter is approximately 1, whereas for the former it is two orders of magnitude higher.

Equation (27) describes a change in K with increasing molecular weight and with the corresponding transformation of the conformation of the molecule from a straight rod into a Gaussian coil. The first term in Eq. (27) represents the contribution to EB of the longitudinal component of the monomer unit's dipole moment and the second term corresponds to its transverse component. The dependences of $K/K_{\infty,\delta=0}$ on x are shown by curves in Fig. 19 for various values of angle δ. The curves differ not only in their limiting values (proportional to $\cos^2 \delta$) but also in their shape. If

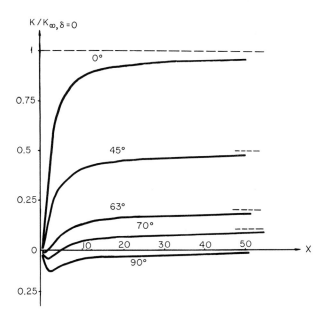

Fig. 19.   Relative Kerr constant $K/K_{\infty,\delta=0}$ versus $x = L/a$ for a rigid-chain polymer. Numbers at curves give angle δ formed by the dipole of the monomer unit and the chain direction.

the dipole μ_0 is perpendicular to the chain ($\delta = 90°$), the Kerr effect is opposite in sign to Δa at all values of x and with increasing x, its absolute value decreases to zero. At all values of δ, the significance of the normal dipole component $\mu_0 \sin \delta$ increases with decreasing x. Consequently, at $\delta > 55°$, EB can change its sign when x is low.

Experimental data (Tsvetkov et al., 1973c, 1974b, 1975b,f) confirm the general type of the dependence of K on x predicted by curves in Fig. 19. This is illustrated in Fig. 20 which shows the dependence of K/K_∞ on x for poly(chlorohexyl isocyanate) (PCHIC), cellulose carbanilate (CC) and ladder poly(chlorophenyl siloxane) (PCPhC). The curves differ not only in absolute value and the sign of the Kerr effect (it is positive for PCHIC and negative for the two other polymers) but also in the steepness of the slope. Experimental data on the relaxational and equilibrium properties of EB for these three polymers permit the determination of the corresponding values of S.

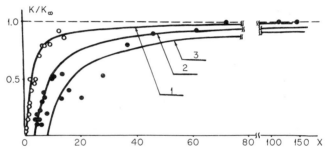

Fig. 20. Relative Kerr constant K/K_∞ versus x for poly(chlorohexyl isocyanate) fractions in tetrachloromethane—curve 1; for cellulose carbanylate in dioxan—curve 2; for cyclolinear chlorophenyl siloxane in benzene—curve 3.

XII. Intramolecular Orientational Order in the Comblike Molecules

A high degree of intramolecular order may exist in chain molecules having a comblike structure. This structure is typical of macromolecules with relatively long (linear) side radicals. In the literature these molecules are usually modeled (Tsvetkov et al., 1964a) by branched chains and for the calculation of their dimensions under θ conditions (Flory, (1953), methods of conformational statistics of noninteracting Gaussian chains are used (Flory, 1969). This modeling and these calculations are permissible if distances between the side groups along the main chain are relatively large or if their length is relatively small.

In contrast, if these conditions are not satisfied, the conformational properties of comblike macromolecules may greatly change due to the interaction of long side groups.

The structure of these molecules is characterized by the fact that although side chains contain the main mass of the molecule, the properties of the molecule as a whole are also determined by the structure of the main chain. In particular, the diphilic nature of the main chain and side chains may exert some effect on the degree of intramolecular order. However, as will be shown below, the length and the chemical structure of side groups are the values essentially governing the intramolecular orientational–axial order of molecules of the comblike type.

The effect of the size and structure of side groups on the conformation of polymer molecules and the equilibrium rigidity of their main chain has been studied systematically for homologous series of poly(alkyl acrylates) (Tsvetkov et al., 1972; Andreeva et al., 1973):

$$-CH_2-CH-$$
$$|$$
$$C=O$$
$$|$$
$$O-C_mH_{2m+1}$$

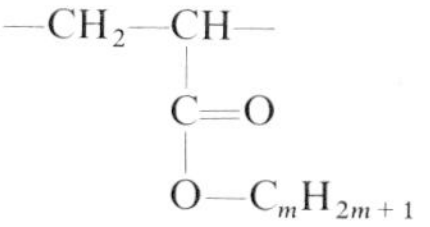

up to poly(octadecyl acrylate), $m = 18$, and of poly(alkyl methacrylates) (Tsvetkov et al., 1969b):

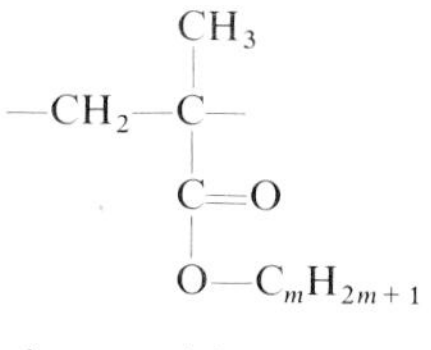

$$CH_3$$
$$|$$
$$-CH_2-C-$$
$$|$$
$$C=O$$
$$|$$
$$O-C_mH_{2m+1}$$

up to poly(cetyl methacrylate), $m = 16$.

The first determinations of the size of comblike molecules by light scattering made for poly(alkylmethacrylates) have shown (Tsvetkov et al., 1964a) that the increase in the length of the ester side group up to poly(octyl methacrylate) inclusive has virtually no effect on the rigidity of the main chain. However, with further increase in the side chain, the size of the molecular coil in solution (the σ^2 parameter, see pp. 58, 59) exhibits a certain increase.

Investigation of translational diffusion and intrinsic viscosity of dilute solutions in homologous series of comblike molecules (Tsvetkov et al., 1972; Andreeva et al., 1973; Tsvetkov et al., 1969b) based on modern theories of hydrodynamic properties of molecules (Stockmayer and Fixman, 1963; Cowie and Bywater, 1965) made it possible to determine the equilibrium rigidity of the main chain of the macromolecule. The experimental material available shows (Table 2) that the increase in the length of the side substituents leads to a two- to fourfold increase in the equilibrium rigidity of the

TABLE 2

Values of the Kuhn Segment A, Segmental Anisotropy $\alpha_1 - \alpha_2$, and the Anisotropy of the Monomer Unit Δa of Some Comblike Macromolecules[a]

Polymer	Monomer unit	m	$A(\text{Å})$	$(\alpha_1 - \alpha_2) \times 10^{25}$ (cm³)	$\Delta a \times 10^{25}$ (cm³)
Esters of polyacrylic acid	$\sim\!-CH_2-CH-\sim$, $C=O$, O, C_mH_{2m+1}	1	21	+16	+1.9
		4	23	−17.8	−1.9
		8	39	−47.9	−3.6
		10	50	−74	−3.7
		12	40	—	—
		16	56	−141	−6.4
		18	72	−190	−6.6
Esters of polymethacrylic acid	CH_3, $-CH_2-C-\sim$, $C=O$, O, C_mH_{2m+1}	1	17	+2	+0.3
		4	17	−14	−2.1
		6	21	−40	−4.6
		8	20	−47	−5.9
		12	30	—	—
		16	44	−160	−8.9

	Structure			
Polycetylvinyl ether	$\sim CH_2-CH-\sim$, $O-C_{16}H_{33}$	60	—	—
Polycholesteryl acrylate		60	-360	-15
Polycetyl ester of n-methacryloyl-oxybenzoic acid		64	-440	-18

^{a} m is the number of carbon atoms in the side radical.

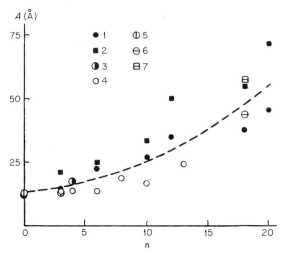

Fig. 21. Values of the Kuhn segment A versus number of bonds in the side chain of poly(alkyl acrylates) (1–3) and poly(alkyl methacrylates) (4–7) according to viscometric data (1, 3–6) and translational friction (2, 7).

main chain. This may be considered as a consequence of the interaction of alkyl side radicals. This result is also clearly seen from Fig. 21 which shows the dependence of the length of the Kuhn segment A on the amount of bonds in the side chain of poly(alkyl acrylates) and poly(alkyl methacrylates) plotted from data on viscometry and translational friction. It is clear that the increase in the length of side groups of the macromolecules investigated is accompanied by a monotonic increase in the equilibrium rigidity of the main chain. The character of this dependence is similar for esters of polyacrylic and methacrylic acids. Data in Table 2 show that this is a common property for a great number of investigated comblike molecules with different structures of side radicals.

Table 2 also shows that the increase in the length of the side radical leads to a decrease in the positive segmental anisotropy $\alpha_1 - \alpha_2$, to a change in its sign, and to an increase in its negative value in passing from methyl to octadecyl esters. Analogous experimental data have been obtained in the series of poly(α-olefins) (Philippoff and Tornqvist, 1968).

It was mentioned on page 50 that the sign of the optical anisotropy of the molecule reflects the microstructure of the polymer chain and of its monomer unit. In this case the change in the anisotropy of the molecules (Table 2) indicates that the side group provides the main contribution to the anisotropy of the molecule and of its monomer unit because its polarizability in the direction normal to the main chain greatly exceeds that in the chain direction.

This change in the optical anisotropy of the molecule may be used to investigate the rigidity and the order of side groups in comblike molecules and to evaluate quantitatively the intramolecular orientational–axial order in these molecules.

Actually, it may be assumed that the monotonic increase in the negative anisotropy of the monomer unit should be slowed down as the side group becomes longer due to its flexibility. Theoretically this regularity was obtained by calculating the differences in the polarizabilities of the side chain $\gamma_\| - \gamma_\perp$ along the axes of its first element (Tsvetkov, 1962, 1965a,b). This calculation, by using the wormlike chain model, leads to the dependence

$$\gamma_\| - \gamma_\perp = \tfrac{1}{3} a\beta(1 - e^{-3x})$$
$$= \beta L[1 - \tfrac{3}{2}x + \tfrac{3}{2}x^2 - \tfrac{9}{8}x^3 + \cdots] \tag{30}$$

where L and β are the length and the anisotropy of the unit length of the wormlike chain, respectively. A comparison of Eq. (30) with Eqs. (7) and (8) for the optical anisotropy of the molecule along the axes of the h vector shows that the trends of dependences $(\gamma_1 - \gamma_2) = f(x)$ and $(\gamma_\| - \gamma_\perp) = f(x)$ are similar: they have a common initial slope in the range of $x \to 0$, but differ in the limiting values at $x \to \infty$ (the limit of the dependence $(\gamma_1 - \gamma_2) = f(x)$ is three times greater).

According to Eq. (30), the contribution of the side radical to the anisotropy of the monomer unit of the comblike molecule is given by

$$\Delta b_L = -\tfrac{1}{2}(\gamma_\| - \gamma_\perp) = -\tfrac{1}{6}a\beta(1 - e^{-6n/v_0}) \tag{31}$$

where n and v_0 are the amounts of valence bonds in the side chain and in its segment, respectively. Equation (31) may be used to analyze conformational properties of some comblike macromolecules.

From experimental data for poly(α-olefins) (Philippoff and Tornqvist, 1968), it is possible to obtain the value of $a\beta = 44 \times 10^{-25}$ cm^3. If the rigidity of the hydrocarbon side chain of poly(α-olefin) is assumed to be equal to that of polyethylene, then, assuming that $v_0 = 16$ (Tsvetkov et al., 1964a), Eq. (31) gives

$$\Delta b_L = -7.3 \times 10^{-25}(1 - e^{-3n/8}) \tag{32}$$

A theoretical curve corresponding to Eq. (32) is shown in Fig. 22 (curve 1). As expected, due to the flexibility of the side chain, with increasing n the anisotropy Δb_L attains saturation determined by the extent of this flexibility. Figure 22 also shows experimental data (Philippoff and Tornqvist, 1968) obtained for a series of poly(α-olefins). They fall on a straight line the slope of which is equal to the initial slope of the curve of Eq. (32); however, in contrast to the theoretical dependence, the experimental curve does not show a tendency toward saturation. This means that the rigidity and the

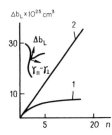

Fig. 22. Contribution Δb_L of the side radical to the anisotropy of the monomer unit of the molecule of poly(α-olefins) versus number of carbon atoms n in the side chain. 1—theory of non-interacting chains that exhibit the flexibility of the polyethylene chain (Eq. (32)). 2—experimental data for Δb_L equal to 2.73×10^{-25} cm^3.

orientational order (determining the value of the anisotropy observed) of side groups in poly(α-olefins) greatly exceed the value that can be obtained from experimental investigations of optical properties of "isolated" hydrocarbon chains in solution.

Quantitative conclusions concerning the rigidity of side groups of comb-like molecules may be obtained from the analysis of experimental data on the optical anisotropy of molecules in a series of poly(alkyl acrylates) (Tsvetkov et al., 1972; Andreeva et al., 1973) and poly(alkyl methacrylates) (Tsvetkov et al., 1969b). The theoretical equation (31) should be compared to experimental values of Δa in the corresponding series of polyacrylates. This comparison is shown in Figs. 23 and 24. The points are experimental

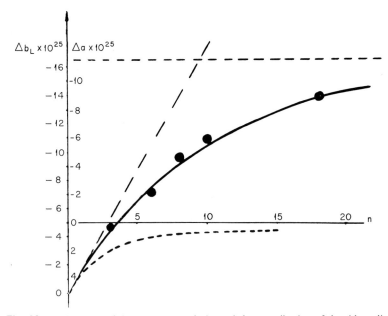

Fig. 23. Anisotropy of the monomer unit Δa and the contribution of the side radical to its anisotropy Δb_L for alkyl esters of polymethacrylic acid versus number of valence bonds in the side chain. Full curve is the theoretical dependence from Eq. (31) at $\beta = 3.3 \times 10^{-25}$ cm^2 and $v_0 = 60$. Dashed curve, from Eq. (31) at $\beta = 3.3 \times 10^{-25}$ cm^2 and $v_0 = 16$.

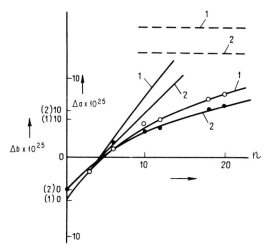

Fig. 24. Anisotropy of the monomer unit Δa (1—toluene, 2—decalin) versus number of valence bonds in the side chain. Full curve is the theoretical dependence from Eq. (31). 1: $\Delta a = 2.6 \times 10^{-25}$ cm^3; $v_0 = 80$; 2: $\Delta a = 2.1 \times 10^{-25}$ cm^3, $v_0 = 80$.

values of the anisotropy of the monomer unit Δa. A comparison of these points to the theoretical dependence in Eq. (31) is made by selecting the values of v_0 and β at which the run of the theoretical curve (31) is in the best agreement with experimental points. Full curve in Fig. 23 gives the value of $v_0 = 60$ for the rigidity of side chains in comblike molecules of polymethacrylic esters. Experimental data on the optical anisotropy in a polymer homologous series of poly(alkyl acrylates) in two solvents (toluene and decalin) give $v_0 = 80$. These values exceed four- to sixfold analogous values of S for free polymethylene chains (e.g., polyethylene) in solution (Tsvetkov et al., 1964a).

The values of the orientational–axial order Q shown in Table 3 correspond to experimental values of the equilibrium rigidity of side chains in comblike

TABLE 3

Degree of Intramolecular Orientational Order Q
in Side Chains of Some Comblike Polymers

Polymer	v_0	x	Q
Poly(cetyl methacrylate)	60	0.66	0.73
Poly(octadecyl acrylate)	80	0.5	0.77
Graft copolymer (main chain: methyl methacrylate; side chain: styrene)	40–80	0.77–0.38	0.69–0.82

molecules. These values of Q are close in their order of magnitude to the corresponding values of Q for the nematic mesophase.

Hence, the experimental data on optical anisotropy of poly(alkyl acrylates) and poly(alkyl methacrylates) indicate that side chains of comblike molecules exhibit an orientational order greatly exceeding the conformational order of flexible polymer chains.

High orientational order in side chains can be attributed to the interaction of alkyl side radicals. The longer are the alkyl side groups, the stronger is this interaction, and it leads both to an increase in the anisotropy of the side chain and to a certain decrease in the equilibrium flexibility of the main chain.

XIII. Intramolecular Order and Equilibrium Rigidity of Macromolecules of Graft Copolymers

Further increase in the length of side chains in comblike molecules (up to hundreds of carbon atoms) may increase the rigidity of the main chain to such an extent that the molecules begin to exhibit properties characteristic of rigid-chain polymers. In particular, as it was pointed out on page 63, that a specific dependence of optical anisotropy on molecular weight appears (Fig. 14). A similar situation is observed for molecules of graft copolymers.

The interest generated by the investigation of structure and physical properties of graft copolymers is quite natural. First, their synthesis provides great possibilities for modifications of physicomechanical properties of high molecular weight materials.

Second, investigations of the size, structure, and configurational properties of macromolecules of graft copolymers may be useful in the study of branched molecules since in principle the method of grafting permits the preparation of required branched structures with a wide variety of types of branching. Actually, by changing the relative lengths of the main and branched chains and the degree of branching, it is possible to synthesize macromolecules with a required distribution of branches.

When molecular structures of this type are investigated, irrespective of the method of their synthesis, it is necessary to determine the type of the copolymer and its composition. The determination of the true structure of its molecules is a more complex problem. Various physical methods have been used to solve it (Tsvetkov *et al.*, 1964a) and the method of flow bire-fringence proved to be the most successful (Tsvetkov *et al.*, 1963, 1964a,b, 1968b; Tsvetkov, 1964). Investigations of optical anisotropy of solutions of graft copolymers provided the only method for the experimental proof of the graft structure of the molecule and for the quantitative determination of the intramolecular orientational–axial order of structures of this type.

Properties of graft copolymers have been studied in greatest detail for model graft copolymers of alkyl methacrylate (the main chain) with styrene (the grafted component). Very peculiar properties of molecules of graft copolymers have been observed in solutions of these graft copolymers.

It has been shown (Tsvetkov *et al.*, 1963, 1964b, 1968b) that the segmental optical anisotropy of molecules of graft copolymer (determined by FB) exceeds many times that of grafted homopolymers. In this case it is positive, i.e., it is opposite in sign to the negative segmental anisotropy of chains of grafted polystyrene although the content of polystyrene in the copolymer exceeds 90%.

Theoretical analysis of experimental data on FB has shown (Tsvetkov *et al.*, 1963, 1964b, 1968b, 1971c) that conformational, optical, and hydrodynamic properties of comblike molecules of graft copolymers may be understood on the basis of the model of a wormlike chain.

The positive sign of the anisotropy of the copolymer molecule $\gamma_1 - \gamma_2$ means that the polarizability of the molecule in the direction of its greatest geometrical length γ_1 (parallel to h) is greater than in the direction normal to h (Fig. 25). High positive value of $\gamma_1 - \gamma_2$ cannot be accounted for by the presence of positively anisotropic chains of poly(alkyl methacrylate) because its anisotropy is too low (Tsvetkov *et al.*, 1964a) and its content in the copolymer is also low ($\sim 10\%$). Evidently, the main contribution to the anisotropy observed is provided by polystyrene because its polarizability b_1 along its chain is much lower than the polarizability b_2 in the direction normal to it. But in this case, if the polarizability γ_1 in the longitudinal

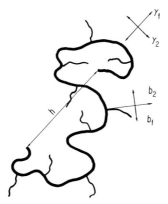

Fig. 25. Configuration of the macromolecule of graft copolymer. γ_1 and γ_2 are main polarizabilities of the macromolecule, b_1 and b_2 are main polarizabilities of the grafted (side) chain.

direction of the macromolecule of the copolymer h is greater, the average direction of the polystyrene chains should be transverse to h. This corresponds to the comblike structure of the graft copolymer.

Just as for comblike molecules described, the optical anisotropy of the molecule of graft copolymer is a tensor sum of the anisotropies contributed by the main chain and side chains of the grafted component. Hence, the anisotropy of grafted chains may be evaluated quantitatively from Eq. (30) along the axes of the first element.

The segmental anisotropy $\alpha_1 - \alpha_2$ of the whole comblike molecule is a sum of the segmental anisotropy of the main chain $(\alpha_1 - \alpha_2)_A$ and the anisotropy provided by all side chains grafted onto a part of the main chain with a length of one segment. Taking this into account, we obtain (Tsvetkov et al., 1963)

$$\alpha_1 - \alpha_2 = (\alpha_1 - \alpha_2)_A - [\Delta b_L S_A z / 2 P_B (1 - z)] \tag{33}$$

where S_A is the number of monomer units in the segment of the main chain of the copolymer (component A), P_B the degree of polymerization of each grafted chain (component B), and z the molar fraction of the grafted component (B) in the copolymer molecule.

Substituting Δb_L from Eq. (31) into Eq. (33) and taking into account that $a\beta = \frac{1}{2} S_B \Delta a_B$, where Δa_B is the anisotropy of the monomer unit in the chain of the grafted component and S_B the number of monomer units in its segment, we finally obtain

$$\alpha_1 - \alpha_2 = S_A \{\Delta a_A - [z \Delta a_B / 12 (1 - z)] f(S_B / P_B)\} \tag{34}$$

where Δa_A is the anisotropy of the monomer unit of the main chain onto which side chains are grafted and the function

$$f(S_B / P_B) = (S_B / P_B)[1 - \exp(-6 P_B / S_B)]$$

It should be noted that Eq. (34) assumes that the macromolecule exhibits a comblike structure, i.e., it is valid under the condition $P_B \ll P_A$ (P_A is the degree of polymerization of the main chain). In passing to the starlike structure ($P_B \gg P_A$) the macromolecule loses the conformation of a linear randomly coiled chain and the h vector is no longer the direction of the preferential orientation of segments. This excludes the validity of concepts on which the derivation of Eq. (34) is based and, consequently, the validity of the equation itself.

In the case of a comblike structure, Eq. (34) may serve to determine S_B / P_B from experimental values of $\alpha_1 - \alpha_2$ if Δa_A, Δa_B, and S_A are known. Hence, this equation allows the evaluation of the degree of intramolecular order of the structure investigated.

Just as for the S_A of the macromolecules considered, the equilibrium rigidity of the main chain may be determined from hydrodynamic properties of graft copolymer molecules (Tsvetkov *et al.*, 1963, 1964a, 1968b). It has been found that the lengths of the Kuhn segment A of the main chain range from 100 to 200 Å for copolymers in which the weight fraction of grafted polystyrene is about 90%.

Hence, values of the equilibrium rigidity of the main chain of graft copolymers are five- to tenfold greater than those for the initial linear homopolymers. The intensity of this phenomenon, which is also observed for comblike macromolecules of the homopolymers, increases with increasing length of grafted chains.

Information about the rigidity of side chains has been obtained (Tsvetkov *et al.*, 1963, 1971c) from analysis of data on FB and by using Eq. (34) and experimental values of S_A.

It was found that the high positive value of segmental anisotropy of graft copolymers is caused by high equilibrium rigidity of side chains: the values of S_B are of the same order of magnitude as the values of the equilibrium rigidity of the main chain S_A. Therefore the conformation of polystyrene side branches (which are much shorter than the main chain) is close to the shape of a slightly curved rod normal to the main chain. This means that a high orientational–axial order is observed in side chains of the graft copolymer; it corresponds to the orientational order of the nematic mesophase (see Table 3).

The sign of the optical anisotropy of the molecule is very important in the analysis of this order. Negative anisotropy of single polystyrene molecules is determined by the preferred orientation of planes of phenyl rings normal to the main chain. Positive anisotropy of the graft copolymer molecule means that the correlation in the orientation of phenyl rings in side groups (Fig. 26) is spread to the whole molecule. The preferred orientation of the ring planes is parallel to the h vector (Fig. 25) and, hence, their contribution to the positive anisotropy of the graft copolymer molecule is very great.

These properties of graft copolymer molecules are observed in cases in which grafted chains are much shorter (ten- to fiftyfold) than the main

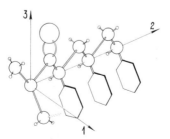

Fig. 26.   Monomer unit of poly(methyl metacrylate) with grafted polystyrene chain. Axis 1 is parallel to the trans chain of poly(methyl metacrylate). (The plane of this chain is the 1, 3 plane.) Axis 2 is parallel to the direction of the trans chain of grafted polystyrene.

chain onto which side chains are grafted, i.e., when the molecule of the graft copolymer is of the comblike type. These properties, moreover, are increasingly pronounced with the degree of grafting, i.e., when the percentage of polystyrene in the copolymer macromolecule is greater. This shows that the interaction of side chains is very important for the appearance of specific properties of graft copolymer molecules, just as in the case of comblike molecules previously considered.

The comblike structure of the molecule of the graft copolymer leads to one more peculiarity of its conformation: high density of segment distribution in the part of the volume adjacent to the main chain. This results in the formation of an uncommon molecular structure of the chain molecule exhibiting high equilibrium rigidity and at the same time forming a random coil that is virtually nondrained by the solvent. When the molecule is modeled by a wormlike chain with the diameter d, the latter property is expressed in high value of d.

The same peculiarity of conformation of the graft copolymer molecule is reflected in a peculiar dependence of FB in its solution on molecular weight (Pavlov et al., 1973; Tsvetkov, 1970). Equilibrium rigidity of graft copolymer chains is high, and therefore they exhibit the dependence of reduced birefringence $[n]/[\eta]$ on the length of the main chain (Fig. 27, curve 1), just as the rigid-chain polymers considered earlier (Fig. 14). Figure 27 also shows initial parts of the dependence of relative values $\Delta \equiv ([n]/[\eta])/([n]/[\eta])_{\infty}$ on x. The peculiarity of graft copolymer molecules is that the slope of the curve $\Delta(x)$ depends on the length of grafted polystyrene branches since this length determines the diameter d of the wormlike chain modeling the molecule. Moreover, values of $[n]/[\eta]$ and, correspondingly, of Δ may

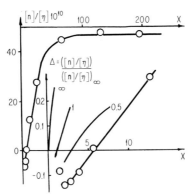

Fig. 27. Values of reduced birefringence $[n]/[\eta]$ and $\Delta \equiv ([n]/[\eta])/([n]/[\eta])_{x}$ versus the x parameter for solutions of fractions of graft copolymer with styrene in bromoform. Numbers at the theoretical curves correspond to different values of the a/d ratio (Tsvetkov, 1970); points correspond to experimental values.

change their sign when x becomes relatively low; this corresponds to changes in the shape of molecules from comblike to starlike structures. The family of curves in Fig. 27 represents theoretical dependences $\Delta(x)$ corresponding to different values of a/d for a wormlike model (a is the persistence length). Their comparison with experimental data permits the evaluation of d and confirms its high value (30–100 Å) for molecules of the graft copolymers investigated.

XIV. Crystallike Molecules with Long-Chain Side Groups

It was shown above that the observed increase in the orientational order in the main and side chains is caused by the interaction of side groups. Conformational properties of comblike polymers, the macromolecules of which combine the main chain of common flexible-chain polymers with long-chain side radicals containing chemical groups favoring the formation of the liquid crystalline phase, may serve as a still more conclusive proof of this concept. One of the outstanding examples of polymers of this type (Tsvetkov *et al.*, 1969c) is offered by poly(phenyl methacrylic) esters of alkoxybenzoic acid (PPhEAA) with the following structure of the monomer unit ($m = 3, 6, 9, 16$)

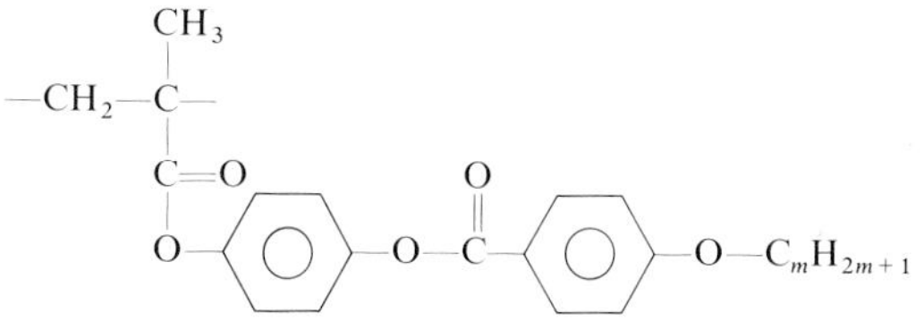

$$-CH_2-\underset{\underset{\underset{O-\langle O\rangle-O-\overset{O}{\overset{\|}{C}}-\langle O\rangle-O-C_mH_{2m+1}}{|}}{\overset{|}{\underset{C=O}{}}}}{\overset{CH_3}{\overset{|}{C}}}-$$

The main part of the side radical consists of a residue of the alkoxybenzoic acid—a compound forming thermotropic liquid crystals. Investigations of the dependence of coefficients of translational diffusion, sedimentation, and intrinsic viscosities on molecular weight (Tsvetkov *et al.*, 1973d) allowed quantitative determination of the number of monomer units in a segment of macromolecules S and was found to be 24. The equilibrium rigidity of their main chain was relatively low and, hence, they exhibited properties typical of flexible-chain polymers. In particular, the size of macromolecules has proven to be very sensitive to the thermodynamic strength of the solvent and may change greatly (fivefold) with changes in the latter (Tsvetkov *et al.*, 1969c, 1973d).

It is very significant that PPhEAA solutions are characterized by high negative FB $[n]/[\eta]$ although the equilibrium rigidity of the main chain is relatively low. Negative segmental anisotropy of the molecule determined by using experimental values of $[n]/[\eta]$ and Eqs. (10) and (14) is shown in Table 4. It exceeds by over an order of magnitude the anisotropy of

TABLE 4

Electrooptical Characteristics of Comblike Polymers, Corresponding Monomers, and Some Nematic Liquid Crystals

	S	$(\alpha_1 - \alpha_2) \times 10^{25}$ (cm^3)	$\Delta a \times 10^{25}$ (cm^3)	$K \times 10^{10}$ $(cm^5 g^{-1}(v/300)^{-2})$	$\tau \times 10^5$ (sec)
1 Poly(phenyl methacrylic) ester of alkoxybenzoic acids (PPhEAA)	24	−3100	−130	−10	5
2 PPhEAA monomer	—	—	—	0.21	—
3 Poly(nonyloxybenzamido styrene) (PNOBS)	20	−2500	−125	−140	0.2

		25	−440	−18	−0.5	—
4	Polycetyl ester of methacryloyl oxybenzoic acid (PCEMA)					
5	PCEMA monomer	—	—	—	0.15	—
6	p-Cyanophenyl ester of nonyloxy benzoic acid	—	—	150	—	—
7	Cetyloxybenzoic acid	—	—	150	—	—

poly(cetyl methacrylate) (see Table 2) and is comparable only to the anisotropy of crystallike molecules such as poly(alkyl isocyanates) (Tsvetkov
et al., 1968a, 1971a, 1974a) or rigid-chain aromatic polyamides (Tsvetkov
et al., 1975d,e, 1976a,b). In this case it has been found that the anistropy
of the monomer Δa exceeds by more than an order of magnitude the anisotropy of the monomer unit of poly(butyl isocyanate) and more than threefold
that of poly(p-benzamide). This means that high optical anisotropy of
PPhEAA molecules is caused by the perfect orientational order in side
groups rather than by the rigidity of their main chains. The degree of order
in side groups is very close to the orientational order in nematic liquid
crystals. Actually, the optical anisotropy of molecules of nematic liquid
crystals (the chemical structure of which is close to that of the PPhEAA
monomer unit), determined from measurements of the anisotropy of molar
refraction of the mesophase taking into account its orientational order Q
(Rjumtsev *et al.*, 1973; Kovshik *et al.*, 1975), has been found to be very close
to the values of Δa for PPhEAA. This is clearly seen from experimental data
in Table 4.

Unique electrooptical properties have been found in solutions of the
polymers investigated; they differ markedly from those of common flexible-
chain polymers. High EB negative in sign indicates that the PPhEAA
macromolecules are not only characterized by the orientational–axial order
but also by the orientational–polar order. The Kerr constants $K_{v=0}$ presented
in Table 4 for the polymer and the corresponding monomer differ not only
in absolute value (fiftyfold) but also in sign. The positive sign of EB of
solutions of the PPhEAA monomer shows that the component of the dipole
moment of the monomer molecule $\mu_{||}$ parallel to the $L_{||}$ axis, to the highest
optical polarizability of the monomer, i.e., parallel to its alkyl chain (Fig. 28a),
is of major significance in this phenomenon. In contrast, in a comblike
polymer molecule (Fig. 28b) in which the main chain is the axis of symmetry,

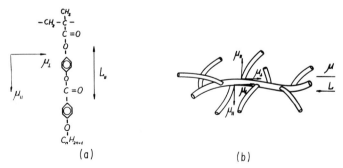

(a)   (b)

Fig. 28. Explanation of the sign and value of EB for monomer (a) and polymers (b) of
phenyl methacrylic esters of alkoxybenzoic acids.

the components of dipole moments $\mu_{\parallel}$ of side radicals normal to the main chain are mutually compensated and in the EB observed experimentally, the components of the monomer $\mu_{\perp}$ parallel to the main chain play a decisive part. These components $\mu_{\perp}$ are normal to the alkyl chain of the monomer but contribute to the component of the dipole moment of the monomer unit $\mu_{0\parallel}$ parallel to the main chain L. Hence, when the molecule is oriented in an electric field, its long-chain side groups are arranged normal to the field direction and this leads to the negative Kerr effect. Thus, molecules of the comblike polymers investigated are characterized by the presence of a polar order in the orientations of $\mu_{0\parallel}$ of longitudinal components of the monomer unit dipoles. As a result, the macromolecule may exhibit a considerable dipole moment ($\mu \approx 100$ D) in the direction of its optical axis, i.e., in the direction of the h vector. A high degree of the orientational–polar order and a relatively high equilibrium flexibility of the main chain ($S = 24$) are combined with very peculiar kinetic properties. For the linear polymers considered (pp. 68–70), these properties provide evidences of kinetic rigidity of macromolecules. In a sinusoidal electric field at frequencies ν up to several thousand hertz, a strong dispersion of EB appears (Fig. 29), which is an indication of the existence of relaxation phenomena and the dipolar character of the macromolecule orientation.

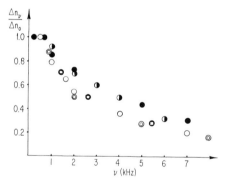

Fig. 29. Relative birefringence versus frequency of the electric field for solutions of some fractions of poly(phenyl methacrylic) ester of cetyloxybenzoic acid of various molecular weight M. ($\bigcirc$) $M = 19.1 \times 10^6$; ($\bigcirc$) $M = 9.9 \times 10^6$; ($\square$) $M = 6.3 \times 10^6$; ($\bullet$) $M = 2.98 \times 10^6$; ($\bullet$) $M = 1.1 \times 10^6$.

Figure 29 shows that the range of relaxation of the Kerr effects has been found to be virtually independent of the molecular weight of the polymer (Tsvetkov et al., 1973d), which is typical of flexible–chain polymers. However, in this case the values of relaxation times τ are 10^{-4}–10^{-5} sec, i.e., they are by five to six orders of magnitude higher than the corresponding values of τ characteristic of flexible-chain polymers. This means that relatively long parts of the macromolecule are oriented as a whole in an electric field (they are intramolecular "domains" containing up to thousands of monomer units—large-scale chain motion). Hence, the kinetic unit of the PPhEAA chain oriented in an electric field greatly surpasses the length of the Kuhn segment characterizing the equilibrium rigidity of the molecule.

Doubtless, these experimental facts are a consequence of pecularities of the chemical structure of long-chain side radicals containing groups capable of forming nematic liquid crystals. They show that the PPhEAA molecules exhibit intramolecular order of the mesomorphic type in which side chains form a mobile liquid crystalline structure. The degree of this order is determined not only by the intramolecular polymer–polymer interactions but also by the polymer–solvent interactions. As already mentioned, the size of PPhEAA molecules is sensitive to the thermodynamic strength of the solvent and can vary greatly when it increases or decreases. Coiling of the main chain in a thermodynamically poor solvent (detected, e.g., from the intrinsic viscosity in various solvents, Table 5) leads to a pronounced increase in the degree of intramolecular orientational order in side chains. This increase is manifested both in the increasing optical anisotropy and in the peculiarities of the Kerr constants of solutions (Table 5). This is presumably due to a decrease in the polymer–solvent interactions that facilitates the formation of the orientational-ordered structure in side groups in a "poor" solvent (tetrachloromethane or a mixture of benzene and heptane) because the intramolecular polymer–polymer interaction increases. In this case, the diphilic nature of molecules that is used sometimes to explain peculiarities of conformational and morphological properties of molecules of comblike

TABLE 5

Intrinsic Viscosity $[\eta]$, Optical Anisotropy
$\alpha_1 - \alpha_2$, and Kerr Constants $K_{v=0}$ of Poly(phenyl methacryloyl)
Ester of Cetyloxybenzoic Acid in Various Solvents

Solvent	$[\eta] \times 10^{-2}$ (cm^3/g)	$(\alpha_1 - \alpha_2) \times 10^{25}$ (cm^3)	$K \times 10^{10}$ $(cm^5 g^{-1}(v/300)^{-2})$
Tetrahydrofuran	3.08	-890	—
Chloroform	2.88	-1400	—
Benzene	2.50	-1600	-2.2
Tetrachloromethane	0.63	-2700	-10
Benzene + heptane (66:34)	0.77	-4200	-18
Benzene + heptane (52:48)	0.54	-4200	-40

polymers cannot be a decisive factor since both the main and side chains are readily dissolved in the solvents used.

These properties are not only typical of the polymer investigated but are also characteristic of other polymers whose macromolecules contain long-chain side radicals with chemical groups favoring the formation of the liquid crystalline state. Actually, similar electrooptical properties have been found in solutions of poly(p-nonyloxybenzamidostyrene) (PNOBS) (Rjumtsev et al., 1976). Main electrooptical characteristics of its molecules are also shown in Table 4.

It is interesting to note that this formation of ordered intramolecular structure is much less pronounced in solutions of the polycetyl ester of methacrylyloyloxybenzoic acid (PCEMA) in which long-chain side radicals contain only one phenyl ring (Table 4). The optical anisotropy of its monomer unit greatly exceeds Δa of molecules of poly(cetyl methacrylate) and almost coincides with the Δa value for polystyrene (Tsvetkov et al., 1964a). This is due to the presence of a benzene ring in the side group. Nevertheless, the value of optical anisotropy of the PCEMA monomer unit is seven- or eightfold lower than that for the poly(phenyl methacrylic) ester of cetyloxybenzoic acid and poly(nonyloxybenzamidostyrene). This shows that side chains of PCEMA do not exhibit the high orientational order characteristic of crystallike molecules of PPhEAA and PNOBS. The absolute values of the Kerr constants $K_{v=0}$ for PCEMA (Table 4) are close to the K value for the corresponding monomer. Nevertheless, for monomer solutions the Kerr constant is positive but in PCEMA solutions it is negative just as for PPhEAA, i.e., it coincides in sign with FB of the same solution. This means that longitudinal axes of side groups of PCEMA are oriented in an electric field normal to it (in contrast to the monomer molecules) and, hence, this orientation is not quite free but is correlated with that of the main chain.

The absence of low-frequency dispersion of the Kerr effect in PCEMA solutions shows that the EB observed should be considered as a result of the "small-scale" (segmental) orientation of the main chain in an electric field caused by the longitudinal component of dipole moments of its segments $\mu_s = \mu_{0\parallel} \cdot S$. In accordance with this, the expression (28) of the Kerr constant for a solution of polymer chains in the Gaussian range of molecular weights may be used to describe this phenomenon quantitatively. Comparison of Eq. (28) with Eqs. (10) and (14) gives

$$\frac{\mu_S^2}{M_S} = \frac{90kTK}{\mathcal{N}_A(\varepsilon + 2)^2([n]/[\eta])} \tag{35}$$

where $M_S = M_0 S$ is the molecular weight of a segment of the macromolecule.

Experimental values of $M_S = 10750$, $K = -0.5 \times 10^{-10}$ cm^5 g^{-1} $(V/300)^{-2}$ and $[n]/[\eta] = -32 \times 10^{-10}$ cm sec^2 g^{-1} and Eq. (35) yield the value of $\mu_S = 7.5\,\text{D}$. This value greatly exceeds the dipole moment that might be expected for the PCEMA monomer unit. This value of μ_S leads to a reasonable value of the longitudinal component of the dipole moment of the monomer unit $\mu_{0\parallel} = \mu_S/S = 0.3\,\text{D}$.

All these results are a direct evidence of the fact that the kinetic unit of PCEMA oriented in an electric field is the chain segment. In contrast to the situation observed in solutions of common flexible-chain polymers, the value of the kinetic unit markedly exceeds the size of the monomer unit. Evidently, this is caused by the orientational interaction between side groups of the chain just as in the PPhEAA and PNOBS molecules. However, in the case of PCEMA the mobility of kinetic units is much higher and their size is much smaller than for PPhEAA and PNOBS since the structure of side groups of the two latter polymers is closer to nematogenic structures (two phenyl rings in the side group (Gray, 1962) compared to one ring for PCEMA). This accounts for the absence of low-frequency dispersion of the Kerr effect in PCEMA solutions; this phenomenon is very pronounced in solutions of PPhEAA and PNOBS.

Hence, the presence of two phenyl rings in the side group is a favorable condition for the formation of the intramolecular liquid crystalline state in solutions of comblike polymers. This conclusion seems valid because the presence of two phenyl rings in the molecule is an important condition for the formation of the thermotropic liquid crystalline state in low molecular weight substances (Gray, 1962).

XV. Random Copolymers of Crystallike Molecules

Strong intramolecular interactions between side chains are apparent in dynamo- and electrooptical properties of solutions of random copolymers of the phenyl methacrylic ester of cetyloxybenzoic acid (PhECA) with cetyl methacrylate (Tsvetkov et al., 1973d). Therefore, it may be assumed that in molecules of these copolymers only the first component actively contributes to the appearance of the intramolecular liquid crystalline order. This makes it possible to investigate main relationships in the formation and changes in the orientational–polar order in macromolecules by varying the composition of the random copolymer.

Figure 30 shows the dependences of reduced flow birefringence $[n]/[\eta]$ and the Kerr constants $K_{v=0}$ on x_A, the molar fraction of PhECA in copolymers. The values of $[n]/[\eta]$ and $K_{v=0}$ increase twenty- and hundredfold, respectively, in passing from poly(cetyl methacrylate) to PPhEAA in accordance with optical and electrooptical properties of homopolymers. Values of optical anisotropy of a copolymer segment $\alpha_1 - \alpha_2$ calculated from experimental values of $[n]/[\eta]$ and Eqs. (10) and (14) are shown as functions of x_A in Fig. 31. It should be noted that the dependences of $[n]/[\eta]$, $\alpha_1 - \alpha_2$, and $K_{\nu=0}$ on x_A markedly deviate from linearity.

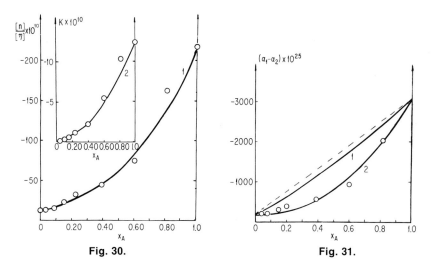

Fig. 30.

Fig. 31.

Fig. 30. Reduced flow birefringence $[n]/[\eta]$ (curve 1) and the Kerr constant K (curve 2) versus molar fraction x_A of phenyl methacryloyl ester of cetyloxybenzoic acid in a random copolymer with cetyl methacrylate.

Fig. 31. Optical anisotropy of the segment $\alpha_1 - \alpha_2$ versus molar fraction x_A of phenyl methacryloyl ester in a random copolymer with cetyl methacrylate. 1 and 2 are the theoretical curves from Eqs. (36) and (37), respectively; points are the experimental data.

In the simplest cases when parts of the molecule of the random copolymer do not exhibit a specific interaction that would lead to the formation of a peculiar intramolecular organization, the character of the dependence of $\alpha_1 - \alpha_2$ on x_A may be represented by the equation (Tsvetkov and Verchotina, 1958; Vitovskaia et al., 1967):

$$\alpha_1 - \alpha_2 = S_1 \Delta a_1 + x_A[S_1 \Delta a_2 + S_2 \Delta a_1 - 2S_1 \Delta a_1]$$
$$+ x_A^2(S_2 - S_1)(\Delta a_1 - \Delta a_2) \tag{36}$$

Here Δa_1 and Δa_2 are the anisotropies of monomer units, S_1 and S_2 the numbers of monomer units in segments of homopolymers of the first and the second components. Equation (36) shows that the deviation from linearity is determined by the second term in the right-hand part of the equation and should appear when both the equilibrium rigidities (S_1 and S_2) and the anisotropies of monomer units (Δa_1 and Δa_2) of both components are different.

The dependence represented by curve 1 in Fig. 31 is readily obtained by using Eq. (36) for the copolymers investigated and experimental values of S and Δa for components shown in Tables 2 and 4.

Although this dependence is curvilinear and the sign of its deviation from linearity (concavity upwards) corresponds to experimental data (points), this curvature is much smaller quantitatively than is actually observed. The reason for this discrepancy between curve 1 in Fig. 31 and experimental data may be related to the fact that Eq. (36) takes into account only the change in the rigidity of the copolymer chain with variations in its composition, but assumes that Δa_1 and Δa_2 are constant at all values of x_A. Nevertheless, as was mentioned earlier, the high value of Δa_1 of the anisotropy of the PPhEAA monomer unit is caused by the interaction of side groups of the macromolecule that form an intramolecular organization with a high degree of orientational order. Obviously, the appearance of these secondary structures is possible only if the concentration (by volume) of side groups in the PhECA chain is relatively high. Consequently, a decrease in the content of PhECA in the copolymer inevitably leads to a decreased probability of the formation of intramolecular order and, hence, to a decrease in the effective value of the anisotropy of the PhECA monomer unit in the copolymer.

If this circumstance is taken into account (Tsvetkov *et al.*, 1973d), this leads to the dependence of $\alpha_1 - \alpha_2$ on x_A determined by the equation

$$\alpha_1 - \alpha_2 = S_1 \Delta a_1 + x_A(S_2 - S_1)\Delta a_1 + x_A{}^2 S_1(\Delta a_2 - \Delta a_1)$$
$$+ x_A{}^3(S_2 - S_1)(\Delta a_2 - \Delta a_1) \tag{37}$$

Curve 2 in Fig. 31 corresponds to Eq. (37), which describes quantitatively the experimental data (points). The result shows directly that in this case the deviation from linearity in the dependence of optical anisotropy of the copolymer on its composition is determined by different anisotropies of monomer units of components rather than by different flexibilities of their main chain; i.e., it is governed by the ability of one of the components to form the intramolecular liquid crystalline order. Probably, the nonlinearity of the dependence of the Kerr constant on the copolymer composition shown in Fig. 30, curve 1 is induced by a similar mechanism.

XVI. Formation of Supermolecular Structures in Solutions of Crystallike Molecules

Peculiarities of intramolecular organization of comblike polymers markedly affect the formation of the supermolecular liquid crystalline state in their solutions. In these systems the properties of both concentrated solutions and polymers in bulk are determined by the supermolecular order formed by side chains of macromolecules. This may be illustrated taking as an example the analysis of changes in electrooptical properties in passing from dilute to concentrated solutions or by investigations of the Kerr effect under conditions close to temperatures of phase separation.

Figure 32 shows the temperature dependence of EB for solutions of phenyl methacrylic ester of hexyloxybenzoic acid in benzene. As was shown before, a drastic increase in the negative effect with decreasing temperature is caused by an increase in the intramolecular orientational–polar order in side groups of the polymer. The situation is markedly changed when supermolecular aggregates are formed in solution.

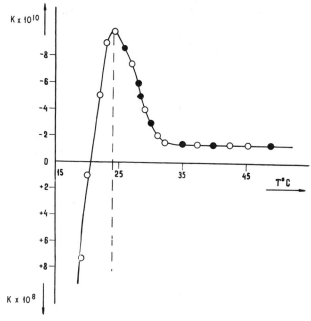

Fig. 32.   The Kerr constant versus temperature for solutions of poly(methacryloyl phenylic) ester of hexyloxybenzoic acid in benzene. Points correspond to polymer concentrations $C = 0.77 \times 10^{-2}$ g cm^{-3} and $C = 1.4 \times 10^{-2}$ g cm^{-3}.

An increase in the polymer concentration or a further decrease in the temperature when it approaches the temperature of the precipitation of the polymer from solution T_0 (10–12° above T_0) leads to the appearance of high positive EB that exceeds by two orders of magnitude the maximum negative value of the Kerr constant.

It is noteworthy that the sign of the supermolecular Kerr effect corresponds to the sign of EB in solutions of the corresponding monomer (Table 4)

and to electrooptical properties of alkoxybenzoic acids (Rjumtsev and Tsvetkov, 1969). This may be attributed to the fact that the formation of supermolecular organization of solution is largely due to the interaction of side
groups not only of one molecule but also of different molecules forming the
supermolecular structure in solution. As a result of this interaction, the
direction of side chains forming a lyotropic liquid crystal becomes the axis
of symmetry, i.e., the axis of the orientational order inside supermolecular
aggregates. In this case the direction of main chains may not be ordered.
The scheme of this arrangement is shown in Fig. 33, which shows that the
supermolecular liquid crystalline organization leads to uniform orientation
not only of longitudinal axes of long-chain side groups but also of longitu-
dinal components of their dipoles $\mu_{||}$ that play an active part in the super-
molecular effect ensuring its positive sign. Normal components $\mu_{\perp}$ do not
form an orientational–ordered system in contrast to what takes place in
a nonaggregated molecule of the polymer considered. It should be emphasized that although signs of EB in a molecular solution and in the aggregate
state are opposite, the mechanisms leading to the properties observed are
similar in both cases. These mechanisms involve interactions of long-chain
side groups of the molecule leading to their parallel orientation both inside
the molecule and on the supermolecular level. This shows that a close
relationship exists between the conformational properties of comblike macromolecules and the lyotropic mesophase appearing in their solutions.

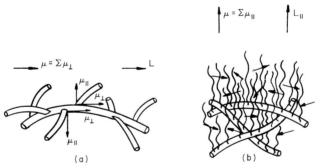

Fig. 33. To the explanation of the sign and value of EB in molecular (a) and aggregated
(b) solutions of poly(phenyl methacrylic) esters of alkoxybenzoic acids.

Hence, the change in the value and sign of the Kerr effect in passing from
monomer solutions to molecular solutions of the polymer and, further, to
its supermolecular structure is related to the peculiarity of the structure of
side groups favoring the formation of lyotropic mesomorphic structures both
on the molecular and on the supermolecular level.

In conclusion it should be noted that the intra and supermolecular liquid crystalline states are very important in processes of polymerization and profoundly affect their rate as well as the structure of the polymers formed (Krentsel and Amerik, 1971; Baturin et al., 1972).

The above experimental material and its evaluation indicate that the degree of ordering of the internal structure in the linear and comblike macromolecules investigated is comparable to the orientational ordering of nematic liquid crystals. It is mainly such crystallike molecular structures that ensure the possibility of the formation of lyotropic mesophases in polymer solutions. Therefore, the first and the most important stage in the study of the liquid crystalline order in macromolecular compounds is the investigation of the orientational order on the intramolecular level.

As shown in this chapter, methods of molecular hydrodynamics and optics have proven to be very effective in these investigations. Among them flow and electric birefringence were the most useful since the optical anisotropy and the dipole moment permit the most sensitive and direct measurements of the orientational (and polar) order in chain macromolecules.

There is no doubt that the synthesis and investigations of the structure of crystallike macromolecules with widely differing chemical constitution will provide a way to obtain polymeric liquid crystals. Such polymers are being more and more widely used in various fields and therefore require further extensive experimental and theoretical investigations.

References

Andreeva, L. N., Gorbunov, A. A., Didenko, S. A., Korneeva, E. V., Lavrenko, P. N., Plate, N. A., and Shibaev, V. P. (1973). *Vysokomol. Soedin., Ser. B* **15**, 209–212.

Andrianov, K. A. (1969). *Vysokomol. Soedin., Ser. A* **11**, 1362–1377.

Andrianov, K. A. (1971). *Vysokomol. Soedin., Ser. A* **13**, 253–265.

Andrianov, K. A., Kurakov, G. A., Sushentsova, F. F., Mjagkov, V. A., and Avilov, V. A. (1965). *Vysokomol. Soedin., Ser. A* **7**, 1477–1488.

Aroney, M., Le Fevre, R. I. W., and Parkins, G. M. (1960). *J. Chem. Soc.* pp. 2890–2895.

Baturin, A. A., Amerik, J. B., Krentsel, B. A., Tsvetkov, V. N., Shtennikova, I. N., and Rjumtsev, E. I. (1972). *Dokl. Akad. Nauk SSSR* **202**, 586–588.

Berry, G., Nomure, H., and Mayhan, K. G. (1967). *J. Polym. Sci., Part A-2* **5**, 1–21.

Black, W. B. (1973). *J. Macromol. Sci., Chem.* 7(1), 3–41.

Burchard, W. (1965). *Makromol. Chem.* **88**, 11–28.

Burchard, W. (1971). *Br. Polym. J.* **3**, 209–213.

Champion, J. W., Dessen, R. A., and Meeten, G. H. (1974). *Polymer* **15**, 301–305.

Cowie, J. M. G., and Bywater, S. (1965), *Polymer* **6**, 197–204.

Crothers, D. M., and Zimm, B. H. (1965), *J. Mol. Biol.* **12**, 525–536.

De Winter, W. (1966). *Rev. Macromol. Chem.* **1**, 329–380.

Doty, P. (1956). *Proc. Natl. Acad. Sci. U.S.A.* **42**, 791–800.

Flory, P. J. (1953). "Principles of the Polymer Chemistry." Cornell Univ. Press, Ithaca, New York.

Flory, P. J. (1969). "Statistical Mechanics of Chain Molecules." Wiley (Interscience), New York.

Freederiksz, V. K., and Tsvetkov, V. N. (1934) *Phys. Z. Sowjet Union* **6**, 490–504.

Frenkel, J. I. (1945). "Kinetic Theory of Liquids." Akad. Nauk SSSR, Moscow.

Garmonova, T. I., Vitovskaia, M. G., Lavrenko, P. N., Tsvetkov, V. N., and Korovina, E. V. (1971). *Vysokomol. Soedin., Ser. A* **13**, 884–891.

Geil, P. H. (1963), "Polymer Single Crystals," Wiley (Interscience), New York.

Gray, G. W. (1962). "Molecular Structure and Properties of Liquid Crystals," Academic Press, New York.

Gray, H. B., Bloomfield, V. A., and Hearst, J. E. (1967). *J. Chem. Phys.* **46**, 1493–1498.

Hearst, J. E. (1963). *J. Chem. Phys.* **38**, 1062–1065.

Hearst, J. E., and Stockmayer, W. H. (1962). *J. Chem. Phys.* **37**, 1425–1433.

Kalmykova, V. D., Kudrjavtsev, G. I., Papkov, S. P., Volokhina, A. V., Iovleva, M. M., Milkova, L. P., Kylichichin, V. G., and Bandurian, S. I. (1971). *Vysokomol. Soedin., Ser. B* **13**, 707–708.

Kirkwood, J., and Riseman, J. (1948). *J. Chem. Phys.* **16**, 565–573.

Kovshik, A. P., Denite, J. I., Rjumtsev, E. I., and Tsvetkov, V. N. (1975). *Kristallografiya* **20**, 861–864.

Krentsel, B. A., and Amerik, J. B. (1971). *Vysokomol. Soedin., Ser. A* **13**, 1358–1374.

Krigbaum, W. R., and Dawkins, I. V. (1974), *In* "Polymer Handbook" (J. Brandrup and E. H. Immergut, eds.), 2nd ed., pp. 319–322, Wiley (Interscience), New York.

Kuhn, W. (1934). *Kolloid-Z.* **68**, 2–31.

Kuhn, W. (1936). *Kolloid-Z.* **76**, 258–276.

Kuhn, W. (1948). *Helv. Chim. Acta* **31**, 1092–1115.

Kuhn, W., and Grün, F. (1942). *Kolloid-Z.* **101**, 248–271.

Kuhn, W., Kuhn, H., and Buchner, P. (1951). *Ergeb. Exakten Naturwiss.* **25**, 1–108.

Langevin, P. (1910). *Radium* (*Paris*) **10**, 249–281.

Lavrenko, P. N., Rjumtsev, E. I., Shtennikova, I. N., Andreeva, L. N., Pogodina, N. V., and Tsvetkov, V. N. (1974). *J. Polym. Sci., Polym. Symp., Part C* **44**, 217–235.

Lawrence, A. S. C. (1938). *Trans. Faraday Soc.* **34**, 660–677.

Le Fevre, C. G., Le Fevre, R. I. W., and Parkins, G. M. (1958). *J. Chem. Soc.* pp. 1468–1476.

Le Fevre, R. I. W., and Sundaram, K. M. S. (1963). *J. Chem. Soc.* pp. 1880–1887.

Lonsdale, K., and Krishnan, K. S. (1936). *Proc. R. Soc., Ser. A* **156**, 597–612.

Maier, W., and Markau, K. (1961). *Z. Phys. Chem.* (*Frankfurt am Main*) **28**, 190–202.

North, A. M. (1972). *Chem. Soc. Rev.* **1**, 49–72.

Ostwald, W. (1931). *Z. Kristallogr., Kristallgeom., Kristallphys., Kristallchem.* **79**, 222–240.

Overberger, C. G., and Moore, J. A. (1970). *Fortschr. Hochpolym-Forsch.* **7**, 113–150.

Papkov, S. P., Iovleva, M. M., Milkova, L. P., Antipova, R. V., Gouchberg, S. S., Kudrjavtsev, G. I., Volokhina, A. V., and Kalmykova, V. D. (1973a). *Vysokomol. Soedin., Ser. B* **15**, 357–360.

Papkov, S. P., Iovleva, M. M., Milkova, L. P., Kalmykova, V. D., Volokhina, A. V., and Kudrjavtsev, G. I. (1973b). *Vysokomol. Soedin., Ser. B* **15**, 757–759.

Papkov, S. P., Bandurjan, S. I., and Iovleva, M. M. (1973c). *Vysokomol. Soedin., Ser. B* **15**, 370–372.

Pauling, L. (1960), "Nature of the Chemical Bond." Cornell Univ. Press, Ithaca, New York.

Pavlov, G. M., Magarik, S. J., Lavrenko, P. N., and Fomin, G. A. (1973), *Vysokomol. Soedin., Ser. A* **13**, 2011–2019.

Philippoff, W., and Tornqvist, E. G. M. (1969). *J. Polym. Sci., Part C* **23**, 881–889.

Porod, G. (1949). *Monatsh. Chem.* **80**, 251–260.

Riseman, J., and Kirkwood, J. (1950). *J. Chem. Phys.* **18**, 512–516.

Rjumstev, E. I., and Tsvetkov, V. N. (1969). *Opt. Spectrosk.* **26**, 607–612.

Rjumtsev, E. I., Kovshik, A. P., Kolomiets, I. P., and Tsvetkov, V. N. (1973). *Kristallografiya* **18**, 1246–1249.

Rjumtsev, E. I., Andreeva, L. N., Aliev, F. M., Kytsenko, L. I., and Tsvetkov, V. N. (1975a). *Vysokomol. Soedin., Ser. A* **17**, 1368–1374.

Rjumtsev, E. I., Aliev, F. M., Vitovskaia, M. G., Yrinov, E. Y., and Tsvetkov, V. N. (1975b). *Vysokomol. Soedin. Ser. A* **17**, 2676–2681.

Rjumtsev, E. I., Aliev, F. M., and Tsvetkov, V. N. (1975c). *Vysokomol. Soedin., Ser. A* **17**, 1712–1725.

Rjumtsev, E. I., Shtennikova, I. N., Pogodina, N. V., Kolbina, G. F., Konstantinov, I. I., and Amerik, J. B. (1976). *Vysokomol. Soedin., Ser. A* **18**, 439–442.

Robinson, C. (1956). *Trans. Faraday Soc.* **52**, 571–592.

Robinson, C. (1961). *Tetrahedron* **13**, 219–234.

Saunders, P. R. (1964). *J. Polym Sci., Part A* **2**, 3765–3770.

Saunders, P. R. (1965). *J. Polym. Sci., Part A* **3**, 1221–1226.

Shmueli, U., Traub, W., and Rosenheck, K. (1969). *J. Polym. Sci., Part A-2* **7**, 515–524.

Shtennikova, I. N. (1973). Thesis, Institute of Macromolecular Compounds Acad. Sci. USSR, Leningrad.

Silberstein, L. (1927a). *Philos. Mag.* **33**, 92–106.

Silberstein, L. (1927b). *Philos. Mag.* **33**, 215–226.

Silberstein, L. (1927c). *Philos. Mag.* **33**, 521–538.

Solc, K., and Stockmayer, W. H. (1971). *J. Chem. Phys.* **54**, 2756–2757.

Stewart, W. G. (1929a). *Phys. Rev.* **31**, 1–10.

Stewart, W. G. (1929b). *Phys. Rev.* **31**, 174–186.

Stewart, W. G. (1929c). *Phys. Rev.* **32**, 153–168.

Stewart, W. G. (1930). *Phys. Rev.* **35**, 726–732.

Stockmayer, W. H., and Fixman, M. (1963). *J. Polym. Sci., Part C* **1**, 137–141.

Stuart, H. A. (1952). "Die Physik der Hochpolymeren." Springer-Verlag, Berlin and New York.

Tsvetkov, V. N. (1939). *Acta Physicochim. URSS* **10**, 555–561.

Tsvetkov, V. N. (1942). *Acta Physicochim. URSS* **16**, 132–147.

Tsvetkov, V. N. (1962). *Vysokomol. Soedin.* **4**, 894–900.

Tsvetkov, V. N. (1964). In "Newer Methods of Polymer Characterization" (B. Ke, ed.), Ch. 14. Wiley (Interscience), New York.

Tsvetkov, V. N. (1965a). *Dokl. Akad. Nauk SSSR* **165**, 360–363.

Tsvetkov, V. N. (1965b). *Vysokomol. Soedin.* **7**, 1468–1475.

Tsvetkov, V. N. (1969). *Eur. Polym. J.* **5**. Suppl., 237–260.

Tsvetkov, V. N. (1970). *Dokl. Akad. Nauk SSSR* **192**, 380–383.

Tsvetkov, V. N. (1972a). *Makromol. Chem.* **160**, 1–26.

Tsvetkov, V. N. (1972b). *Dokl. Akad. Nauk SSSR* **205**, 328–331.

Tsvetkov, V. N. (1974). *Vysokomol. Soedin., Ser. A* **16**, 944–965.

Tsvetkov, V. N. (1976). *Eur. Polym. J.* **12**, 867–871.

Tsvetkov, V. N., and Frisman, E. V. (1944). *Acta Physicochim. URSS* **19**, 323–327.

Tsvetkov, V. N., and Sosnovski, A. N. (1943). *Acta Physicochim. URSS* **18**, 358–369.

Tsvetkov, V. N., and Verchotina, L. N. (1958). *Zh. Tekh. Fiz.* **28**, 97–108.

Tsvetkov, V. N., Magarik, S. J., Klenin, S. I., and Eskin, V. E. (1963). *Vysokomol. Soedin.* **5**, 3–10.

Tsvetkov, V. N., Eskin, V. E., and Frenkel, S. Y. (1964a). "Structure of Macromolecules in Solutions." Nauka, Moscow.

Tsvetkov, V. N., Klenin, S. I., and Magarik, S. Y. (1964b). *Vysokomol. Soedin.* **6**, 400–405.

Tsvetkov, V. N., Andreeva, L. N., and Kvitchenko, L. N. (1965). *Vysokomol. Soedin.* **7**, 2001–2010.

Tsvetkov, V. N., Garmonova, T. I., and Vitovskaia, M. G. (1966). *Vysokomol. Soedin.* **8**, 980–986.

Tsvetkov, V. N., Shtennikova, I. N., Rjumtsev, E. I., Getmanchuk, Y. P., Spirin, Y. L., and Drjagileva, R. I. (1968a). *Vysokomol. Soedin., Ser. A* **10**, 2132–2144.

Tsvetkov, V. N., Magarik, S. Y., Kadyrov, T., and Andreeva, G. A. (1968b). *Vysokomol. Soedin., Ser. A* **10**, 943–954.

Tsvetkov, V. N., Andrianov, K. A., Vinogradov, E. L., Shtennikova, I. N., Jakyshkina, S. E., and Pahomov, V. I. (1969a). *J. Polym. Sci., Part C* **23**, 385–391.

Tsvetkov, V. N., Hardy, D., Shtennikova, I. N., Korneeva, E. V., Pirogova, G. F., and Nitrai, K. (1969b). *Vysokomol. Soedin., Ser. A* **11**, 349–358.

Tsvetkov, V. N., Shtennikova, I. N., Rjumtsev, E. I., Kolbina, G. F., Konstantinov, I. I., Amerik, Y. B., and Krentsel, B. A. (1969c). *Vysokomol. Soedin. Ser. A* **11**, 2528–2536.

Tsvetkov, V. N., Shtennikova, I. N., Rjumtsev, E. I., and Getmanchuk, Y. P. (1971a). *Eur. Polym. J.* **7**, 767–774.

Tsvetkov, V. N., Andrianov, K. A., Okhrimenko, G. J., and Vitovskaia, M. G. (1971b). *Eur. Polym. J.* **7**, 1215–1230.

Tsvetkov, V. N., Andreeva, L. N., Magarik, S. Y., Lavrenko, P. N., and Fomin, G. A. (1971c). *Vysokomol. Soedin., Ser. A* **13**, 2011–2019.

Tsvetkov, V. N., Andreeva, L. N., Korneeva, E. V., and Lavrenko, P. N. (1972). *Dokl. Akad. Nauk SSSR* **205**, 895–898.

Tsvetkov, V. N., Rjumtsev, E. I., Kolomiets, I. P., Kovshik, A. P., and Gantseva, N. L. (1973a). *Opt. Spektrosk.* **35**, 880–883.

Tsvetkov, V. N., Rjumtsev, E. I., Shtennikova, I. N., Peker, T. V., and Tsvetkova, N. V. (1973b). *Eur. Polym. J.* **9**, 1–6.

Tsvetkov, V. N., Andrianov, K. A., Rjumtsev, E. I., Shtennikova, I. N., Vitovskaia, M. G., and Makarova, N. N. (1973c). *Eur. Polym. J.* **9**, 27–34.

Tsvetkov, V. N., Rjumtsev, E. I., Shtennikova, I. N., Korneeva, E. V., Krentsel, B. A., and Amerik, Y. B. (1973d). *Eur. Polym. J.* **9**, 481–492.

Tsvetkov, V. N., Rjumtsev, E. I., Aliev, F. M., and Shtennikova, I. N. (1974a). *Eur. Polym. J.* **10**, 55–60.

Tsvetkov, V. N., Rjumstev, E. I., Andreeva, L. N., Pogodina, N. V., Lavrenko, P. N., and Kutsenko, L. G. (1974b). *Eur. Polym. J.* **10**, 563–570.

Tsvetkov, V. N., Rjumtsev, E. I., and Pogodina, N. V. (1975a). *Dokl. Akad. Nauk SSSR* **224**, 112–115.

Tsvetkov, V. N., Rjumtsev, E. I., Pogodina, N. V., and Shetennikova, I. N. (1975b). *Eur. Polym. J.* **11**, 37–42.

Tsvetkov, V. N., Zakcharova, E. N., and Mikchaylova, N. A. (1975c). *Dokl. Akad. Nauk SSSR* **224**, 1365–1368.

Tsvetkov, V. N., Kudrjavtsev, G. I., Shtennikova, I. N., Peker, T. V., Zakcharova, E. N., Kalmykova, V. D., and Volokhina, A. V. (1975d). *Dokl. Akad. Nauk SSSR* **224**, 1126–1129.

Tsvetkov, V. N., Kudrjavtsev, G. I., Rjumtsev, E. I., Nikolaev, V. J., Kalmykova, V. D., and Volokhina, A. V. (1975e). *Dokl. Akad. Nauk SSSR* **224**, 398–401.

Tsvetkov, V. N., Andrianov, K. A., Rjumtsev, E. I., Shtennikova, I. N., Pogodina, N. V., Kolbina, G. F., and Makarova, N. N. (1975f). *Eur. Polym. J.* **11**, 771–780.

Tsvetkov, V. N., Shtennikova, I. N., Peker, T. V., Kudrjavtsev, G. I., Volokhina, A. V., and Kalmykova, V. D. (1976a). *Eur. Polym. J.* **12**, 517–520.

Tsvetkov, V. N., Kudrjavtsev, G. I., Shtennikova, I. N., Peker, T. V., Kalmykova, V. D., and Volokhina, A. V. (1976b). *Vysokomol. Soedin., Ser. A* **18**, 2212–2217.

Vitovskaia, M. G., and Tsevtkov, V. N. (1976). *Eur. Polym. J.* **12**, 251–257.

Vitovskaia, M. G., Tsvetkov, V. N., Godunova, L. I., and Sheremeteva, T. V. (1967). *Vysokomol. Soedin., Ser. A* **9**, 1682–1687.

Vitovskaia, M. G., Astapenko, E. P., Nikolaev. V. J., Didenko, S. A., and Tsvetkov, V. N. (1976). *Vysokomol. Soedin., Ser. A* **18**, 691–695.

Watson, J. D., and Crick, F. H. (1953). *Nature (London)* **171**, 737–738.

Yahindra, N., and Rao, V. S. R. (1970a). *J. Polym. Sci. Part A-2* **8**, 2033–2034.

Yahindra, N., and Rao, V. S. R. (1970b). *Biopolymers* **9**, 783–790.

Yahindra, N., and Rao, V. S. R. (1971a). *J. Polym. Sci., Part A-2* **9**, 1149–1151.

Yahindra, N., and Rao, V. S. R. (1971b). *Biopolymers* **10**, 1891–1900.

Yahindra, N., and Rao, V. S. R. (1972). *J. Polym. Sci., Part A-2* **10**, 1369–1382.

Yamakawa, H. (1971). "Modern Theory of Polymer Solutions." Harper, New York.

3

Liquid Crystalline Order in Polymers with Mesogenic Side Groups

Alexandre Blumstein
Edward C. Hsu

Department of Chemistry
Polymer Program
University of Lowell
Lowell, Massachusetts

I. Introduction

The realization that synthetic polymers are capable of mesophase formation goes back to the time when birefringence of solutions of rigid, helical macromolecules of biological origin was first described in detail (Elliott and Ambrose, 1950). A few years later Paul Flory established the theoretical foundation for the formation of nematic mesophases from solutions of rigid, rodlike particles from purely geometrical requirements (Flory, 1956).

Following the discovery of anionic initiation and interfacial polycondensation, two other classes of polymers, in addition to polypeptides, were found to display mesomorphic behavior. Block and graft copolymers with long sequences of monomer units of dissimilar polarity were synthesized and found to display a variety of lyotropic mesophases, not unlike the classical low molecular amphiphiles (Ricuaros and Szwarc, 1959; Skoulios *et al.*, 1960). This development led to new and important materials: micro-phase composites combining extreme properties, unachievable by statistical copolymerization. At about the same time, a class of rigid macromolecules prepared by condensation was developed to enhance the heat stability of materials for the requirements of space technology. Some of these rigid polymers were shown to form anisotropic solutions in agreement with the earlier predictions of Flory.

Anisotropic solutions of aromatic polyamides when spun give highly oriented high strength, high modulus fibers of increasing technological importance. The recent industrial development of such fibers has brought the field of nematic solutions of rigid chain macromolecules into sharp focus (Preston, 1975). While in the case of block copolymers the driving force for the formation of microdomains of various molecular organizations is the segregation between two long and dissimilar monomer sequences, the for-mation of the nematic phases of rigid, rodlike macromolecules stems from the geometrical anisotropy of the macromolecule as a whole and is assisted by molecular attraction forces.

During the last decade it became apparent that similar phenomena can also be associated with the formation of ordered structures by polymers in which the elements of mesogenic structures are built not into the main chain, as for example in aromatic polyamides, but into the side group. This led to the development of a third class of organized polymers: polymers with side group structure leading to mesomorphic behavior. The structure and properties of such polymers are dominated by the nature and structure of the side group, which in turn depends on the nature of the parent monomer.

The purpose of this chapter is to review the field of polymers with side group structure leading to mesomorphic behavior. For the sake of brevity such side groups will subsequently be called "mesomorphic" or "mesogenic" although the parent monomer may or may not display mesomorphism.

Formation of ordered domains in polymers is related to the phenomenon of local order in liquids, in which "oriented aggregates" of variable size and shape can exist. The driving force for aggregation stems from van der Waals forces. If the geometry of molecules is such as to favor extensive intermolec-ular contacts, ordered domains will persist even at high levels of thermal motion and even in the isotropic melt. For *n*-alkanes it has been shown that a nearly hexagonal packing of the cylindrical molecules in their melt does exist even though there is no evidence for long range ordering (Golik *et al.*, 1967). X-ray patterns of *n*-paraffins containing terminal bromine atoms give a small angle reflection that could well be due to clustering of such molecules (Golik *et al.*, 1967). Clustering is also observed in aliphatic alcohols and acids (Stewart and Marrow, 1927; Marrow, 1928).

Aggregation in isotropic liquids is due to molecular interactions (Frenkel, 1945). The amphiphilic nature of the above mentioned molecules is responsible for the formation of ordered aggregates (micellization) even though the size of such clusters is much below the size of anisotropic domains observed in liquid crystals. Clustering in isotropic liquids generated by thermotropic liquid crystals (i.e., above the nematic–isotropic transition) was demonstrated by V. N. Tsvetkov who has investigated the birefringence of such phases as a function of temperature in the presence of magnetic, hydrodynamic, and acoustic fields (Tsvetkov, 1944; Tsvetkov and Krozer, 1958).

More recently, de Vries was able to demonstrate the existence of pre-transitional or "cybotactic" states in low molecular nematic phases by analyzing low angle X-ray diffraction patterns obtained from oriented spec-imens of nematic fluids (de Vries, 1970).

The experimental evidence for the existence of oriented aggregates in isotropic or nematic phases of interacting anisotropic molecules is accumu-lating. Polymerization of such molecules will bring the anisotropic moieties into close proximity, introduce orientational restrictions, and reduce thermal motion. Thus we can logically expect polymerization to promote the devel-opment of mesomorphic order, although the order in the polymer may not be the same as that of the parent monomer.

II. Monomers

In the past ten years a large number of mesomorphic (or potentially mesomorphic) monomers has been synthesized. These monomers are mainly mono- or divinyl derivatives of compounds that display thermotropic liquid crystalline phases. Polymerization can be achieved by means of free radical initiation, including high energy irradiation. Tables 1–4 give a list of four groups of monomers roughly classified by organic functional group.

Derivatives of benzoic acid are listed in Table 1, Schiff base derivatives in Table 2, derivatives of steroids in Table 3, and miscellaneous monomers in Table 4. Monomers with long linear hydrocarbon chains such as α-olefins, alkyl ethers, and esters have not been included as they very rarely display mesomorphism, in contrast with the corresponding polymers. Phase tran-sitions are listed using the linear notation method introduced by Verbit (Verbit, 1971). For example, K 50 S 100 N 150 I represents

$$\text{crystal} \xrightarrow{50^\circ C} \text{smectic} \xrightarrow{100^\circ C} \text{nematic} \xrightarrow{150^\circ C} \text{isotropic}$$

K 100 I (80 C 60 S 40 K) represents

$$\text{crystal} \xrightarrow{100^\circ C} \text{isotropic}$$
$$40^\circ C \uparrow \qquad\qquad \downarrow 80^\circ C$$
$$\text{smectic} \xleftarrow{60^\circ C} \text{cholesteric}$$

TABLE 1

Derivatives of Benzoic Acid

No.	Molecular structure	Transition temperatures (°C)	Organization of polymer	References
1	$CH_2=CHCOO$—⬡—$COOH$ 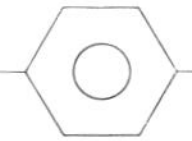	K 201 I	Crystalline	Blumstein et al. (1976)
2	$CH_2=CH-CH_2-O$—⬡—$COOH$ 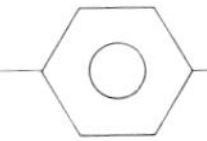	K 160 I		Tanaka (1976)
3	$CH_2=C(CH_3)COO$—⬡—$COOH$ 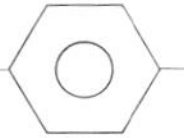	K 182 I	Crystalline and smectic	Amerik and Krentsel (1967); Blumstein et al. (1971, 1974, 1975, 1976)
4	$CH_2=C(CH_3)CH_2O$—⬡—$COOH$ 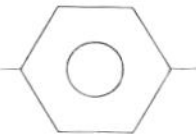	K 129 I		Tanaka (1976)
5	$CH_2=CHOCH_2CH_2O$—⬡—$COOH$ 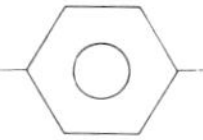	K 165 S 174 N 178 I		Tanaka (1976)
6	$CH_3(CH_2)_7CH=CH(CH_2)_5O$—⬡—$COOH$ 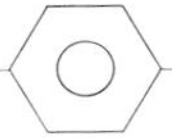	K 65 I		Tanaka (1976)

The derivatives of benzoic acid (Table 1) are generally synthesized from *p*-hydroxybenzoic acid and the appropriate acid chloride by a Shotten–

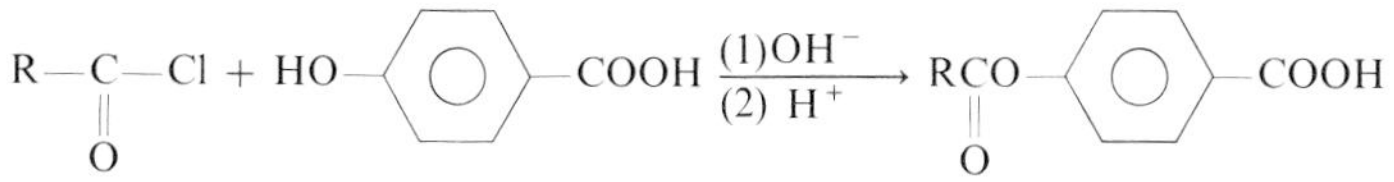

$$R-\overset{\overset{\displaystyle O}{\|}}{C}-Cl + HO-\bigcirc-COOH \xrightarrow[(2)\ H^+]{(1)OH^-} RCO-\bigcirc-COOH$$
$$\overset{\|}{O}$$

(I)

Baumann method **(I)**. The Schiff-base monomers (Table 2) are prepared by condensation of the appropriate aldehyde and amine according to the sequence depicted in schemes **(II)** and **(III)**:

(1)   Styrene derivatives:

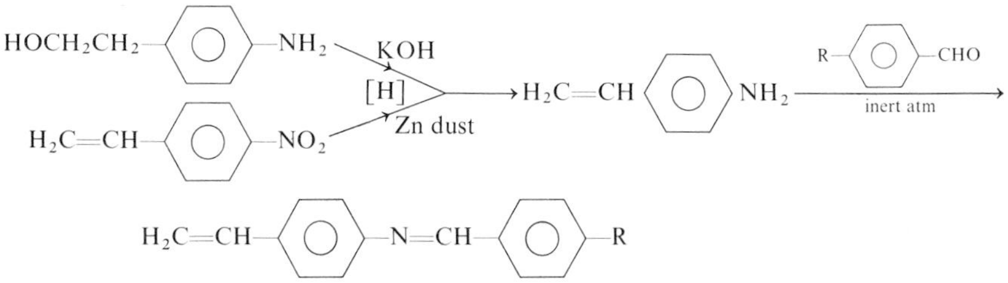

(II)

(2)   Acryloyl or methacryloyl derivatives are prepared in several steps **(III)**:

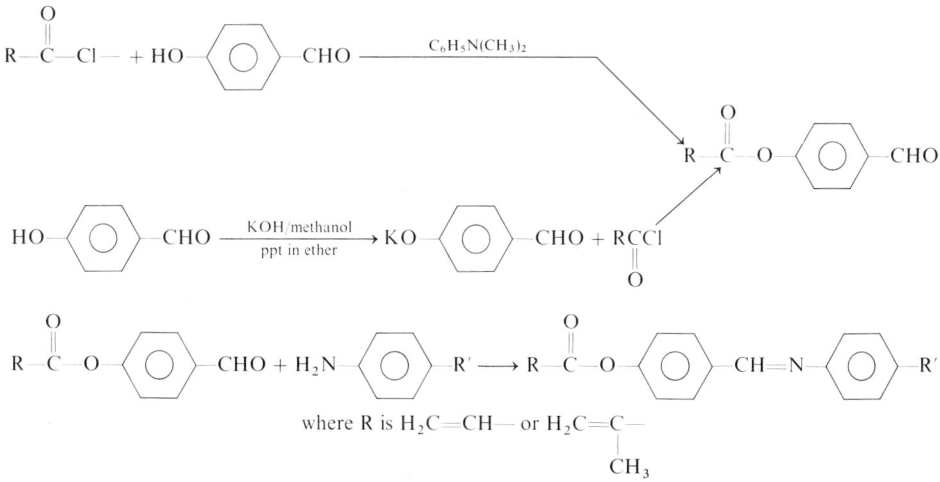

where R is $H_2C=CH-$ or $H_2C=\underset{\underset{\displaystyle CH_3}{|}}{C}-$

(III)

TABLE 2

Schiff-Base Derivatives

No.	Molecular structures	Transition temperatures (°C)	Organization of polymer	References
A. Styrene derivatives				
1	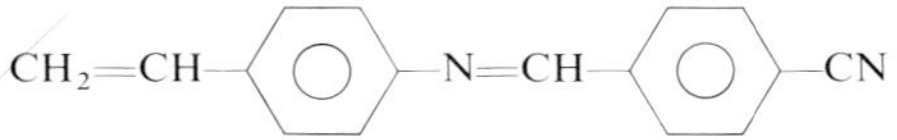 CH$_2$=CH— ⬡ —N=CH— ⬡ —CN	K 113.8 N 140. 5 I	Nematic	Hsu and Blumstein (1977)
2	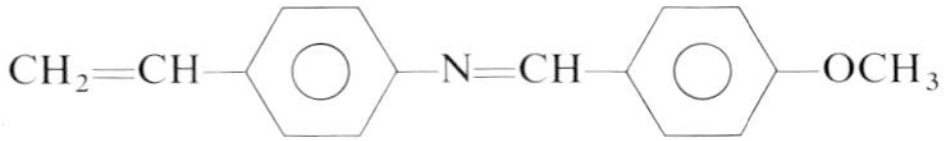 CH$_2$=CH— ⬡ —N=CH— ⬡ —OCH$_3$	K 97.3 N 110.6 I		Paleos et al. (1968)
3	 CH$_2$=CH— ⬡ —N=CH— ⬡ —O-n-C$_4$H$_9$	K 88.3 N 120.6 I	Smectic intermediate	Paleos et al. (1968); Blumstein et al. (1975)
4	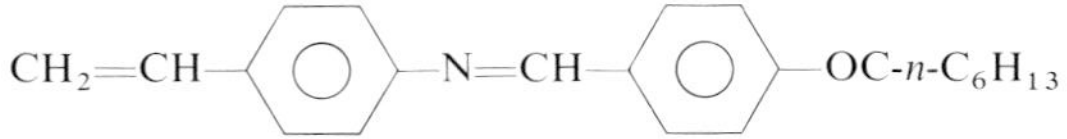 CH$_2$=CH— ⬡ —N=CH— ⬡ —OC-n-C$_6$H$_{13}$	K 94 S 97.5 N 116 I	Smectic intermediate	Hsu et al. (1977)
5	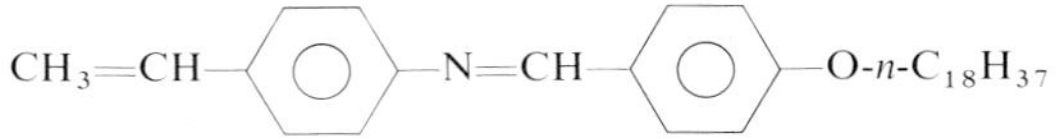 CH$_3$=CH— ⬡ —N=CH— ⬡ —O-n-C$_{18}$H$_{37}$	K 98 S 104 I		Paleos et al. (1968)
6	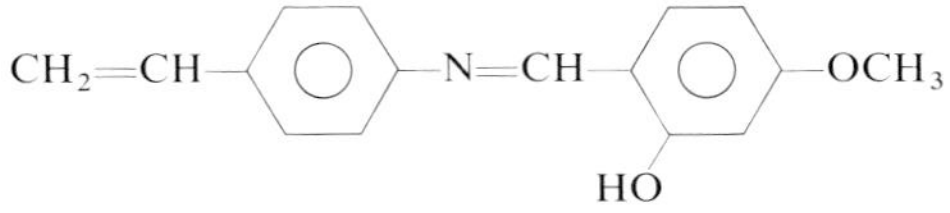 CH$_2$=CH— ⬡ —N=CH— ⬡ —OCH$_3$ (HO)	K 111 N 124 I		Paleos and Labes (1970)
7	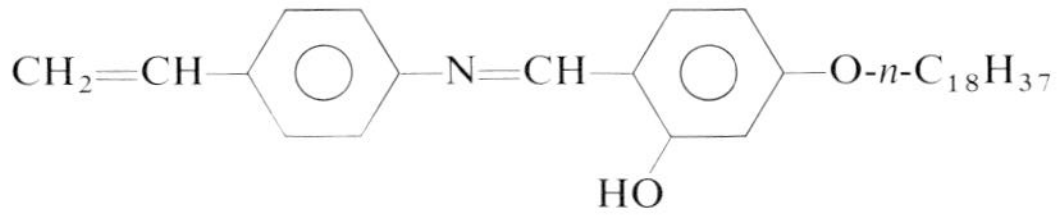 CH$_2$=CH— ⬡ —N=CH— ⬡ —O-n-C$_{18}$H$_{37}$ (HO)	K 93.5 S 108 I		Paleos and Labes (1970)

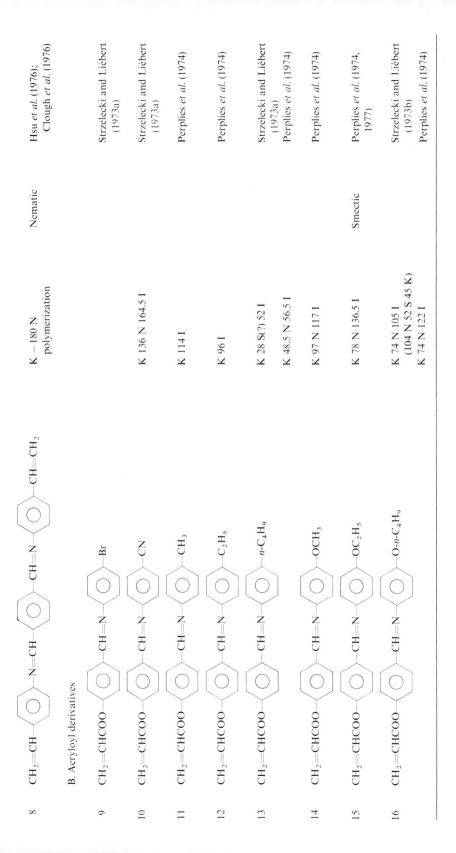

No.	Structure	Transition temperatures		Reference
8	CH_2=CH—⬡—N=CH—⬡—CH=N—⬡—CH=CH_2	K – 180 N polymerization	Nematic	Hsu et al. (1976); Clough et al. (1976)

B. Acryloyl derivatives

No.	R	Transition temperatures		Reference
9	CH_2=CHCOO—⬡—CH=N—⬡—Br			Strzelecki and Liébert (1973a)
10	CH_2=CHCOO—⬡—CH=N—⬡—CN	K 136 N 164.5 I		Strzelecki and Liébert (1973a)
11	CH_2=CHCOO—⬡—CH=N—⬡—CH_3	K 114 I		Perplies et al. (1974)
12	CH_2=CHCOO—⬡—CH=N—⬡—C_2H_5	K 96 I		Perplies et al. (1974)
13	CH_2=CHCOO—⬡—CH=N—⬡—n-C_4H_9	K 28 S(?) 52 I / K 48.5 N 56.5 I		Strzelecki and Liébert (1973a) / Perplies et al. (1974)
14	CH_2=CHCOO—⬡—CH=N—⬡—OCH_3	K 97 N 117 I		Perplies et al. (1974)
15	CH_2=CHCOO—⬡—CH=N—⬡—OC_2H_5	K 78 N 136.5 I	Smectic	Perplies et al. (1974, 1977)
16	CH_2=CHCOO—⬡—CH=N—⬡—O-n-C_4H_9	K 74 N 105 I (104 N 52 S 45 K) / K 74 N 122 I		Strzelecki and Liébert (1973b) / Perplies et al. (1974)

TABLE 2 *(continued)*

No.	Molecular structures	Transition temperatures (°C)	Organization of polymer	References
17	$CH_2=CHCOO$—⟨ring⟩—$CH=N$—⟨ring⟩—O-n-C_8H_{17}	K 70 S 122 I		Perplies *et al.* (1977)
18	$CH_2=CHCOO$—⟨ring⟩—$CH=N$—⟨ring⟩—O-n-$C_{12}H_{25}$	K 87 S 108 I		Perplies *et al.* (1977)
19	$CH_2=CHCOO$—⟨ring⟩—$CH=N$—⟨ring⟩—$CH=CHCOO$—CH_3	K 111 S_A 148 N 209 I	Smectic	Strzelecki *et al.* (1975)
20	$CH_2=CHCOO$—⟨ring⟩—$CH=N$—⟨ring⟩—$CH=CHCOO$—C_2H_5	K 80 S_A 130 N 169 I	Smectic	Strzelecki *et al.* (1975)
21	$CH_2=CHCOO$—⟨ring⟩—$CH=N$—⟨ring⟩—$CH=CHCOO$—n-C_3H_7	K 90 S_A 134 N 164 I	Smectic	Strzelecki *et al.* (1975)
22	$CH_2=CHCOO$—⟨ring⟩—$CH=N$—⟨ring⟩—$CH=CHCOO$—n-C_4H_9	K 77 S_A 125 N 139 I	Smectic	Strzelecki *et al.* (1975)
23	$CH_2=CHCOO$—⟨ring⟩—$CH=N$—⟨ring⟩—$CH=CHCOO$—n-C_5H_{11}	K 59 S_A 125 N 140 I	Smectic	Strzelecki *et al.* (1975)
24	$CH_2=CHCOO$—⟨ring⟩—$CH=N$—⟨ring⟩—$CH=CHCOO$—n-C_6H_{13}	K 61 S_A 117 N 128 I	Smectic	Strzelecki *et al.* (1975)
25	$CH_2=CHCOO$—⟨ring⟩—$CH=N$—⟨ring⟩—$CH=CHCOO$—n-C_7H_{15}	K 65 S_A 120 N 129 I	Smectic	Strzelecki *et al.* (1975)

No.	Structure	Transitions	Phase	Reference
26	CH₂=CHCOO—⟨ring⟩—CH=N—⟨ring⟩—CH=CHCOO—n-C$_8$H$_{17}$	K 78 S$_A$ 114 N 124 I	Smectic	Strzelecki et al. (1975)
27	CH₂=CHCOO—⟨ring⟩—CH=N—⟨ring⟩—CH=CHCOO—n-C$_9$H$_{19}$	K 66 S$_A$ 124 N 127 I	Smectic	Strzelecki et al. (1975)
28	CH₂=CHCOO—⟨ring⟩—CH=N—⟨ring⟩—CH=CHCOO—n-C$_{10}$H$_{21}$	K 74 S$_A$ 125 I	Smectic	Strzelecki et al. (1975)
29	CH₂=CHCOO—⟨ring⟩—CH=N—⟨ring⟩—CH=CHCOO—n-C$_{11}$H$_{23}$	K 72 S$_A$ 124 I	Smectic	Strzelecki et al. (1975)
30	CH₂=CHCOO—⟨ring⟩—CH=N—⟨ring⟩—COCH$_3$	K 110 I (108 N 86 K)		Strzelecki and Liébert (1973b)
31	CH₂=CHCOO—⟨ring⟩—CH=N—⟨ring⟩—COOH	K 218 S → polymerization K 280 (decomposition)		Strzelecki and Liébert (1973b) Perplies et al. (1977)
32	CH₂=CHCOO—⟨ring⟩—CH=N—⟨ring⟩—COOC$_2$H$_5$	K 73 I (60 N 30 K)		Strzelecki and Liébert (1973b) Perplies et al. (1977)
33	CH₂=CHCOO—⟨ring⟩—CH=N—⟨ring⟩—COO-n-C$_4$H$_9$	K 76 I		Strzelecki and Liébert (1973b)
34	CH₂=CHCOO—⟨ring⟩—CH=N—⟨ring⟩—NHCOCH$_3$	K 197 I		Strzelecki and Liébert (1973b)
35	CH₂=CHCOO—⟨ring⟩—CH=N—⟨ring⟩—N=N—⟨ring⟩	K 118 N 196 I		Strzelecki and Liébert (1973b)

TABLE 2 (continued)

No.	Molecular structures	Transition temperatures (°C)	Organization of polymer	References
36	$CH_2=CHCOO$—⬡—$CH=N$—⬡—$N=CH$—⬡—$OOCCH=CH_2$	K 180 S → polymerization	Smectic and nematic	Strzelecki and Liébert (1973a)
37	$CH_2=CHCOO$—⬡—$CH=N$—⬡(Cl)—$N=CH$—⬡—$OOCCH=CH_2$	K 185 S 189 I		Strzelecki and Liébert (1973a)
38	$CH_2=CHCOO$—⬡—$CH=N$—⬡(SO_3Na)	K 300 (decomposition)		Perplies et al. (1977)
39	$CH_2=CHCOO$—⬡—$CH=N$—$N=CH$—⬡—$OCOCH=CH_2$	K 140 N → polymerization	Nematic	Strzelecki and Liébert (1973b)
	C. Methacryloyl derivatives			
40	$CH_2=C(CH_3)COO$—⬡—$CH=N$—⬡—CH_3	K 68 I		Perplies et al. (1974)
41	$CH_2=C(CH_3)COO$—⬡—$CH=N$—⬡—C_2H_5	K 62 I		Perplies et al. (1974)
42	$CH_2=C(CH_3)COO$—⬡—$CH=N$—⬡—$n\text{-}C_4H_9$	K 58 I		Perplies et al. (1974)
43	$CH_2=C(CH_3)COO$—⬡—$CH=N$—⬡—OCH_3	K 130.5 I		Perplies et al. (1974)

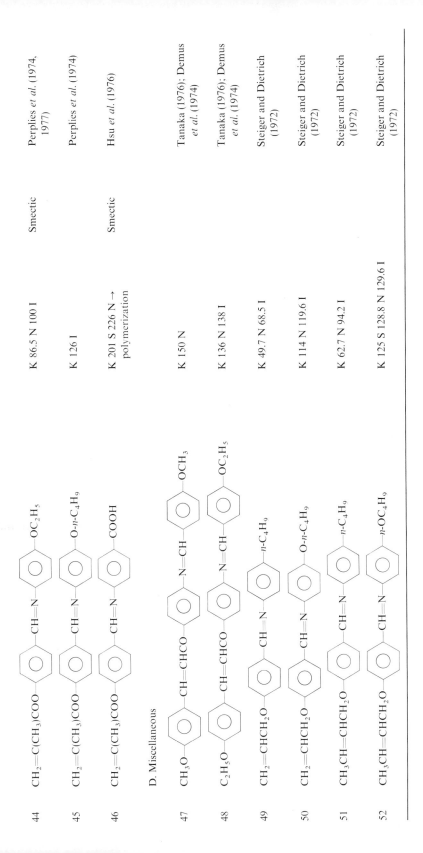

No.	Structure	Transition	Type	Reference
44	$CH_2=C(CH_3)COO$–⟨⟩–$CH=N$–⟨⟩–OC_2H_5	K 86.5 N 100 I	Smectic	Perplies et al. (1974, 1977)
45	$CH_2=C(CH_3)COO$–⟨⟩–$CH=N$–⟨⟩–O-n-C_4H_9	K 126 I		Perplies et al. (1974)
46	$CH_2=C(CH_3)COO$–⟨⟩–$CH=N$–⟨⟩–$COOH$	K 201 S 226 N → polymerization	Smectic	Hsu et al. (1976)

D. Miscellaneous

No.	Structure	Transition	Type	Reference
47	CH_3O–⟨⟩–$CH=CHCO$–⟨⟩–$N=CH$–⟨⟩–OCH_3	K 150 N		Tanaka (1976); Demus et al. (1974)
48	C_2H_5O–⟨⟩–$CH=CHCO$–⟨⟩–$N=CH$–⟨⟩–OC_2H_5	K 136 N 138 I		Tanaka (1976); Demus et al. (1974)
49	$CH_2=CHCH_2O$–⟨⟩–$CH=N$–⟨⟩–n-C_4H_9	K 49.7 N 68.5 I		Steiger and Dietrich (1972)
50	$CH_2=CHCH_2O$–⟨⟩–$CH=N$–⟨⟩–O-n-C_4H_9	K 114 N 119.6 I		Steiger and Dietrich (1972)
51	$CH_3CH=CHCH_2O$–⟨⟩–$CH=N$–⟨⟩–n-C_4H_9	K 62.7 N 94.2 I		Steiger and Dietrich (1972)
52	$CH_3CH=CHCH_2O$–⟨⟩–$CH=N$–⟨⟩–n-OC_4H_9	K 125 S 128.8 N 129.6 I		Steiger and Dietrich (1972)

TABLE 3

Derivatives of Steroids

No.	Molecular structures	Transition temperatures (°C)	Organization of polymer	References
1	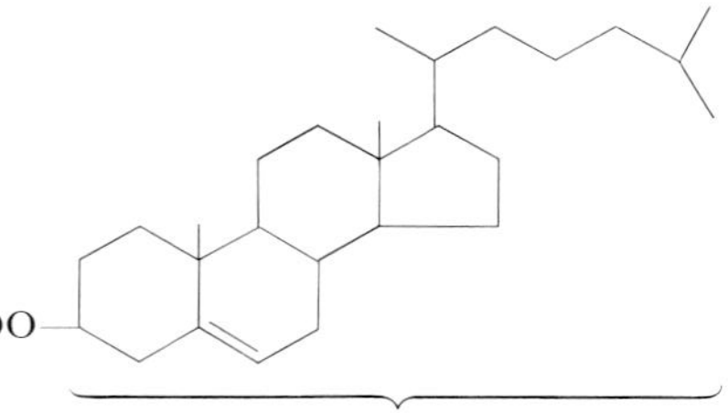 CH₂=CHCOO— (steroid structure, C₂₇H₄₅)	K 118.7 N 125.8 I		Hardy et al. (1969a, 1970)
		K 127 I (90 C)		Toth and Tobolsky (1970)
		K 122.5 C 125 I		de Visser et al. (1970, 1971, 1972)
		K 118 C 126 I		Tanaka et al. (1973)
		K 127 I (78 C)		Tanaka (1976)
2	CH₂=C(CH₃)COO—C₂₇H₄₅	K 108 C 112 I		Tanaka et al. (1972)
		K 109 I (103 C 85 K)		Saeki et al. (1972)
		K 114 I (111.5 C 92 K)		de Visser et al. (1972)
		K 103 C 112 I		Tanaka et al. (1973)
3	trans-CH₃CH=CHCOO—C₂₇H₄₅	K 127 C 162 I		Tanaka (1976); Demus et al. (1974)
4	CH₃C≡CCOOC₂₇H₄₅	K 115 C 127 I		Tanaka (1976); Demus et al. (1974)
5	CH₂=CHCH₂CH₂COO—C₂₇H₄₅	K 74 C 95 I		Tanaka (1976); Demus et al. (1974)
6	CH₃CH=CHCH=CHCOO—C₂₇H₄₅	K 127 C 168 I		Tanaka (1976); Demus et al. (1974)
7	CH₂=CHOOCCH₂CH₂COO—C₂₇H₄₅	K 82 C 89 I		Nyitrai et al. (1976a,b)

No.	Structure	Transition temperatures	Mesophase	References
8	CH₂=CHCOO—⟨C₆H₄⟩—COO—C₂₇H₄₅	K 128 C → polymerization	Smectic	Hsu et al. (1977)
9	cis-CH₃(CH₂)₇CH=CH(CH₂)₇COO—C₂₇H₄₅	K 50.5 I (47.5 C 42 S)		Tanaka (1976); Demus et al. (1974)
10	trans-CH₃(CH₂)₇CH=CH(CH₂)₇COO—C₂₇H₄₅	K 58 C 65.5 I		Tanaka (1976); Demus et al. (1974)
11	CH₃(CH₂)₇CH=CH(CH₂)₁₁COO—C₂₇H₄₅	K 26 C 41 I		Tanaka (1976); Demus et al. (1974)
12	CH₃(CH₂)₄CH=CHCH₂CH=CH(CH₂)₇COO—C₂₇H₄₅	K 42 I (36.5 C 34 S)		Tanaka (1976); Demus et al. (1974)
13	CH₃(CH₂CH=CH)₃(CH₂)₇COO—C₂₇H₄₅	K 35.5 I (29 C 27.5 S)		Tanaka (1976); Demus et al. (1974)
14	CH₂=C(CH₃)CO(CH₂)₁₀COO—⟨C₆H₄⟩—CH=CHCOO—C₂₇H₄₅	K 105 N 145 I		Tanaka (1976)
15	CH₂=C(CH₃)CONH(CH₂)ₙCOO—C₂₇H₄₅ (n = 2, 5, 6, 8, 10, 11)		Smectic	Shibaev et al. (1976)

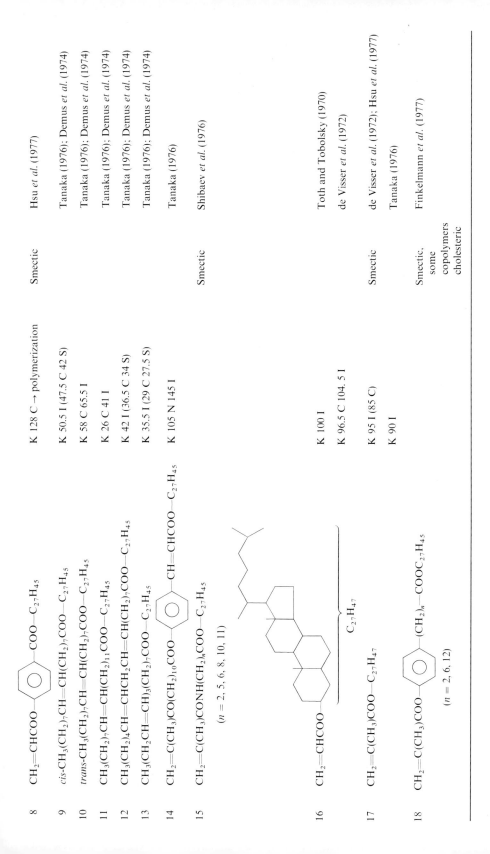

$C_{27}H_{47}$

No.	Structure	Transition temperatures	Mesophase	References
16	CH₂=CHCOO—C₂₇H₄₇	K 100 I		Toth and Tobolsky (1970); de Visser et al. (1972)
17	CH₂=C(CH₃)COO—C₂₇H₄₇	K 96.5 C 104.5 I (85 C)	Smectic	de Visser et al. (1972); Hsu et al. (1977); Tanaka (1976)
18	CH₂=C(CH₃)COO—⟨C₆H₄⟩—(CH₂)ₙ—COOC₂₇H₄₅ (n = 2, 6, 12)	K 90 I	Smectic, some copolymers cholesteric	Finkelmann et al. (1977)

TABLE 4

Miscellaneous Monomers

No.	Molecular structure	Transition temperatures (°C)	Organization of polymer	References
1	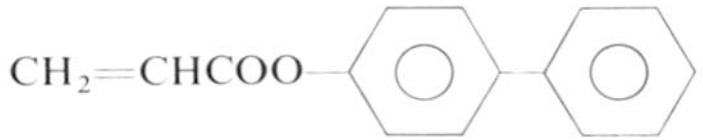 CH_2=CHCOO—⟨biphenyl⟩	K 65 I	Smectic	Baccaredda et al. (1971) Ceccarelli et al. (1975)
2	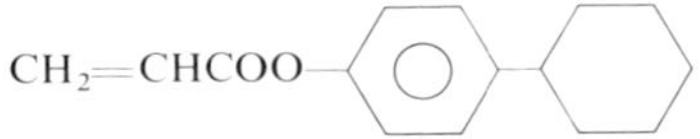 CH_2=CHCOO—⟨phenyl-cyclohexyl⟩	K 62.5 I	Smectic	Magagnini et al. (1974)
3	C_2H_5—CH=CHCH=CHCOOH	K 44 N 46 I		Markan and Maier (1962)
4	n-C_4H_9—CH=CHCH=CHCOOH	K 23 N 53.5 I		Markan and Maier (1962)
5	n-C_6H_{13}—CH=CHCH=CHCOOH	K 32 N 62.5 I		Markan and Maier (1962)
6	$CH_3(CH_2)_7$CH=CH$(CH_2)_7$COOCH=CH_2	K 45 S 18 I K 36 S 8 I		Amerik and Krentsel (1967) Hardy et al., (1976)
7	n-CH_2=CHOC$_{16}$H$_{33}$	K 2.5 S 16.5 I		Hardy et al. (1969b)
8	 CH_2=CHCOO—⟨phenyl⟩C≡C—⟨phenyl⟩—OOCCH=CH_2	K 135 N 141 I		Strzelecki and Liébert (1973b)
9	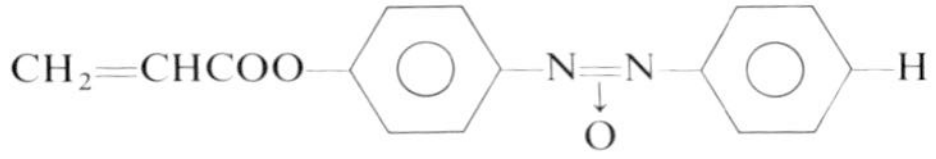 CH_2=CHCOO—⟨phenyl⟩N=N(→O)—⟨phenyl⟩—H	K 83 I		Cser et al. (1976)
10	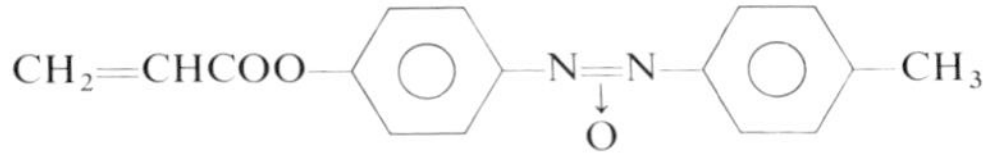 CH_2=CHCOO—⟨phenyl⟩N=N(→O)—⟨phenyl⟩—CH_3	K 83 N 89 I		Nyitrai et al. (1976a) Cser et al. (1976)

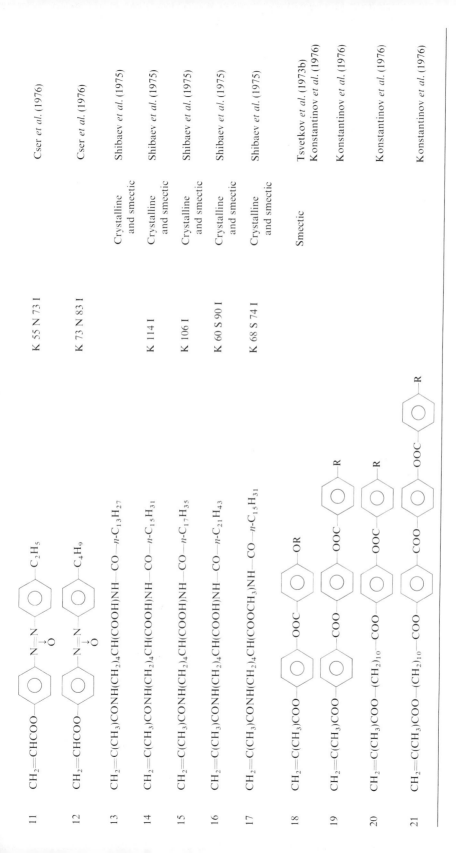

No.	Structure	Phase behaviour	Reference
11	$CH_2=CHCOO$—⬡—$N=N(\rightarrow O)$—⬡—C_2H_5	K 55 N 73 I	Cser et al. (1976)
12	$CH_2=CHCOO$—⬡—$N=N(\rightarrow O)$—⬡—C_4H_9	K 73 N 83 I	Cser et al. (1976)
13	$CH_2=C(CH_3)CONH(CH_2)_4CH(COOH)NH$—$CO$—$n\text{-}C_{13}H_{27}$	Crystalline and smectic	Shibaev et al. (1975)
14	$CH_2=C(CH_3)CONH(CH_2)_4CH(COOH)NH$—$CO$—$n\text{-}C_{15}H_{31}$	K 114 I; Crystalline and smectic	Shibaev et al. (1975)
15	$CH_2=C(CH_3)CONH(CH_2)_4CH(COOH)NH$—$CO$—$n\text{-}C_{17}H_{35}$	K 106 I; Crystalline and smectic	Shibaev et al. (1975)
16	$CH_2=C(CH_3)CONH(CH_2)_4CH(COOH)NH$—$CO$—$n\text{-}C_{21}H_{43}$	K 60 S 90 I; Crystalline and smectic	Shibaev et al. (1975)
17	$CH_2=C(CH_3)CONH(CH_2)_4CH(COOCH_3)NH$—$CO$—$n\text{-}C_{15}H_{31}$	K 68 S 74 I; Crystalline and smectic	Shibaev et al. (1975)
18	$CH_2=C(CH_3)COO$—⬡—OOC—⬡—OR	Smectic	Tsvetkov et al. (1973b); Konstantinov et al. (1976)
19	$CH_2=C(CH_3)COO$—⬡—COO—⬡—OOC—⬡—R		Konstantinov et al. (1976)
20	$CH_2=C(CH_3)COO$—$(CH_2)_{10}$—COO—⬡—OOC—⬡—R		Konstantinov et al. (1976)
21	$CH_2=C(CH_3)COO$—$(CH_2)_{10}$—COO—⬡—COO—⬡—OOC—⬡—R		Konstantinov et al. (1976)

TABLE 4 (continued)

No.	Molecular structure	Transition temperatures (°C)	Organization of polymer	References
22	$CH_2{=}CHCOO{-}$⟨ring⟩$-N{=}N-$⟨ring⟩$-OCH_3$	K 97 N 117 I		Lecoin et al. (1975b)
23	$CH_2{=}CHCOO{-}$⟨ring⟩$-N{=}N-$⟨ring⟩$-OC_2H_5$	K 89 N 137 I		Lecoin et al. (1975b)
24	$CH_2{=}CHCOO{-}$⟨ring⟩$-N{=}N-$⟨ring⟩$-n{-}OC_4H_9$	K 88 N 123 I		Lecoin et al. (1975b)
25	$CH_2{=}CHCOO{-}$⟨ring⟩$-N{=}N{\rightarrow}O-$⟨ring⟩$-OCH_3$	K 87 N 147 I		Hochapfel et al. (1976)
26	$CH_2{=}CHCOO{-}$⟨ring⟩$-N{=}N{\rightarrow}O-$⟨ring⟩$-OC_2H_5$	K 114 N 160 I		Hochapfel et al. (1976)
27	$CH_2{=}CHCOO{-}$⟨ring⟩$-N{=}N{\rightarrow}O-$⟨ring⟩$-O{-}n{-}C_4H_9$	K 65 N 145 I		Hochapfel et al. (1976)
28	$CH_3{-}(CH_2)_3{-}CH{=}CH{-}CH{=}CH{-}COOH$	K 20 N 49 I		Lecoin et al. (1975a)
29	$CH_2{=}CHCOO{-}$⟨ring⟩$-COO{-}$⟨ring⟩$-O{-}n{-}C_6H_{13}$		Smectic	Lorkowski and Reuther (1976)
30	$CH_2{=}C(CH_3)COO{-}(CH_2)_n{-}O{-}$⟨ring⟩$-COO{-}$⟨ring⟩$-R$		Smectic	Finkelmann et al. (1977)

Derivatives of steroids (Table 3) are generally prepared by reaction of the appropriate acid chloride and steroid (cholesterol or cholestanol) in a suitable solvent. For most steroid derivatives, it is very difficult to obtain a state of high purity by conventional recrystallization. It is therefore not surprising that different phase transition temperatures are reported by different investigators.

III. Liquid Crystalline Order in Tightly Cross-Linked Networks

In 1933 Vorländer, one of the pioneers in the field of liquid crystals, had mentioned the possibility of preparing liquid crystalline resins and lacquers (Vorländer, 1933). In 1963, Herz succeeded in "locking in" the organization of the neat phase of the 11-sodium styrylundecanoate water system (Herz et al., 1963). Polymerization was carried out in the presence of a small amount of divinyl benzene used to cross-link the structure and to permanently lock into the polymer the molecular organization of a neat soap. The presence of the cross-linking agent insured the stability of the neat phase of the polysoap well beyond the range of temperatures and compositions of thermodynamic stability for this phase. Poly(11-sodiumstyrylundecanoate) was the first representative of a new class of synthetic polymeric materials: polymers with locked-in molecular organization, frequently but improperly referred to as "solid liquid crystals."

Locking in of the two-dimensional organization of a monolayer of methyl methacrylate molecules was reported by Blumstein et al. (1969). Here again a cross-linking agent (tetraethylene glycol dimethacrylate) was used to stabilize the two-dimensional arrangement of monomer molecules. Other experiments have been performed on monolayers of monomers in order to retain their two-dimensional organization. In all cases polyfunctional monomers were present in large proportions (Bresler et al., 1941; Dubault et al., 1973). The two-dimensional order of the monomeric film was retained through cross-linking, but the molecular organization of the polymers was not investigated in enough detail to warrant any conclusion as to their intrinsic ability to organize.

In a series of recent papers Liébert and Strzelecki have reported locking in the organization of a large number of mesomorphic monomers (Strzelecki and Liébert, 1973a,b; Liébert and Strzelecki, 1973a,b; Bouligand et al., 1974; Strzelecki et al., 1975). Synthesis and bulk polymerization of a series of monomeric Schiff base derivatives of acrylic acid and para-substituted diphenyl Schiff bases were described (Table 2: monomers 9, 10, 13, 31, 36, and 37). Organization of the polymers and that of the monomers before polymerization was studied by means of polarizing microscopy. A perfect retention of the nematic, smectic, and cholesteric organization of the monomeric

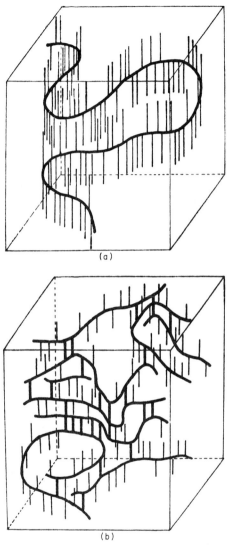

Fig. 1. Polymer with nematic disposition of side groups (Strzelecki and Liébert, 1973a). (a) Linear chain, (b) cross-linked network.

phase can be achieved provided large amounts of cross-linking agents are added. The monomer forms a mixed mesophase that must contain at least 30% by weight of a divinyl monomer of chemical structure similar to that of the monofunctional monomer. Figures 1–3 show models of nematic, smectic, and cholesteric arrangements of side groups in linear and cross-linked

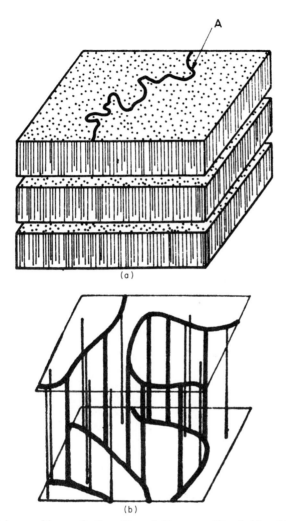

Fig. 2.   Polymer with smectic disposition of side groups (Strzelecki and Liébert, 1973a). (a) Linear chain, (b) cross-linked network.

polymers (Strzelecki and Liébert 1973a). The polymer of *N*-(*p*-acryloyloxy-benzylidene)-*p*-aminobenzoic acid is an exceptional case in which cross-linking is due to hydrogen bonding between benzoic acid moieties. In all other instances addition of a cross-linking agent was required to prevent precipitation and relaxation of the nascent macromolecules. The resulting polymers were insoluble, infusible glasses with liquid crystalline organization. It is remarkable that the authors succeeded in locking in, not only the overall

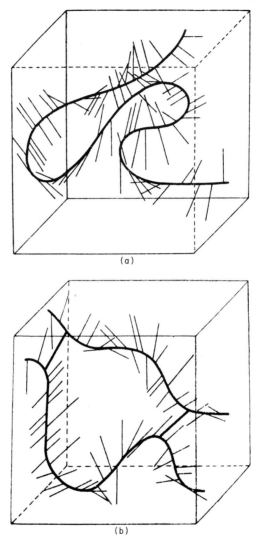

Fig. 3. Polymer with cholesteric disposition of side groups (Sterzelecki and Liébert, 1973a). (a) Linear chain, (b) cross-linked network.

organization of their monomers, but also their characteristic textures. This was demonstrated by conoscopic observation of thin, polished specimens. Nematic phases were locked into cross-linked networks derived from mono-mers 9, 10, 13, 16, 35, and 39 (Table 2), nematic and smectic phases were obtained with polymers derived from monomer 36 (Table 2), smectic phases were obtained with polymers derived from monomers 19–29 (Table 2), and cholesteric or twisted nematic phases were obtained from terpolymers of mono- and diacrylic Schiff bases and cholesteryl acrylate. The amount of cholesteryl acrylate determines the chirality or the pitch of the helix in the polymeric glass. A typical composition of seven parts of cyanoacrylic Schiff base (Table 2, monomer 10), two parts of di-(*n*-*p*-acryloyloxybenzylidene)-*p*-diaminobenzene (Table 2, monomer 36), and one part of cholesteryl acrylate (Table 3, monomer 1) gives a helix pitch of 15,800 Å. Changing the composition to 7:2.75:0.25 of these components increases the pitch to 44,000 Å (Liébert and Strzelecki, 1973a).

"Monocrystals" or uniformly oriented films of copolymers of diacrylic and monoacrylic Schiff bases were obtained by polymerization of thin (0.1 cm) films of monomer in a magnetic field of 5000–8000 Oe. Such films when polished and examined in convergent laser light display interference figures characteristic of nematics aligned parallel to the surface of the film (Liébert and Strzelecki, 1973b). A detailed microscopic study of copolymers of mono- and diacrylic Schiff bases and of terpolymers with cholesteryl acrylate revealed a far reaching fixation of monomer textures without the complication of surface defects common in mesophases. Such defects can be simply removed by careful polishing of the specimen (Bouligand *et al.*, 1974). This technique provides physicists with an additional tool for the study of various textures of mesophases. Crystal to mesophase transition temperatures of the Schiff base monomers described by Liébert and Strzelecki are high, typically above 200°C. Bulk polymerization in this range of temperatures is difficult to control. Fixation of structure can also take place at lower temperatures using monomers for which transition temperatures are significantly lower. Alkyl *N*-(*p*-acryloyloxybenzylidene)-*p*-aminocinnamates have a melting range much below 200°C (Table 2, monomers, 19–29). Most give smectic A structures at temperatures below 100°C and nematic structures below 170°C (Strzelecki *et al.*, 1975). Locking in of the mesomorphic order can be carried out by means of ultraviolet radiation (Liébert *et al.*, 1975). The cinnamic monomers have the advantage over the simpler, higher melting Schiff bases of being able to generate both smectic and nematic phases, the former of remarkable stability. Organization of the alkyl (*N*-*p*-acryloyloxybenzylidene)-*p*-aminocinnamate in both phases can be locked in the polymer without incorporation of a cross-linking agent. It is assumed by the authors that cross-linking takes place through the opening of the cinnamic double bond.

The study of Liébert, Strzelecki, and co-workers has been carried out on well-defined liquid crystalline textures. It shows without doubt that faithful locking in of mesomorphic textures down to the smallest regions is possible if care is being taken to prevent segmental relaxation. It provides means of obtaining a "profile" of the molecular organization of the liquid crystal, a technique complementary of rapid quenching of mesophases below their glass transition temperatures (Chistyakov, 1964; Kessler and Lydon, 1974) with the added advantage of permanency at room temperature and above.

IV. Mesomorphic Order in Linear Polymers

Development of mesomorphic order and its persistence in the polymer are not limited to insoluble and infusible networks such as in the previous section. A great number of monomers of the type $H_2C{=}CRR'$ or $H_2C{=}CHR$, where R is a bulky, anisotropic side group, give soluble polymers in which a nematic, smectic, and possibly cholesteric organization of side groups prevails. The monomers listed in Tables 1–4 are not necessarily mesomorphic, but the lathlike or disklike shape of the R group and the presence of highly polarizable moieties and suitably oriented dipoles insure coupling between neighboring side groups and lead to the development of intra- and intermolecular order in the polymer. This ordering can only be achieved at levels of segmental motion that are compatible with the existing strength of interaction between the side groups. If the level of thermal energy exceeds the energy of coupling between side groups, the intra- and intermolecular order of such macromolecules is destroyed. However, orientation of side groups in such polymers is often cooperative especially in molecules where the side groups are bearers of dipoles and polarizable groups. The orientation is then spontaneous and difficult to destroy.

A. Intramolecular Order

The first step in the study of such orientation effects is the investigation of the intramolecular ordering in polymers (Tsvetkov, 1965, 1969; Tsvetkov et al., 1972, 1973a,b; Korneeva et al., 1975). Tsvetkov and co-workers have extensively studied the dilute solution behavior of macromolecules containing long aliphatic side groups and side groups containing, in addition to an alkyl chain, a rigid moiety composed of two benzene rings joined in the para position by an ester linkage. In the former group of macromolecules, side group interactions occur through dispersion forces, in the latter through induction and dipolar forces (Tsvetkov, 1969).

By a combination of methods including studies of streaming birefringence, diffusion and sedimentation, Kerr effect, intrinsic viscosity, and light scattering, the macromolecular parameters of such chains in a variety of solvents were established. These parameters included among others segmental anisotropy of polarizabilities $\alpha_1 - \alpha_2$ and Kuhn's statistical segment length A (or v number of monomer units encompassed by a statistical segment). The macromolecular backbones remain essentially flexible, although ordering of the side groups does increase the backbone rigidity to some extent. This is reflected in an enhanced value of A and becomes evident as one passes from polyalkyl esters in which the number of carbon atoms n in the side chain is smaller than 8 (no interactions) to esters in which n exceeds 8 (presence of (interactions) (Tsvetkov, 1969; see also Chapter 2, Table 2). This phenomenon is even more pronounced in a series of poly(p-phenylmethacrylic esters of p-n-alkyloxybenzoic acid) (Table 4, monomer 18) in which the side group is characterized by a strongly mesogenic chemical structure. Even so the rigidity of such macromolecules is far below the rigidity of stiff linear chains of the aramid type. The value of $\alpha_1 - \alpha_2$ in this series of polymers was shown to be of the order of several thousand (Tsvetkov, 1969; Tsvetkov *et al.*, 1973b), comparable in magnitude to the optical anisotropies of highly rigid aromatic polyamides. This indicates a high degree of side group ordering. The resulting dipole moment in the direction of the macromolecular backbone was found to be unusually high, in comparison with other flexible macromolecules, indicating a high degree of dipole alignment. The combination of orientational (axial) and dipolar orders is a unique characteristic of flexible macromolecules with mesogenic side groups. Recent studies of copolymers of p-phenylmethacrylic esters of the p-cetyloxybenzoic acid and cetyl methacrylate in the realm of respective compositions from 1:1 to 7:3 (mole ratios) have shown that in this interval of compositions the rigidity and conformation of such macromolecules remain unchanged (Korneeva *et al.*, 1975). The intramolecular order in polymers is extensively described in Chapter 2 of this book.

B. Intermolecular Order

1. Polymers with Linear Paraffinic Side Groups

A mesomorphic, layered structure was proposed in 1961 for polymers with long linear paraffinic side groups (Brownawell and Feng, 1961).

Among the typical compounds of this kind are the higher homologues of poly(olefins) (Turner-Jones, 1964), poly(n-acrylates), poly(vinyl n-alkyl ethers), poly(vinyl n-alkyl esters), and poly(acylstyrenes) (Greenberg and Alfrey, 1954; Shibaev *et al.*, 1968; Jordan, 1971). They all have in common a long paraffinic side group that becomes dominant when the number n of methylene units exceeds 8. The side group crystallizes easily. Above the crystalline melting point a large degree of order is still preserved and layered liquid crystalline structures are formed. The corresponding monomers however are isotropic in their molten state, with the exception of vinyl oleate that has been found to be mesomorphic (Amerik and Krentsel, 1967; Farinha-Martins, 1969).

The field of polymers with long n-paraffinic side groups was extensively reviewed by Platè and Shibaev who have also studied different properties of these comblike structures (Platè and Shibaev, 1974). X-ray and electron diffraction studies performed on poly(n-alkylacrylates), poly(n-alkyl methacrylates), and poly(n-alkylvinyl esters) above the crystalline melting point indicate that two important spacings appear (Shibaev *et al.*, 1969; Platè *et al.*, 1971). One, the wide angle spacing d' is independent of the length of the side group and only slightly dependent on the rigidity of the bridge between the backbone of the polymer and the alkyl moiety of the side group. The spacing d', located between 4.6 and 5.0 Å, is attributed to the distance between side groups. The second spacing d'' is at small angles and increases regularly with the length of the alkyl moiety in the side group. It is attributed to the distance between the strata that confine the macromolecular backbones. According to Platè and Shibaev this regular variation of $d''(n)$ can be expressed for a variety of comblike polymers by a linear equation: $d'' = d_0 + an$, where d_0 is the diameter of the macromolecular backbone augmented by the attachment bridge and a gives the side chain length increment for each additional CH_2 group, varying from 0.98 to 1.76. Both parameters depend strongly on the degree of flexibility of the alkyl moiety, which in turn depends on the freedom of C–C rotations allowed by the attachment bridge. Thus if the flexibility is high, as in poly(acrylates) or poly(vinyl esters), the alkyl moiety takes on a number of kinked conformations leading to an overall shortening of d'' and values of a smaller than 1.22 Å (for an extended C–C zigzag, $a = 1.27$ Å). For polymers with somewhat higher rigidity such as poly(methacrylates), the value of a equals 1.27 Å, indicating a zigzag conformation of the paraffinic side group. The values of d'' are in agreement with a model in which backbones are confined to layers spaced by distances roughly equal to the length of the side chain. The model for such ordering is therefore a smectic layered arrangement of randomly coiled backbones with a single layer of side groups perpendicular to the lamellar planes containing the backbone.

Less understood is the case of polymers in which the alkyl group is attached to the backbone through highly rigid and polar bridges, for example, in poly(p-phenyl methacrylic esters) of p-n-alkoxybenzoic acids or poly(n-alkylmaleinimides). The values of d'' and a are abnormally high and correspond to distances considerably in excess of the length of one side group. It is possible that double layers of side groups are introduced between the lamellar planes containing these backbones, as in bilayered forms of lamellar crystals of low amphiphilic molecular compounds. In such coumpounds (ethyl esters of fatty acids, n-aliphatic acids, alcohols), bilayered structures form, as a rule, for strongly interacting molecules and single layers form for weakly interacting species (Platè and Shibaev, 1974). The layers may be tilted. Thus the study of low molecular weight analogs of such polymers could be very instructive. Platè and Shibaev in reviewing the results of investigation of low molecular amphiphiles and comparing them to the results obtained with polymers with long aliphatic moieties state that the "backbone chain is an additional structural factor promoting layered ordering observed even with poly(α-olefins) containing no polar groups."

It is thus established that the structure of mesomorphic polymers with long aliphatic side groups is characterized by a layered ordering of the methylene side chains.

2. Polymers with Rigid and Anisotropic Side Groups

a. Acidic Side Groups. Because of strong hydrogen bonding between carboxylic groups, acids often form thermotropic liquid crystals. A necessary condition is that the COOH group be attached to a rigid moiety, usually aromatic. Aliphatic carboxy compounds, with the possible exception of alkanoic acids containing conjugated double bonds, are isotropic above their melting points. A benzoic acid moiety attached to its neighbor by hydrogen bonds provides a sequence of three coplanar hexagonal rings that strongly enhances rigidity and anisotropy of the molecule (**IV**).

(**IV**)

Thus *p-n*-alkoxybenzoic acids are mesogenes starting with the *n*-propyloxybenzoic acid. The mesogenic potential increases with the length of the alkoxy group (Gray, 1962). These acids form strongly tilted smectic C structures in which the angle of tilt of the molecule with respect to the plane of the layers is of the order of 45° (Chistyakov and Chaikovsky, 1969). There are indications that even the nematic phases of *p-n*-alkoxybenzoic acids display a layered disposition of molecules within small aggregates of molecules above the nematic–isotropic transition (Blumstein *et al.*, 1977a). The most extensively studied vinyl polymers containing the benzoic acid moiety are the polymers of methacryloyloxybenzoic acid and acryloyloxybenzoic acid. The monomers (Table 1, monomers 1 and 3) are not mesomorphic and melt sharply at 182 and 201°C, respectively. They display strong interactions in the isotropic phase and orient easily when dissolved in stable liquid crystalline phases (Blumstein *et al.*, 1974). Such compounds are sometimes called "potentially mesogenic." As can be seen from Table 1, both monomers when polymerized lead to spontaneous development of layered arrangements. Mesomorphic layered structures develop in both polymers regardless of polymerization conditions (Blumstein *et al.*, 1975, 1976). The backbone is confined to lamellar planes while the interacting aromatic moieties are oriented parallel to each other and tilted by an angle of 50–60° with respect to the plane of the layers.

The layered structure forms spontaneously from solution upon evaporation of solvent and persists in the solid state until thermal degradation of the polymer. Abrupt precipitation of the polymer from dilute solution leads to an amorphous structure. In addition to a layered disposition of main chains, both polymers display side group crystallization that, in the case of poly-(acryloyloxybenzoic acid), may reach 40% (Blumstein *et al.*, 1976). The derivatives of acrylic acid are more prone to crystallize than that of methacrylic acid presumably because of higher backbone flexibility.

This flexibility also plays an important part in the crystallization of long chain polyacrylates as compared to long chain polymethacrylates, favoring a more stable two-layer structure in the former case (Shibaev *et al.*, 1970; Platè and Shibaev, 1974). For poly(acryloyloxybenzoic acid two crystalline forms have been observed. The more extensively crystallized form obtained on slow polymerization in bulk was indexed to a monoclinic unit cell (Blumstein *et al.*, 1976). The less extensively crystallized form was classified as related to the smectic E_t (Clough *et al.*, 1977).

It is remarkable that smectic structures of poly(methacryloyloxybenzoic acid) and poly(acryloyloxybenzoic acid) as well as their crystalline domains are formed by atactic polymers (Blumstein *et al.*, 1975, 1976). Stereoregularity does not appear to be a necessary condition for crystallization in polymers with strongly interacting side groups. This extends the realm of atactic crystalline polymers beyond the example of polymers with long linear paraffinic side groups (Platè and Shibaev, 1974).

Hydrogen bonding plays a most important part by maintaining a rigid planar three-ring structure that favors strong lateral interactions. Methylation leads to the collapse of order and crystallinity in both poly(methacryloyloxybenzoic acid) and poly(acryloyloxybenzoic acid) (Blumstein *et al.*, 1976). Poly[*N*-(*p*-methacryloyloxybenzylidene)-*p*-aminobenzoic acid] has also been studied (Blumstein *et al.*, 1975). The monomer is mesogenic and has both a smectic and a nematic phase (Table 2, monomer 46).

b. Schiff Bases. As can be seen from the great variety of Schiff base derivatives listed in Table 2, this group of monomers has been the most extensively studied. The monomers are generally nematic in their liquid crystalline state sometimes preceded by a smectic organization. Polymerization generates macromolecules that often are more highly organized than the parent monomer. In Table 2 monomers 1, 8, and 39 are nematic and their polymers also give a nematic organization, while monomers 3, 4, 15, 19–29, 44, and 46 polymerized in their nematic or isotropic state give polymers with a layered organization.

From a phenomenological point of view one can qualitatively explain these changes by extending to the case of polymers some simple structural considerations initially developed for low molecular thermotropic liquid crystals by Gray (1962). Nematic phases are as a rule obtained through strong terminal and weak lateral interactions between molecules. Smectic phases are the result of weak terminal and stronger lateral attraction forces. Thus one can state very schematically that dipoles operating across the long molecular axes (transverse dipoles) enhance smectogenic tendencies, while dipoles operating along these axes weaken these tendencies through mutual electrostatic repulsion. However, the precise part of a given dipole in establishing the mesophase is not simple and depends on the overall balance of terminal to lateral interactions. A terminal dipole acting at an angle can have a nematogenic effect by strengthening the terminal attraction between molecules. Removed from the end of the molecule, the same dipole will strengthen the lateral attractions. This is the pattern of behavior of many homologous series in which lower members display nematic behavior, whereas the higher members (long paraffinic chains) are smectogenic. If a structural element weakens the cohesion of molecules in the smectic layers, a nematic or isotropic phase results. Thus branching weakens lateral cohesion and also weakens the smectogenic tendencies. These considerations are still qualitatively valid in the case of polymers in which the mesomorphic side groups are placed in the vicinity of each other by their attachment to the backbone of the macromolecule.

It is apparent from Table 2 that the presence of two strong transverse dipoles leads in the overwhelming majority of cases to polymers with a well-defined lamellar (smectic) organization. Cinnamic ester derivatives of Schiff bases (Table 2, monomers 19–29) and diacrylic derivatives of Schiff bases (Table 2, monomer 36) are good examples of this tendency. In the latter case the existence of a smectic and a nematic phase allows to prepare the polymer from both phases and to show that its organization is smectic and independent of the phase in which the polymerization took place (Clough *et al.*, 1976). Conversely, the absence of strong transverse dipoles and the presence of one or several strong longitudinal dipoles in a rigid and polarizable side group will decrease the tendency of the polymer to organize into well-defined layered arrangements by pushing the macromolecular backbone out of the confining planes. Poly[N-(p-cyanobenzylidene)-p-aminostyrene] and poly-[p-phenylene-bis(N-methylene-p-aminostyrene)] (Table 2, monomers 1 and 8) are examples of such structures that are exclusively nematic.

In all cases the presence in the side group of long linear paraffinic chain dramatically increases the tendency to form layered structures analogous to those discussed in Section III.B.2.a.

Some polymers given in Table 2 develop arrangements of side groups intermediate between a well-defined lamellar organization and a nematic organization. Monomers 3, 4, and 44 (Table 2) give polymers characterized by such an intermediate structure. The X-ray patterns of this group of polymers have in common a rather broad but intense low angle reflection and a diffuse ring at 4–5 Å. This is compatible with a more fluid layered structure than the well-defined smectic arrangements of polymers with strong lateral coupling of side groups such as polymers from acrylic and methacrylic derivatives of benzoic acid (Blumstein *et al.*, 1975, 1976). They can perhaps be considered as a special case of nematic arrangements of higher order (de Vries, 1974). It is possible to see from Table 2 that monomers from which such polymers are obtained have a structure intermediate between the structure of smectic and exclusively nematic monomers. For example, the replacement of the strong longitudinal terminal C≡N dipole in monomer 1 (Table 2) by the weak, transversal dipole of an alkoxy group in monomer 3 produces a polymer with such an intermediate structure. Similarly, the replacement of one or more transversal dipoles (carboxy group) by a longitudinal dipole impairs the well-defined smectic organization in the polymer.

Monomer 39 (Table 2) gives a polymer with nematic organization of side groups, in spite of two strong transversal dipoles. This apparent discrepancy can be rationalized. In the molecule of di-(*N-p*-acryloyloxybenzylidene) hydrazine (Table 2, monomer 39), the relatively free rotation around the N–N hydrazine bond unsettles the rigidity of the side group. This last point was confirmed experimentally by X-ray measurements of the average distance between the side groups in this polymer that was found substantially larger than in polymers with layered organization (Clough *et al.*, 1976).

These last examples show clearly the danger of predicting the nature of the order in polymers with mesogenic side groups from qualitative structural considerations. The combined influence of geometry, polarizability of structural elements, and distribution of polar interactions in the side group on the balance of lateral terminal interactions is often difficult to estimate. One can state however that polymerization of di- and monofunctional monomers in their mesomorphic state does not necessarily lead to the "locking in" of molecular organization of the mesophase, but develops consistently the highest order compatible with the levels of interactions and the geometry of the side group.

c. Ester Side Groups. As can be seen from Table 4, a number of monomeric esters do not display any mesomorphism but the polymers of these esters organize into liquid crystalline arrangements.

A series of acrylic esters have been investigated (Magagnini *et al.*, 1974). The nature of the ester group has been changed and the side group geometry, rigidity, and anisotropy were varied systematically. The following monomers and their polymers were studied: *p*-biphenyl acrylate (Table 4, monomer 1), *p*-biphenyl methacrylate, vinyl-*p*-phenyl benzoate, *p*-phenylbenzyl acrylate, *p*-benzylphenyl acrylate, and *p*-cyclohexylphenyl acrylate (Table 4, monomer 2). None of these monomers exhibited mesomorphism. Among polymers only poly(*p*-biphenyl acrylate) and poly(*p*-cyclohexylphenyl acrylate) display smectic organization. The X-ray data are compatible with a double layer of side groups oriented perpendicularly to the flexible backbone located in the plane of the strata (Newman *et al.*, 1975).

That poly(*p*-biphenyl acrylate) was found to display mesomorphic order is understandable as it is well known that 4 and 4-4' substituted biphenyls provide a rich source of liquid crystals (Gray, 1974). The rigid, anisotropic, and highly polarizable diphenyl moiety attached to the chain through a bridge provided with a strong transverse dipole is enough to induce strong lateral interactions between neighboring side groups and lead to a layered organization quite similar to the one described for poly(*p*-methacryloyloxy-benzoic acid) and poly(*p*-acryloyloxybenzoic acid) (Blumstein *et al.*, 1975, 1976). Introduction of a methylene group between the backbone and the diphenyl moiety increases the flexibility of the side group, decreases its linearity, and weakens the lateral cohesion between side groups. The introduction of a methyl group into the backbone that decreases the flexibility of the main chain also upsets the lateral cohesion between the acryloyloxydiphenyl side groups. In both cases amorphous polymers result.

More difficult to explain is the destruction of smectic arrangements of side groups upon inversion of the order of atoms in the ester linkage, from (**V**) in

(**V**)   (**VI**)

poly(biphenyl acrylate) to (**VI**) in poly(vinyl-*p*-phenyl benzoate). This effect can perhaps be rationalized by a better alignment of carboxy dipole bearing groups along the macromolecular backbone for poly(biphenyl acrylate). The ester linkage of poly(vinyl-*p*-phenyl benzoate) provides more flexibility and perhaps weaker coupling between the transverse carboxylic dipoles. It is interesting to note that the replacement of one phenyl ring by a cyclohexane ring can be tolerated by the polymer without collapse of the smectic organization. Such replacements usually decrease the stability of liquid crystalline structures due to the unfavorable geometry of the puckered cyclohexane ring and a decrease in polarizability anisotropy (Gray, 1974). It seems however that flexibility of a cyclohexane ring is severely reduced upon binding to a phenyl ring (Magagnini et al., 1974). Replacement of the phenyl ring by an azophenyl group also leads to highly organized polymers (Magagnini et al., 1974; Balloni, 1972).

The copolymers of p-biphenyl acrylate and p-cyclohexylphenyl acrylate are characterized by the same type of layered morphology as the homopolymers (Newman et al., 1975). On the other hand copolymers of p-biphenyl acrylate with n-vinyl carbazole are amorphous though both homopolymers are ordered (Magagnini et al., 1974). It was also shown that the poly(p-biphenyl acrylate) displaying layered order was atactic (Ceccarelli et al., 1975). Interestingly, isotactic poly(p-biphenyl acrylate) prepared by means of anionic initiators displayed a layered organization identical to the order observed in the atactic polymer (Ceccarelli et al., 1975). This is a significant result indicating that interaction forces between side groups prevent the backbone of the isotactic poly(p-biphenyl acrylate) from taking on helical conformations.

Other well-studied polymers are the polymerized methacryloyloxyphenyl esters of p-n-alkoxybenzoic acids (Table 4, monomer 18). The synthesis and mesomorphic properties of their saturated analogs were described (Konstantinov et al., 1974). Because of several transverse dipoles and a linear paraffinic terminal group, nearly all of these compounds are smectic with a few displaying a limited range of nematic behavior at high temperatures.

The polymers of methacrylic esters of p-n-alkoxybenzoic acids are characterized by a high degree of intramolecular order in dilute solution and a layered (smectic) organization in bulk (Petrukhin et al., 1971). The polymer has a smectic organization with presumably two tilted layers of side groups between the lamellar strata containing the backbone in a "wormlike" conformation. Poly(methacryloyloxyphenyl) esters of p-n-alkoxybenzoic acid are intriguing not only because of their high degree of organization in bulk and solution, but also because of rather small differences in density between the crystalline monomer and the highly ordered polymers (Baturin et al., 1972; Amerik, 1976). Lorkowsky and Reuther (1976) have prepared oriented, optically monoaxial polymers of p-n-hexyloxyphenylester of acryloyloxybenzoic acid (Table 4, monomer 29) from aligned nematic phases of the monomer. The polymer displays a double-layered alignment of side groups.

A new type of mesogenic ester has been recently described by Finkelmann et al. (1977) (Table 4, monomer 30) and Shibaev et al. (1976). In these types of monomeric esters, the mesogenic group is connected to the backbone through a sequence of methylene units. The introduction of a flexible spacer lowers the glass transition temperature of the polymer with a concomitant "decoupling" of the mesogenic moiety. The polymer thus displays on heating a true liquid crystalline behavior with first order transitions between mesophases. The glass transition temperature of ordered polyesters has been also drastically reduced through copolymerization of mesogenic esters with nonmesogenic long chain methacrylate and acrylate esters (Shibaev *et al.*, 1977; Blumstein *et al.*, 1977b). The liquid crystalline order in the copolymer is, however, critically dependent on the geometry of the nonmesogenic side group which interferes with the packing of mesogenic side groups (Blumstein *et al.*, 1977b).

    d.   Cholesteric Ester Side Groups.   Derivatives of cholesterol are playing a central part in the investigation of cholesteric phases due to the natural abundance and biological importance of cholesterol and its derivatives. However in the words of Gray:

> any compound consisting of geometrically anisotropic molecules that are disposed to arrange themselves parallel to one another without the ends of the molecules lying in planes in the crystal exhibit nematic properties on melting. If such a compound possesses a center of asymmetry in the molecule and is optically active, then the liquid crystal formed will be a twisted nematic, i.e., cholesteric (Gray, 1974, p. 142).[†]

Depending on the molecular context in which an anisotropic, chiral moiety of cholesterol is included, one can distinguish twisted nematic phases and twisted smectic phases, the latter occuring when the ratio of terminal to lateral interactions becomes low. Cholesterol itself is isotropic and only replacement of the hydroxyl with an appropriate substituent in the sterol can induce liquid crystalline behavior. The tendency of the cholesteric moiety to stimulate liquid crystalline behavior in spite of a structure that does not entirely satisfy the conditions of mesophase formation is somewhat of a puzzle.

    Mesomorphic order in polymers containing a cholesterol moiety has not been extensively studied. The two simplest monomers with chiral moieties are cholesteryl acrylate and cholestanyl acrylate (Table 3, monomers 1 and 16). It must be pointed out that controversy exists in the literature concerning the location and nature of the mesomorphic transitions of cholesteryl acrylate. Hardy claims that cholesteryl acrylate is in the smectic state between -7 and $118.7°C$ and a nematic state between 118.7 and $125.8°C$ (Hardy *et al.*, 1969a, 1970). Toth and Tobolsky claim a single monotropic transition at approximately $90°C$ from isotropic to cholesteric (Toth and Tobolsky, 1970). De Visser claims an enantiotropic phase between $122.5–125°C$ (de Visser *et al.*, 1970, 1971, 1972). Finally Tanaka reported an enantiotropic cholesteric mesophase between 118 and $126°C$ (Tanaka *et al.*, 1973). Whatever its true location, the mesophase must be of cholesteric (twisted nematic) structure since it is being frequently used to induce chirality in nematic monomers (Strzelecki and Liébert, 1973a; Liébert and Strzelecki, 1973a; Liébert *et al.*, 1975).

[†] From "Liquid Crystals and Plastic Crystals, Volume 1   Preparation, Constitution and Applications," by G. W. Gray and P. A. Winsor (Editors), Ellis Horwood, Publishers, Chichester.

Introduction of a cholesteric moiety into an already mesogenic side group usually does not destroy the mesogenic potential of the molecule but introduces a twist into the liquid crystalline phases. In polymers, however, chirality is frequently lost unless the cholesteric structure of the monomeric phase is "locked in" by the use of cross-linking agents (Strzelecki and Liébert, 1973a; Liébert and Strzelecki, 1973a; Strzelecki *et al.*, 1975; Hsu *et al.*, 1977).

The polymer of cholesteryl *p*-acryloyloxybenzoate displays a highly ordered lamellar organization without the selective reflections of light characteristic of the chiral monomer. Similarly, the polymer of cholesteryl methacrylate considered amorphous in the literature revealed a high degree of lamellar order (Hsu *et al.*, 1977). Thermal polymerization of cholesteryl methacrylate in its monotropic, twisted nematic phase leads to a polymer with no chirality present (Saeki *et al.*, 1972). Attachment of the cholesteric moiety to the end of a long, flexible side group through esterification with cholesterol of *N*-methacryloyl-ω-amino carboxylic acids (Table 3, monomer 15) leads to the formation of a smectic organization of remarkable stability (Shibaev *et al.*, 1976). In a recent development, Finkelmann *et al.* (1977) succeeded in preparing an uncross-linked cholesteric copolymer of two monomers in which the cholesteric moiety was placed at the end of flexible spacers of widely different lengths. Specifically, the composition of the copolymer was approximately equimolar with respect to comonomers which contained 2 and 12 methylene units in their respective spacers. The mismatch between the length of the adjacent side groups favored formation of a (twisted) nematic order.

Recent activity in the field of study of liquid crystalline order in polymers has brought to light a new class of polymers with nematic, smectic, or cholesteric organization of side groups. These polymers often display a high degree of intramolecular as well as intermolecular organization and help bridge the gap between the highly organized biological macromolecules and the less ordered simpler, synthetic polymers on which most structure–property correlations have hitherto been studied. The structure and properties of this group of polymers are dominated by the side group, its geometry, rigidity, polarizability and polarity. The side group imposes on the polymer its local organization and determines its dynamic and equilibrium properties. From the very scant data available, it appears that tacticity in such polymers does not play its otherwise determinant role as a factor influencing supermolecular structure and crystallinity (Saeki *et al.*, 1972; Platè and Shibaev, 1974; Blumstein *et al.*, 1975, 1976; Ceccarelli *et al.*, 1975; Blumstein *et al.*, 1977b).

The properties of such polymers are determined by the anisotropy of their structure, size of the oriented domains, and the strength of the forces determining domain stability. With the exception of polymers with linear paraffinic side groups for which the thermal, mechanical, and dielectric properties have been determined and extensively reported (Platè and Shibaev, 1974), only very scant data are available on properties of other polymers of this group. Alignment of anisotropic domains to form anisotropic glasses has been successfully achieved by polymerizing a nematic monomer in a magnetic field (Liébert and Strzelecki, 1973b; Clough et al., 1976). Determination of thermal expansion of such glasses indicates a high degree of anisotropy of thermal expansion that, in the oriented "smectic" polymeric glass, follows the pattern of expansion of the corresponding smectic phases of oriented low molecular compounds (Clough et al., 1976). Investigation of the mechanical and viscoelastic properties of such polymers as well as their morphology is complicated by very high viscosity, high T_g, and poor solubility. Nevertheless, strong evidence exists from the study of electrooptical properties of polymer solutions of varying concentrations (see Chapter 2 of this book), measurement of relaxation times in a sinusoidal electric field (Tsvetkov et al., 1973a), that ordered aggregation of such macromolecules can take place under suitable conditions, and this in turn means that polymeric materials such as films or fibers with oriented anisotropic domains can in principle be obtained from solutions and melts of such macromolecules.

The use of liquid crystalline phases as media for polymerization has been somewhat disappointing from the point of view of topochemical and topotactic effects, in spite of high hopes formulated for these systems (de Gennes, 1969; Blumstein, 1970; Blumstein et al., 1971). Useful developments came however from the study of polymers with structural elements related to mesomorphic behavior and these polymers can be expected to play an increasingly important part in the technology of new materials.

Acknowledgment

Partial support of the National Science Foundation under the Research Grant DMR 75–17397 is gratefully acknowledged.

References

Amerik, Y. B. (1976). Personal communication.
Amerik, Y. B., and Krentsel, B. A. (1967). J. Polym. Sci., Part C **16**, 1383.
Baccaredda, M., Magagnini, P. L., Pizzirani, G., and Giusti, P. (1971). J. Polym. Sci., Part B **9**, 303.
Balloni, P. (1972). Thesis, Univ. of Pisa, Pisa.
Baturin, A. A., Amerik, Y. B., Krentsel, B. A., Tsvetkov, V. N., Shtenikova, I. N., and Riumtsev, E. I. (1972). Dokl. Akad. Nauk SSSR **202**, 586.
Blumstein, A. (1970). Adv. Macromol. Chem. **2**, 123.
Blumstein, A., Blumstein, R. B., and Vanderspurt, T. H. (1969). J. Colloid Interface Sci. **31**, 236.
Blumstein, A., Kitagawa, N., and Blumstein, R. B. (1971). Mol. Cryst. Liq. Cryst. **12**, 215.

Blumstein, A., Billard, J., and Blumstein, R. (1974). *Mol. Cryst. Liq. Cryst.* **25**, 83.

Blumstein, A., Blumstein, R. B., Clough, S. B., and Hsu, E. C. (1975). *Macromolecules* **8**, 73.

Blumstein, A., Clough, S. B., Patel, L., Blumstein, R. B., and Hsu, E. C. (1976). *Macromolecules* **9**, 243.

Blumstein, A., Patel, L., and Skoulios, A. (1977a). *Symp. Liquid Crystals and Oriented Fluids* Amer. Chem. Soc. Meeting, Chicago, August.

Blumstein, A., Osada, Y., Clough, S. B., Hsu, E. C., and Blumstein, R. B. (1977b). *Polymer Prepr., Am. Chem. Soc., Div. Polym. Chem.* **18** (*2*), 14.

Bouligand, Y., Cladis, P. E., Liébert, L., and Strzelecki, L. (1974). *Mol. Cryst. Liq. Cryst.* **25**, 233.

Bresler, S., Youdin, M., and Talmud, D. (1941). *Acta Physicochim. URSS* **14**, 71.

Brownawell, D., and Feng, I. M. (1961). *J. Polym. Sci.* **60**, S19.

Ceccarelli, G., Frosini, V., Magagnini, P. L., and Newman, B. A. (1975). *J. Polym. Sci., Polym. Lett. Ed.* **13**, 109.

Chistyakov, I. G. (1964). *J. Struct. Chem. (USSR)* **5**, 507.

Chistyakov, I. G., and Chaikovsky, W. M. (1969). *Mol. Cryst. Liq. Cryst.* **7**, 279.

Clough, S. B., Blumstein, A., and Hsu, E. C. (1976). *Macromolecules* **9**, 123.

Clough, S. B., Blumstein, A., and de Vries, A. (1977). *Polymer Prepr., Am. Chem. Soc., Div. polym. Chem.* **18** (*2*), 1.

Cser, F., Nyitrai, K., Seyfried, E., and Hardy, G. (1976). *Magy. Kem. Polym. J.* (*Hung.*) **82**, 207.

de Gennes, P. G. (1969). *Phys. Lett. A* **28**, 725.

Demus, D., Demus, H., and Zaschke, H. (1974). "Flussige Kristalle in Tabellen," p. 97, VEB Dtsch. Verlag, Leipzig.

de Visser, A. C., Feyen J., De Groot, K., and Banties, A. (1970). *J. Polym. Sci., Part B* **8**, 805.

de Visser, A. C., De Groot, K., Feyen, J., and Banties, A. (1971). *J. Polym. Sci., Part A-1* **9**, 1893.

de Visser, A. C., De Groot, K., Feyen, J., and Banties, A. (1972). *J. Polym. Sci., Part B* **10**, 851.

de Vries, A. (1970). *Mol. Cryst. Liq. Cryst.* **10**, 219.

de Vries, A. (1974). Personal communication.

Dubault, A., Veyssie, M., Liébert, L., and Strzelecki, L. (1973). *Nature (London), Phys. Sci.* **245**, 94.

Elliott, A., and Ambrose, E. J. (1950). *Discuss. Faraday Soc.* **9**, 246.

Farinha-Martins, A. (1969). *C. R. Acad. Sci., Ser. B* **268**, 1731.

Finkelmann, H., Ringsdorf, H., and Wendorff, J. H. (1977). *Commun. Symp. on Mesomorphic Order in Polymers and 'Polymerization in Liquid Crystalline Media,* Amer. Chem. Soc. Meeting, Chicago, August.

Flory, P. J. (1956). *Proc. R. Soc., Ser. A* **234**, 73.

Frenkel, Y. I. (1945). "Kinetic Theory of Liquids." Akad. Nauk SSSR, Moscow.

Golik, A. Z., Skrshevsky, A. F., and Adamenko, I. I. (1967). *Zh. Strukt. Khim.* **8**, 105.

Gray, G. W. (1962). "Molecular Structure and Properties of Liquid Crystals." Academic Press, New York.

Gray, G. W. (1974). *In* "Liquid Crystals and Plastic Crystals, Volume 1 Preparation, Constitution and Applications" (G. W. Gray and P. A. Winsor, eds.), pp. 103–152. Ellis Horwood, Chichester, England.

Greenberg, S. A., and Alfrey, T. (1954). *J. Am. Chem. Soc.* **76**, 6280.

Hardy, G., Nyitrai, K., and Cser, F. (1969a). *Prepr. IUPAC, Int. Symp. Macromol. Chem., Budapest* **4**, 121.

Hardy, G., Nyitrai, K., Cser, F., Cselik, G., and Nagy, I. (1969b). *Eur. Polym. J.* **5**, 133.

Hardy, G., Cser, F., Kallo, A., Nyitrai, K., Bodor, G., and Lengyel, M. (1970). *Acta Chim. Acad. Sci. Hung.* **65**, 287.

Hardy, G., Cser, F., Fedorova, N., and Batky, M. (1976). *Magy. Kem. Polym. J.* (*Hung.*) **82**, 191.

Herz, J., Reiss-Husson, F., Rempp, P., and Luzzati, V. (1963). *J. Polym. Sci.* **4**, 1275.

Hochapfel, A., Lecoin, D., and Viovy, R. (1976). *Mol. Cryst. Liq. Cryst.* **37**, 109.

Hsu, E. C., and Blumstein, A. (1977). *J. Polym. Sci., Polym. Lett. Ed.* **15**, 129.

Hsu, E. C., Lim, L. K., Blumstein, R. B., and Blumstein, A. (1976). *Mol. Cryst. Liq. Cryst.* **33**, 35.

Hsu, E. C., Clough, S. B., and Blumstein, A. (1977). *J. Polym. Sci., Polym. Lett. Ed.* **15**, 545.

Jordan, E. F., Jr. (1971). *J. Polym. Sci., Part A-1* **9**, 3367.

Kessler, J. O., and Lydon, J. E. (1974). *In* "Liquid Crystals and Ordered Fluids" (J. F. Johnson and R. S. Porter, eds), Vol. 2, pp. 331–339. Plenum. New York.

Konstantinov, I. I., Amerik, Y. B., and Krentsel, B. A. (1974). *Mol. Cryst. Liq. Cryst.* **29**, 1.

Konstantinov, I. I., Baturin, A. A., Minaeva, V. S., Amerik, Y. B., and Krentsel, B. A. (1976). *Abstr. Int. Liq. Cryst. Conf. 6th* p. G-4.

Korneeva, E. B., Lavrenko, P. N., Shtennikova, I. N., Konstantinov, N. A., Polotskii, A. E., Baturin, A. A., Amerik, Y. B., Krentsel, B. A., and Tsvetkov, V. N. (1975). *Vysokomol. Soedin., Ser. A* **17**, 2582.

Lecoin, D., Hochapfel, A., and Viovy, R. (1975a). *J. Chim. Phys.* **72**, 1029.

Lecoin, D., Hochapfel, A., and Viovy, R. (1975b). *Mol. Cryst. Liq. Cryst.* **31**, 233.

Liébert, L., and Strzelecki, L. (1973a). *Bull. Soc. Chim. Fr.* p. 603.

Liébert, L., and Strzelecki, L. (1973b). *C. R. Acad. Sci., Ser. C* **276**, 647.

Liébert, L., Strzelecki, L., and Vacogne, D. (1975). *Bull. Soc. Chim. Fr.* p. 2073.

Lorkowski, H. J., and Reuther, F. (1976). *Plaste Kautschuk* **2**, 81.

Magagnini, P. L., Marchetti, A., Matera, F., Pizzirani, G., and Turchi, G. (1974). *Eur. Polym. J.* **10**, 585.

Markan, K., and Maier, W. (1962). *Chem. Ber.* **95**, 889.

Marrow, R. M. (1928). *Phys. Rev.* **31**, 10.

Newman, B. A., Frosini, V., and Magagnini, P. L. (1975). *J. Polym. Sci., Polym. Phys. Ed.* **13**, 87.

Nyitrai, K., Cser, F., Lengyel, M., Seyfried, E., and Hardy, G. (1976a). *Magy. Kem. Polym. J.* (*Hung.*) **82**, 195.

Nyitrai, F., Cser, F., Bui, D. N., and Hardy, G. (1976b). *Magy. Kem. Polym. J.* (*Hung.*) **82**, 210.

Paleos, C. M., and Labes, M. M. (1970). *Mol. Cryst. Liq. Cryst.* **11**, 385.

Paleos, C. M., Laronge, T. M., and Labes, M. M. (1968). *Chem. Commun.* p. 1115.

Perplies, E., Ringsdorf, H., and Wendorff, J. H. (1974) *Makromol. Chem.* **175**, 553.

Perplies, E., Ringsdorf, H., and Wendorff, J. H. (1977). *In* "Polymerization of Organized Systems" (Hans-Georg Elias, ed.), pp. 149–164. Gordon & Breach, New York.

Petrukhin, B. S., Shibaev, V. P., Renio, M., and Platè, N. A. (1971). *Vysokomol. Soedin., Ser. B* **13**, 405.

Platè, N. A., and Shibaev, V. P. (1974). *J. Polym. Sci., Macromol. Rev.* **8**, 117.

Platè, N. A., Shibaev, V. P., and Petrukhin, P. S. (1971). *Vysokomol. Soedin., Ser. B* **13**, 757.

Preston, J. (1975). *Polym. Eng. Sci.* **15**, 199.

Ricuaros, D. M., and Szwarc, M. (1959). *Trans. Faraday Soc.* **55**, 1644.

Saeki, H., Iimura, K., and Takeda, M. (1972). *Polym. J.* **3**, 414.

Shibaev, V. P., Petrukhin, B. S., Zoubov, Y. A., Platè, N. A., and Kargin, V. A. (1968). *Vysokomol. Soedin., Ser. A* **10**, 216.

Shibaev, V. P., Petrukhin, B. S., and Zoubov, Y. A. (1969). *Symp. Macromol. Chem. USSR, 17th, Prepr., Moscow* p. 77.

Shibaev, V. P., Petrukhin, B. S., Platè, N. A., and Kargin, V. A. (1970). *Vysokomol. Soedin., Ser. A* **12**, 140.

Shibaev, V. P., Talrose, R. V., Karakhanova, F. I., Kharitonov, A. V., and Platè, N. A. (1975). *Dokl. Akad. Nauk SSSR* **225**, 632.

Shibaev, V. P., Freidzon, J. S., and Platè, N. A. (1976). *Dokl Akad. Nauk SSSR* **227**, 1412.

Skoulios, A., Finaz, G., and Parod, J. (1960). *C. R. Acad. Sci.* **251**, 739.

Steiger, E. L., and Dietrich, H. J. (1972). *Mol. Cryst. Liq. Cryst.* **16**, 279.
Stewart, G. W., and Marrow, R. M. (1927). *Phys. Rev.* **31**, 10.
Strzelecki, L., and Liébert, L. (1973a). *Bull. Soc. Chim. Fr.* p. 597.
Strzelecki, L., and Liébert, L. (1973b). *Bull. Soc. Chim. Fr.* p. 605.
Strzelecki, L., Liébert, L., and Keller, P. (1975). *Bull. Soc. Chim. Fr.* p. 2750.
Tanaka, Y. (1976). *Org. Synth. Chem. (J.)* **34**, 2.
Tanaka, Y., Kabaya, S., Shimura, Y., Okada, A., Kuribata, Y., and Sakakibara, Y. (1972). *J. Polym. Sci., Part B* **10**, 261.
Tanaka, Y., Shiozaki, H., and Shimura, Y. (1973). *Chem. Abstr.* **78**, 136543.
Toth, W. J., and Tobolsky, A. V. (1970). *J. Polym. Sci., Part B* **8**, 289.
Tsvetkov, V. N. (1944). *Zh. Eksp. Teor. Fiz.* **14**, 35.
Tsvetkov, V. N. (1965). *Vysokomol. Soedin.* **7**, 1468.
Tsvetkov, V. N. (1969). *Vysokomol. Soedin.* **11**, 132.
Tsvetkov, V. N., and Krozer, S. P. (1958). *Zh. Tekh. Fiz* **28**, 1444.
Tsvetkov, V. N., Riumtsev, E. I., Konstantinov, I. I., Amerik, Y. B., and Krentzel, B. A. (1972). *Vysokomol. Soedin., Ser. A* **14**, 67.
Tsvetkov, V. N., Riumtsev, E. I., Shtennikova, I. N., Konstantinov, I. I., Amerik, Y. B., and Krentsel, B. A. (1973a). *Vysokomol. Soedin., Ser. A* **15**, 2270.
Tsvetkov, V. N., Riumtsev, E. I., Shtennikova, I. N., Korneeva, E. V., Krentsel, B. A., and Amerik, Y. B. (1973b). *Eur. Polym. J.* **9**, 481.
Turner-Jones, A. (1964). *Makromol. Chem.* **71**, 1.
Vervit, L. (1971). *Mol. Cryst. Liq. Cryst.* **15**, 89.
Vorländer, D. (1933). *Trans. Faraday Soc.* **29**, 907.

4

Synthesis and Properties
of Rodlike Condensation Polymers

J. Preston

Monsanto Textiles Company
Monsanto Triangle Park Development Center, Inc.
Research Triangle Park, North Carolina

I. Introduction

In recent years the phenomenon of liquid crystalline behavior has received considerable attention, as witness the present volume and some excellent reviews of this subject. However, the study of condensation polymers as they relate to the liquid crystalline state has received relatively little attention until recently, possibly because of a lack of demonstrated utility. Earlier, the use of liquid crystalline solutions of poly(γ-benzyl-L-glutamate) made it possible to spin this polymer more rapidly than isotropic solutions (Ballard, 1963; Ballard and Griffiths, 1964). Further study has been neglected largely because the tensile properties of these fibers are not unusual vis-à-vis numerous other fibers, including some common commercial ones. With the discovery of ultrahigh-strength/high-modulus fibers, renewed interest has been shown in spinning from liquid crystalline solutions.

In this chapter attention will be given to the ultrahigh-strength/high-modulus fibers prepared from rodlike polymers spun from isotropic and anisotropic solutions. It should be noted, however, that not all of the rodlike polymers can be spun from anisotropic solutions because such solutions cannot always be prepared due to insufficient solubility of the polymers in the organic solvents presently available or because of insufficient stability of certain of these polymers in potent inorganic solvents. Nevertheless, certain rodlike polymers yield ultrahigh-strength/high-modulus fibers without use of anisotropic spinning solutions. It may be inferred that if more powerful organic solvents or nondegrading acidic or basic solvents can be found, even better fiber properties may be obtained by spinning from anisotropic solutions. Although some salient features will be presented, a detailed discussion of the physical state that exists in the spinning solutions of rodlike polymers will be left for others to discuss. Black (1975) and Carter and Schenk (1975) have recently reviewed the literature concerning the use of anisotropic dopes for spinning ultrahigh-strength/high-modulus fibers. These reviews are recommended reading in this area as is the review of solutions of extended chain polymers in the chapters by Baird and Tsvetkov *et al.* in this volume. Here only the synthesis, spinning, and tensile properties will be stressed.

The only commercially available ultrahigh-strength/high-modulus fiber is an aramid, a generic term adopted in 1974 by the United States Federal Trade Commission for designating fibers of the aromatic polyamide type and defined as "Aramid—A manufactured fiber in which the fiber-forming substance is a long-chain synthetic polyamide in which at least 85% of the amide (—CO—NH—) linkages are attached directly to two aromatic rings." The first aramid fiber of high strength and exceptionally high initial modulus was introduced (Hannell, 1970) by the du Pont Company in 1970 for use in tires under the experimental name of Fiber B. Later, another aramid fiber, also called Fiber B, was marked (Wilfong and Zimmerman, 1973) having a much higher strength (by nearly twofold) than that of the early Fiber B. An aramid fiber of comparable strength to the second Fiber B but having a low elongation-to-break and a much higher initial modulus was introduced under the experimental name of PRD-49. These products, once introduced commercially, were marketed under the trademarks Kevlar and Kevlar-49, respectively. Production from a semiworks facility in 1975–1976 reportedly was 6 million pounds per year and the projected annual capacity for Kevlar is approximately 50 million pounds per year. Although the introductory price of Kevlar tire cord was $2.85/lb, the price was increased to $3.15/lb in 1975 and more recently to $3.95/lb. PRD-49 was introduced at a cost of about 100/lb but the price has fallen to $18–25/lb and is expected to fall even further with high volume production. It can be appreciated that while the annual production of these fibers is not great, the total value of the product in dollars is quite significant commercially.

II. Syntheses

Polymers that have been used to prepare ultrahigh-strength/high-modulus fibers consist of the same types of units commonly observed in certain simple aromatic compounds that exhibit the liquid crystalline phenomenon. Typical of such units are para-oriented aromatic rings

and ester ($-CO-O-$) and amide ($-CO-NH-$) linkages. The polymers[†] are of the wholly aromatic type and are polyamides, polyamide–hydrazides, polyhydrazides, and polyester. Each of these types will be discussed in turn.

A. Polyamides

Aromatic polyamides are formed by reactions that lead to the formation of amide linkages between aromatic rings. In practice this generally means the reaction of aromatic diamines and aromatic diacid chlorides in an amide-type solvent at low temperatures. Aramid fibers of exceptional tensile strength and modulus are spun from solution, either that in which the polymer is formed, or from various acids, e.g., sulfuric acid.

When the low temperature polycondensation route employing diacid chlorides and diamines is used to make the rodlike aromatic polyamides, they are prepared in highest molecular weight when mixed solvents are employed. In a typical example, poly(p-phenylene terephthalamide) (PPD-T) is best prepared in a 1:3 molar ratio of hexamethylphosphoramide (HPT)[‡] to N-methylpyrrolidone (NMP) or in a 2:3 molar ratio of dimethylacetamide (DMAc) to HPT (Federov *et al.*, 1973). Also, polymer solids should be in the range of 6–7% to achieve the highest level of molecular weight (Preston, 1975; Bair *et al.*, 1976). At this concentration it is possible to obtain anisotropic solutions and it is quite possible that the growing polymer chains, closely packed in the anisotropic state, may become inaccessible to the monomers, thereby limiting molecular weight. The lower polymer molecular weight attained at higher concentrations may merely be a consequence of more rapid rates of gelation (Bair *et al.*, 1976).

[†] Some other types of polymers, polyimides and polyureas, have been spun to fibers of moderately high tensile properties but will not be discussed here. Also, fibers from linear polyethylene have been described (Blades and White, 1963) having tensile strengths as high as 23 gpd and initial modulus as high as 429 gpd at an elongation-to-break of 5.9%. However, such materials are outside the scope of this chapter. Another type of polymer reported very recently (Morgan, 1976a), polyazomethine, has been melt spun to fibers which, after heat treatments to increase the molecular weight of the polymer, have very high strength and initial modulus.

[‡] HPT has been found to cause cancer in laboratory animals. Inhalation of the vapors and skin contact with the liquid should be avoided (Zapp, 1975).

The simplest type of rodlike polyamide is of the —AB— type and is poly-(p-benzamide) (PPB). The first version of Fiber B appears to have been based on this composition. PPB may be prepared (Eq. (1)) in high molecular weight from an acid chloride–amine hydrochloride monomer or in moderate molecular weight directly from p-aminobenzoic acid using the phosphorylation reaction (Yamazaki *et al.*, 1975). In the former case the reaction is carried out in an amide-type solvent or in *N,N,N',N'*-tetramethylurea (TMU). The PPB thus formed can be spun directly to fiber from the solution obtained.

$$\text{HCl·NH}_2\text{—}\langle\bigcirc\rangle\text{—CO—Cl} \xrightarrow[-\text{HCl}]{\text{TMU}} \left[\text{NH—}\langle\bigcirc\rangle\text{—CO}\right] \tag{1}$$

PPB

PPB obtained via the phosphorylation reaction (Eq. (2)) must be isolated, dried, and redissolved in either an organic solvent (e.g., DMAc and NMP,

$$\text{NH}_2\text{—}\langle\bigcirc\rangle\text{—CO—OH} \xrightarrow[\text{pyridine}]{(\text{Ph—O})_3\text{P}} \text{PPB} \tag{2}$$

preferably containing dissolved lithium chloride) or a strong acid, e.g., sulfuric acid, for spinning to fiber. Spinning from organic solvents can be carried out either by the wet or dry spinning methods. Of course, only the wet spinning method may be used for spinning from sulfuric acid solutions.

A typical rodlike polyamide is that used to make Kevlar, poly(p-phenylene terephthalamide) (PPD-T). In the synthesis (Eq. (3)) of PPD-T, the mono-

$$\text{NH}_2\text{—}\langle\bigcirc\rangle\text{—NH}_2 + \text{Cl—CO—}\langle\bigcirc\rangle\text{—CO—Cl} \xrightarrow{\text{NMP/HPT}}$$

$$\left[\text{NH—}\langle\bigcirc\rangle\text{—NH—CO—}\langle\bigcirc\rangle\text{—CO}\right] \tag{3}$$

PPD-T

mers are mixed in solution, but on the attainment of high molecular weight the polymer precipitates, resulting in a gel structure (Kwolek, 1972; Blades, 1973). High shearing action is required in order to obtain high molecular weight. In the laboratory high shearing action may be obtained by using a food blender operated at high speed. Commercially this may be done using a special device such as that described in a recent du Pont patent (Fitzgerald and Likhyani, 1974).

PPB-T polymer is precipitated into water, washed free of solvent, dried, and then redissolved in acid solvents (e.g., sulfuric acid, chlorosulfuric acid, fluorosulfuric acid, or hydrofluoric acid) for the preparation of fibers by wet spinning methods (Kwolek, 1972; Blades, 1973). Other —AA—BB— polyamides (Tables I, II) are prepared and spun to fiber in similar fashion.

Rodlike polyamides may be prepared (Eqs. (4), (5)) from intermediates having preformed amide linkages and/or containing benzheterocyclic groups.

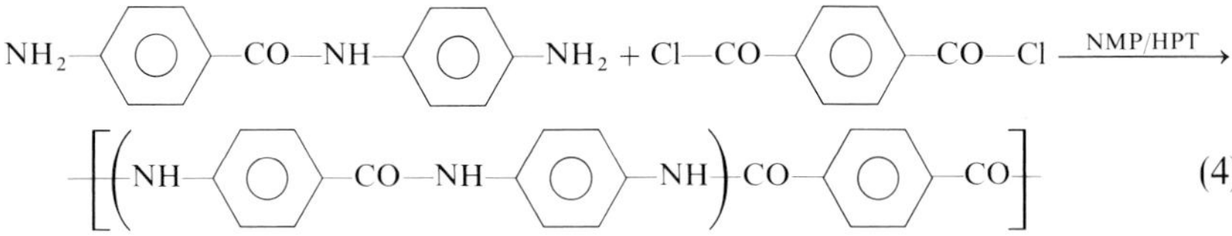

$$NH_2-\langle O \rangle-CO-NH-\langle O \rangle-NH_2 + Cl-CO-\langle O \rangle-CO-Cl \xrightarrow{NMP/HPT}$$

$$\left[\left(NH-\langle O \rangle-CO-NH-\langle O \rangle-NH\right)CO-\langle O \rangle-CO\right] \qquad (4)$$

PB-T

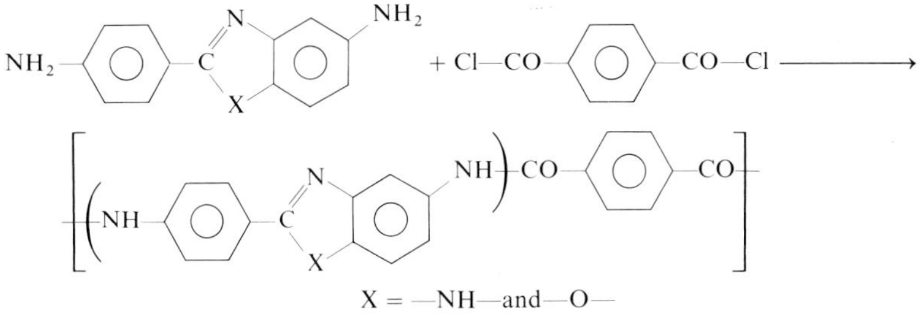

$$NH_2-\langle O \rangle-C\langle^N_X\rangle\langle O \rangle^{NH_2} + Cl-CO-\langle O \rangle-CO-Cl \longrightarrow$$

$$\left[\left(NH-\langle O \rangle-C\langle^N_X\rangle\langle O \rangle^{NH}\right)CO-\langle O \rangle-CO\right] \qquad (5)$$

$$X = -NH- \text{ and } -O-$$

Some very excellent fibers (Table I) have been spun from sulfuric acid using the polyterephthalamide of 4,4'-diaminobenzanilide, PB-T. Polymer and co-polymers with PPD-T of the type produced according to Eq. (5) appear to have been employed by Soviet workers to make high-strength/high-modulus fibers.

B. Polyamide–Hydrazides

Polyamide–hydrazides (Table III) are prepared (Eq. (6)) by low temperature polycondensation in similar fashion to aromatic polyamides. Mixed

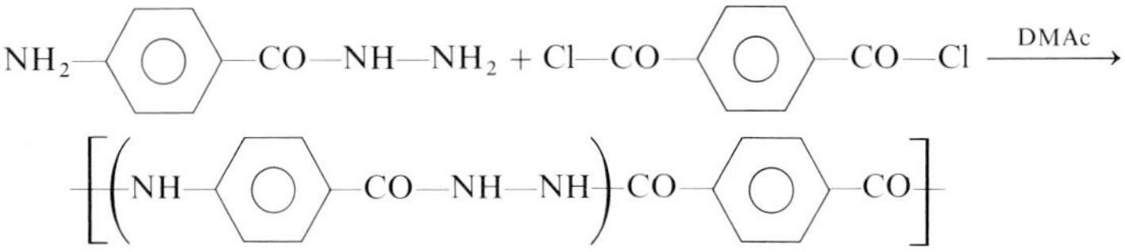

$$NH_2-\langle O \rangle-CO-NH-NH_2 + Cl-CO-\langle O \rangle-CO-Cl \xrightarrow{DMAc}$$

$$\left[\left(NH-\langle O \rangle-CO-NH-NH\right)CO-\langle O \rangle-CO\right] \qquad (6)$$

PABH-T

solvents may be employed but apparently are not required. Dissolved salts, e.g., LiCl, may be added at the start of the polymerization to aid in the solubility of the monomers and the polymers produced. Fibers are usually spun from the solutions in which the polymer is formed.

Although dimethyl sulfoxide (DMSO) is not a suitable solvent for polymerization of polyamide–hydrazides, it is a polymer solvent and can be used

TABLE I

Ultrahigh-Strength/High-Modulus Fibers from Aromatic Polyamides[a]

No.	Structure	η_{inh}[b]	Denier[c]	$T/E/M_i$[d]	W-T-B[e]
1	$\left[\text{NH}-\bigcirc-\text{CO}\right]$	3.7	1.0	19/4.0/570	0.41
2	$\left[\text{NH}-\bigcirc-\text{NH}-\text{CO}-\bigcirc-\text{CO}\right]$	4.8	1.3	24/5.0/484	—
3	$\left[\text{NH}-\bigcirc-\text{NH}-\text{CO}-\bigcirc\bigcirc-\text{CO}\right]$	3.6	2.0	17/4.9/470	0.40
4	$\left[\text{NH}-\bigcirc-\text{NH}\left(\text{CO}-\bigcirc_{\text{Cl}}-\text{CO}\right)\right]$	3.7	1.9	21/4.8/640	0.54

5		5.3	3.5	18/5.8/470	0.57
6		3.1	2.1	16/2.5/780	0.22
7		3.1	3.9	18/6.5/370	0.61
8		4.3	5.6	17/3.9/620	0.39

[a] Data taken from Blades (1973); all of the fibers were spun from anisotropic solutions in sulfuric acid by means of the dry jet—wet spinning method. The tensile properties are for *as-spun* fibers.

[b] Determined at 30°C on 0.5 g of fiber dissolved in 100 ml of sulfuric acid.

[c] Denier per filament.

[d] T = tenacity (g/den); E = elongation-to-break (%); M_i = initial modulus (g/den).

[e] $W\text{-}T\text{-}B$ = work-to-break (g cm/den cm).

TABLE II

Ultrahigh-Modulus Fibers from Aromatic Polyamides[a]

No.	Structure	M_i[b]	Reference
1	$-[-NH-\text{C}_6\text{H}_2(\text{Cl})(\text{Cl})-NH-CO-\text{C}_6\text{H}_4-CO-]-$	490	Kwolek (1972)
2	$-NH-\text{C}_6\text{H}_4-CO-NH-\text{C}_6\text{H}_4-NH-CO-[-\text{C}_6\text{H}_4-CO-NH-\text{C}_6\text{H}_4-NH-CO-\text{C}_6\text{H}_4-CO-]-$	583	Kwolek (1972)
3	$-NH-\text{C}_6\text{H}_4-NH-CO-\text{C}_6\text{H}_3(\text{Cl})-CO-NH-[-\text{C}_6\text{H}_4-NH-CO-\text{C}_6\text{H}_4-CO-]-$	763	Kwolek (1972)
4	$-NH-\text{C}_6\text{H}_3(\text{CH}_3)-CO-NH-\text{C}_6\text{H}_4-N=N-\text{C}_6\text{H}_3(\text{CH}_3)-NH-CO-[-\text{C}_6\text{H}_4-NH-CO-\text{C}_6\text{H}_4-CO-]-$	587	Bach and Hinderer (1970)
5	$-(-NH-\text{C}_6\text{H}_4-CO-NH-)-(-NH-\text{C}_6\text{H}_4-CO-)-$	707	Kwolek (1972)

[a] All of these fibers were hot-drawn; tensile strengths were not particularly high in relation to the initial module.
[b] Initial modulus (g/den).

TABLE III Ultrahigh-Strength/High-Modulus Fibers from Aromatic Polyamide–Hydrazides[a]

No.	Comonomer A	Molar ratio A/PABH	η_{inh}[b]	$T/E/M_i$[c]	$W\text{-}T\text{-}B$[d]	Reference
1	None	0	—	18.4/3.7/718	—	Preston (1975)
2	NH_2–⟨⟩–CH_2–⟨⟩–NH_2	9:1	4.8	14.0/5.8/443	0.538	Preston et al. (1973b)
3	NH_2–NH–CO–⟨⟩–CO–NH–NH_2	7:3	4.5	10.1/5.1/637	0.162	Preston et al. (1973b)
4	NH_2–NH–CO–⟨⟩(⟨⟩)–CO–NH–NH_2	7:3	4.8	12.8/3.0/541	0.235	Preston et al. (1973b)
5	benzoxazole–⟨⟩–benzoxazole with NH_2	9:1	7.6	12.8/3.0/572	0.223	Preston et al. (1973b)
6	NH_2–⟨⟩–CO–NH–⟨⟩–NH_2	1:1	5.1	18.1/3.1/776	0.385	Preston et al. (1973b)
7	NH_2–NH–CO–⟨⟩–CO–NH–NH_2(0.2) + NH_2–⟨⟩–NH_2(0.8)	1:1	5.3	11.6/2.8/536	—	Tohyama et al. (1976)

[a] All of the fibers were *hot-drawn*.
[b] Determined at 30°C in an organic solvent, e.g., DMSO or DMAcLiCl on 0.5 g polymer dissolved in 100 ml of solvent.
[c] T = tenacity (g/den); E = elongation-to-break (%); M_i = initial modulus (g/den).
[d] $W\text{-}T\text{-}B$ = work-to-break (g cm/den cm).

for spinning (Preston *et al.*, 1973a; Alfonso *et al.*, 1977) of the poly(tereph-thalamide–hydrazide) of *p*-aminobenzhydrazide, PABH-T. Sulfuric acid and methanesulfonic acid are solvents for the aromatic polyamide–hydrazides but these solvents cause extensive polymer degradation and hence are not suitable spinning solvents. Althouth intensive studies have been conducted by Ciferri (1975), anisotropic solutions of PABH-T suitable for spinning dopes have not been prepared in organic solvents even in the presence of dissolved salts.

C. *Polyhydrazides*

In general, simple polyhydrazides are prepared (Eq. (7)) in similar fashion to polyamide–hydrazides. Because of the low solubility of hydrazide mono-

$$NH_2-NH-CO-\langle O \rangle-CO-NH-NH_2 + Cl-CO-\langle O \rangle-CO-Cl \xrightarrow{NMP}$$

$$\left[-NH-NH-CO-\langle O \rangle-CO-\right] \tag{7}$$

TDH-T

mers, salts, e.g., LiCl, are added to the solvents at the start of polyconden-sation to enhance both monomer and polymer solubility. Fibers may be spun directly from the solutions in which the polymers are produced. However, fibers from solution of polyhydrazides,e.g., poly(terephthalhydrazide) (TDH-T), in organic ammonium bases and in sulfuric acid, respectively, are claimed to produce improved results (Hartzler, 1976; Morgan and Hartzler, 1977).

Very good results have been obtained by the use of copolyhydrazides (Eq. (8)) to prepare ultrahigh-strength/high-modulus fibers. No evidence of aniso-tropic solutions would be expected at the low solids level ($\sim 4\%$) reported (Dobinson and Pelezo, 1973) in the preparation of fibers from such copoly-mers.

$$\left.\begin{array}{c} NH_2-NH-CO-\langle O \rangle-CO-NH-NH_2 \\ + \\ NH_2-NH-CO-CO-NH-NH_2 \end{array}\right\} + 2\ Cl-CO-\langle O \rangle-CO-Cl \xrightarrow{DMAc/LiCl}$$

$$\left[\left(NH-NH-CO-\langle O \rangle-CO\right) + \right.$$

$$\left.\left(NH-NH-CO-CO-NH-NH-CO-\langle O \rangle-CO\right)\right] \tag{8}$$

D. Polyesters

Aromatic polyesters may be made (Schaefgen et al., 1975) in a melt from the reaction (Eq. (9)) of diacids with bis-acetates of diphenols and substituted hydroquinones. In order to suppress the melting points of these polymers

$$CH_3-CO-O-Ar-O-CO-CH_3 + HO-CO-R-CO-OH \xrightarrow[-CH_3-CO-OH]{\Delta}$$

$$\left[O-Ar-O-CO-R-CO \right]$$

Ar = (image of substituted benzene rings with Cl, CH₃, CH₃; CH₃, CH₂—CH₃ substituents)

R = (image of aromatic and cyclic ring structures)

$$-C_6H_4-O-CH_2-CH_2-O-C_6H_4-,$$

$$-C_6H_4-O-C_6H_4-$$

$$(9)$$

below the decomposition point of rodlike aromatic polyesters having high symmetry, unsymmetrical monomers, and/or random copolymers are employed for the melt spinning of fibers.

It is interesting to note that the polymer melts exhibit liquid crystalline behavior within certain temperature ranges and spinning must be carried out within this range. A temperature must be selected sufficiently above the melting point that melt viscosity is not too high. However, at too high a temperature above the melt, the anisotropic state may be lost.

After spinning, the reaction (Eq. (9)) in the solid state (i.e., in the fiber) is carried further and the molecular weight of the polyesters in increased with attendant increase in tensile strength and modulus of the fiber.

III. Solution Properties

Aside from the great interest in anisotropic solutions for the preparation of fibers having a high level of tensile properties, the study of solutions of rodlike polymers is of considerable theoretical interest. Flory (1956), in a classic paper, predicted the solution properties of rodlike polymers and Hermans (1962) experimentally confirmed the theory for poly(γ-benzyl-L-glutamate), which is rodlike in solution because of helix formation. With the advent of aromatic polyamides based on para-oriented rings, an entirely new type of polymeric liquid crystalline solution was observed which forms not because of helix formation but because of inherently rigid, extended chain structures.

The very high inherent viscosity values for the rodlike polymers and the a constant approaching 2 in the Mark–Houwink relationship are indicative of the extended chains of such polymers. For polyamides of $\bar{M}_w < 12,000$ a indeed is 1.7 for PPB and PPD-T, but for $\bar{M}_w > 12,000$, i.e., in the range suitable for spinning useful fibers, a is 1.08 and 1.06, respectively (Schaefgen et al., 1976). It is of interest to note that an a value for the polyamide–hydrazide PABH-T was also found to be 1.06 (Burke, 1973) although liquid crystalline solutions suitable for spinning fibers have not been reported. The persistence length found by Arpin and Strazielle (1976) for PPB in sulfuric acid is $\simeq 400$ Å while Schaefgen et al. (1976) found a somewhat smaller value, 200 Å, and Papkov et al. (1974) found a higher value, 1000 Å. For PABH-T, a persistence length of only 75 Å was found (Alfonso et al., 1977), which may explain why anisotropic solutions have not been reported for this polymer.

Presumably the rodlike molecules in the liquid crystalline state have the short-range order typical of nematic mesophases. The liquid crystals can be oriented readily under shear at the spinneret and for a short while immediately after extrusion. Coagulation of the fiber in the spin bath "sets" the orientation of the fibers, resulting in fibers having unusually high strength and high initial modulus, along with high "toughness" without the necessity of hot-drawing to develop orientation and a high level of tensile properties. When hot-drawing is required to obtain orientation, as is the case for fibers spun from isotropic solutions, excessive crystallinity is imparted and the resulting fibers develop a higher modulus, with a corresponding reduction in elongation-to-break and toughness (i.e., work-to-break).

For poly(p-phenylene terephthalamide) (PPD-T), the critical concentration for transition from the isotropic phase to an anisotropic phase in sulfuric acid (100.2–100.5%) decreases, in qualitative agreement with Flory's theory, as polymer molecular weight increases (Schaefgen et al., 1976). The critical concentration of PPD-T in sulfuric acid is considerably higher than in an organic solvent system (e.g., HPT/NMP/LiCl) for a polymer of a given η_{inh} value. This is attributed to a lower effective axial ratio for PPD-T in sulfuric acid than in the amide solvent/salt system.

Because poly(p-benzamide) (PPB) is considerably more soluble than PPD-T, it has been studied more in organic solvent/salt systems. The critical concentration point at which bulk viscosity decreases varies considerably with solvent, polymer molecular weight, and temperature, as pointed out by Kwolek *et al.* (1976) and by Papkov *et al.* (1974). Using the solvent TMU–LiCl, 14% polymer concentration can be reached before there is an abrupt drop in bulk viscosity as the solution changes from isotropic to a mixture of isotropic and anisotropic and finally totally anisotropic (Kwolek *et al.*, 1976). With DMAc–4% LiCl, the critical concentration point occurs at about 5% polymer concentration (Kwolek *et al.*, 1976) while the critical concentration point in NMP–5% LiCl is ~6% polymer concentration (Preston and Hofferbert, 1977). The two phases (isotropic and anisotropic) that form at the critical concentration point can frequently be separated by allowing the solutions to stand or by centrifuging them. Kwolek *et al.* (1976) have shown that the anisotropic phase from a solution of PPB in TMU–LiCl has the higher polymer molecular weight, density, concentration, volume, and lower bulk viscosity. Although fiber data are often cited from spinning of these separated phases to show the superiority of use of the anisotropic solution, it should be recalled, as pointed out by Kwolek *et al.* (1976), that the anisotropic phase does contain higher molecular weight polymer and a higher level of polymer solids. Hence, fiber properties should be somewhat better for these reasons alone.

Qualitatively, liquid crystalline solutions of rodlike aromatic polyamides can be recognized visually by a hazy quality at rest and opalescence under low shear, e.g., as that achieved by stirring a solution with a rod. They also depolarize plane-polarized light, showing birefringent domains in the polarizing microscope. Thick samples of the pure nematic solution of low molecular weight PPB have been shown (Panar and Beste, 1976) to relax to a transparent state that has random "nematic" (threadlike) lines running through it. When such a sample is placed in a magnetic field of a few thousand gauss, the lines align in the direction of the field and slowly disappear. Thus, the original depolarizing solution takes on the characteristics of a uniaxial birefringent crystal. Panar and Beste (1976) have made the very interesting observation that an anisotropic solution of low molecular weight PPB (20% solids in DMAc/LiCl) can be converted to the cholesteric form by the addition of an optically active solute, (+)1-methylcyclohexanone, which associates sufficiently with backbone groups to impart a preferred chirality to the entire molecule. Parallel lines typical of poly(γ-benzylglutamate) solutions are formed.

The ease of attaining orientation for rodlike polyamides can be demonstrated by smearing a liquid crystalline solution of PPB on a microscope slide. Thus, Morgan (1976b) reports that an electron micrograph of the dried and washed film shows ~ 77% of the polymer chains in a crystalline state aligned within 10% of the long axes of the fibrils.

Because PPB and PPD-T chains are already extended, no stiffening in solution due to a polyelectrolyte effect should occur. Schaefgen *et al.* (1976)

found this to be the case for polymers of PPD-T up to an inherent viscosity of 2.5. However, at higher values of inherent viscosity a noticeable polyelectrolyte effect is seen in strong acids.

Although phase diagrams for PPB in DMAc/LiCl have been reported by Kwolek (1972), Panar and Beste (1976), and Papkov et al. (1974), significantly different results have been obtained by Salaris et al. (1976). For concentrated solutions used as spinning dopes, especially when sulfuric acid or HF are the solvent, the diagrams according to Kwolek are suitable guides for predicting solution behavior. Ciferri (1975) has studied the phase diagrams for PABH-T in DMSO and concluded that in the usual range of solid concentrations used for spinning, it is not possible to employ anisotropic solutions.

IV. Fiber Production

With the exception of the melt-spun fibers from aromatic polyesters, all of the fibers reported here are spun from solutions of the polymers. Dry spinning techniques have been reported to yield fibers of high initial modulus but in general high strength fibers have not been obtained by this route, probably because of heat-induced crystallinity in the fibers before high orientation is attained. Wet spinning in the conventional manner of immersing the jet in the coagulation bath can be employed, but superior results apparently can be achieved by means of the dry jet or so-called "air-gap" method of spinning (Blades, 1973; Morgan, 1968, 1972), which leaves the jet dry above the coagulation bath. Both organic and inorganic solvents may be used for wet spinning. With some inorganic solvents, e.g., sulfuric acid, dry spinning would be impossible. Organic solvents are generally rendered more potent by the addition of soluble salts, e.g., lithium chloride.

The wet spinning process lends itself well to continuous in-line solvent removal by washing, drying on heated rollers, and hot-drawing. The latter operation is usually required to obtain a high strength and high initial modulus fiber when isotropic rather than anisotropic solutions are employed for spinning. Hot-drawn fibers from rodlike polymers are generally more brittle or less "tough" (i.e., have lower work-to-break values) than fibers which are spun from anisotropic dopes. Not all of the rodlike polymers can be dissolved to yield the anisotropic solutions, however, because sufficiently powerful and nondegrading solvents are not known for them. Certain of the rodlike polyamides can be dissolved in potent solvents, such as sulfuric acid, to reach the high solids level to lead to the formation of liquid crystals, i.e., close-packed aggregates of the rodlike molecules mutually aligned within a packet. Hot-drawing of fibers spun from anisotropic solutions is not necessary for the attainment of high strength, but the initial modulus can be increased as much as twofold without an increase in tenacity, albeit with a sacrifice in elongation-to-break (compare Kevlar with Kevlar-49). It is interesting to note that heating or "annealing" of fibers of PPB, PPD-T, and PABH-T leads to an appreciable increase in initial modulus and a slight increase in the tenacity of these fibers. These data may be taken as indirect evidence of the extended chain crystal morphology existing in these fibers. Fibers from polymers that chain fold generally relax under heat annealing and have decreased strength and modulus with an increase in elongation-to-break.

Recently, anisotropic spinning solutions employing an organic base have been reported (Hartzler, 1976) for polyhydrazides and superior fibers have been reported by a wet spinning method compared to the same compositions spun from an organic solvent by a dry spinning method. If anisotropic solutions could be formed from the polyamide–hydrazides and hydrazide copolymers that have yielded fibers of high tensile strength and modulus from organic solvents, it seems likely that further increases in strength and "toughness" without the embrittlement caused by hot-drawing could be obtained.

Probably one of the more remarkable reports of spinning ultrahigh-strength/high-modulus fibers concerns the spinning of rodlike aromatic polyesters from an anisotropic melt (Schaefgen *et al.*, 1975). The as-spun fibers are heated further to increase the molecular weight of the polymers. Surprisingly, the elongation-to-break of the fibers *increases* along with tensile strength and modulus.

An important feature of the ultrahigh-strength/high-modulus fibers is that the fibers are of fine diameter, usually less than 6 denier per filament and generally, more ~1.2–2 denier per filament. This situation is similar to that for glass fibers that must be of very small cross-sectional area in order to exhibit good mechanical properties in flexure.

Attainment of a high initial modulus for even a crudely spun fiber usually means that the polymer is capable of yielding high-strength/high-modulus fibers. High strength, largely a function of the degree of perfection of a fiber (i.e., lack of voids, fissures, etc.), can often be developed on optimization of the spinning process, including proper choice of solvent, spinning method, coagulation conditions, etc.

V. Fiber Properties

The fibers from rodlike polymers generally have rather high melting points and in the case of those fibers from polymers having amide-type linkages, they do not melt in the conventional sense because decomposition occurs simultaneously with melting. Nevertheless, an endothermic peak in the differential thermal analysis (DTA) of aramid, polyamide–hydrazide, and polyhydrazide fibers can be obtained and the values observed are generally above

400°C and as high as 550°C. Glass transitions for the fibers range from about 300°C to above 400°C. Despite the high melting points of several fibers from rodlike polymers, they are not particularly useful as heat resistant fibers because of their rather low elongations-to-break that decrease after exposure to an elevated temperature thereby becoming too brittle.

Most of the fibers of the ultrahigh-strength/high-modulus type have been reported to be either highly crystalline or crystallizable. For a discussion of apparent crystallite size, lateral crystallite order, longitudinal order along the fiber axis, and molecular orientation along the fiber axis in PPB and PPD-T, the reader is referred to a paper by Ballou (1976). Densities for these fibers generally are high, 1.4–1.5 g/cm^3. Small angle X-ray diffraction studies of the fibers from rodlike polymers show no evidence of chain-folding.

Most of the rodlike polyamide fibers, with the notable exception of those containing hydrazide linkages, characteristically burn only with difficulty and upon burning, produce chars. Limiting oxygen index (LOI) values, generally accepted as a measure of flame resistance for polymers, are quite high, i.e., greater than 28, for the aramid fibers from rodlike polymers. Certain of the all para-oriented aramid fibers, e.g., Kevlar, have high dimensional stability (i.e., resistance to shrinkage) in a flame. Addition of 1% phosphorus to a fiber from poly (p-phenylene terephthalamide) raises the LOI value from ∼ 28 to 40–42.

The chief interest in fibers from the rodlike polymers has been in their ultrahigh-strength/high-modulus properties. These fiber properties are given in Tables I–X. The rodlike aramid polymers used to prepare the fibers of Table I are of the —AB—, —AA—BB—, ordered copolymer and copolymers of limited order types. Terlon, Vnivlon, and SVM are high-strength/high-modulus fibers developed in the Soviet Union; the properties of these fibers would suggest that they are spun from, respectively, an —AB— type polyamide (poly-p-benzamide) spun from an organic solvent, an —AA—BB— type polyamide [poly(p-phenylene terephthalamide)] spun from sulfuric acid, and a copolyamide of limited order containing benzheterocyclic groups spun sulfuric acid (Sokolova et al., 1974; Kudryavtsev et al., 1974).

Fibers of poly(p-benzamide), PPB, spun from organic solvents, even when anisotropic dopes were employed, have inferior properties to those given in Table I because lower molecular weight polymers and lower solids levels have been used. Thus, Preston and Hofferbert (1977), and Kwolek et al. (1976) using a wet spinning method, have reported fibers spun from PPB of η_{inh} of = 1.5–1.6. As-spun tensile properties reported by these authors, respectively, were $T/E/M_i$ = 10.4/4.1/483 and 8.6/3.0/470. It is important to note that as-spun elongation-to-break values and initial modulus values were comparable whether PPB fibers were spun from an organic solvent or from sulfuric acid (Table I). The tensile strength, however, for the fiber spun from the former solvent, was less than one-half that for the fiber spun from the latter solvent.

Probably the highest tensile strength, 37 gpd, for any of these fibers is that claimed (Blades, 1973) by spinning of poly(p-phenylene terephthalamide), PPD-T, from sulfuric acid, the highest initial modulus of $1,700 \times 10^9$ dyne/cm^2 is that for PPB claimed (Alfonso et al., 1977) by spinning from an organic solvent.

The initial modulus of some miscellaneous polyamides that have been spun to fibers having high initial moduli but not particularly high strength are given in Table II. As noted earlier, strength can probably be developed by use of better solvents and spinning processes.

The polyamide–hydrazide that has been studied most is that based on the polyterephthalamide of p-aminobenzhydrazide, PABH-T. The tensile properties of fibers from this composition and from selected copolymers with PABH-T are given in Table III. The data for PABH-T given in Table III are for fibers spun from polymer of moderate molecular weight. Fiber (Table IV) from higher molecular weight PABH-T shows considerably higher strength (but *not* elongation-to-break and initial modulus) and is intermediate between Kevlar and Kevlar-49 as regards elongation-to-break and initial modulus. Tensile strength, however, is lower for both yarn and single filaments, particularly so for the latter.

TABLE IV

Comparison of PABH-T Fibers Spun from Organic Solvent with PPD-T Spun from H_2SO_4[a]

	X-500 (PABH-T) 1.5 (den/fil)	Kevlar (PPD-T) 1.4 (den/fil)	Kevlar-49 (PPD-T) 1.3 (den/fil)
η_{inh}			
DMAc–5% LiCl	12.3	Insoluble	Insoluble
Conc. sulfuric	Decomposed	5.5	6.1
Yarn			
T (g/den)	18.3	17.8^b–21.5	—
E (%)	2.9	3.5^b	2.1
M_i (g/den)	849		—
Single filament			
T (g/den)	21.8(24.2)	24.9^b–27.2$(31.2)^c$	29.8
E (%)	3.7	5.1^b–6.3	3.0
M_i(g/den)	691	437^b–517	1,014

[a] Data taken from Preston and Black (1976).

[b] Data from a sample which showed relatively low tensile properties; thus the range is shown for a number of specimens sampled over a 2-year period.

[c] Data in parentheses indicate individual high values for a given series of breaks.

TABLE V

Fibers from Polyhydrazides Spun from Organic Solvents

Structure	$T/E/M_i$[a]	Reference
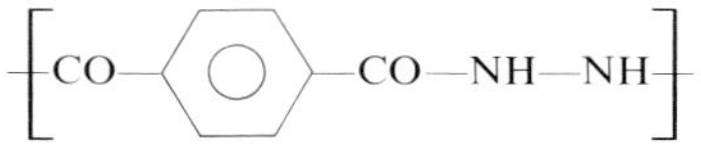	6.0/1.3/490[b]	Frazer (1972)
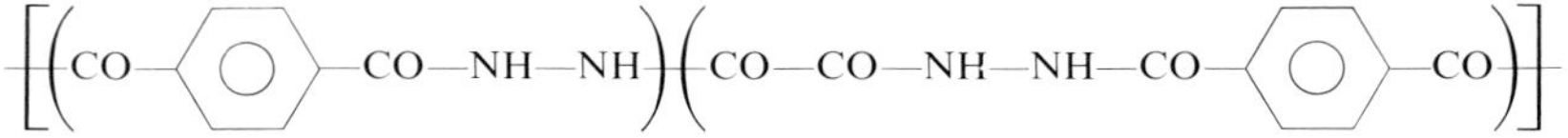	15.6/5.1/483[c]	Dobinson and Pelezo (1973)

[a] T = tenacity (g/den); E = elongation-to-break (%); M_i = initial modulus (g/den).
[b] Spun from DMSO using the dry spinning method.
[c] Spun from DMAc–LiCl using the dry jet–wet spinning method.

TABLE VI

Fibers from Organic Base Solutions of Poly(terephthaloylhydrazide)[a]

Solvent	% Polymer solids	η_{inh}	$T/E/M_i$[b]	Denier
10.0% N(Me)₄OH	9.1	4.7	10.7/7.5/320	0.75
16.7% N(Me)₄OH	10.0	4.6	7.9/11.2/206	1.6
10.0% N(Et)₄OH	9.1	5.1	10.5/13.7/220	2.04
1.1 × draw @ 200°C			12.7/7.4/363	1.96

[a] Data taken from Hartzler (1976); data are for *as-spun* fibers unless otherwise indicated.
[b] T = tenacity (g/den); E = elongation-to-break (%); M_i = initial modulus (g/den).

The polyhydrazide fiber that has been studied most is that based on the polyterephthalhydrazide, TDH-T. The tensile properties of fibers from this polymer are given in Tables V and VI. The improvement in tensile properties obtained by spinning polyhydrazides from an anisotropic spinning solution using an aqueous organic base (Table VI) versus spinning from an isotropic spinning solution using an organic solvent should be noted. Copolymers of TDH-T with the polymer from the polyterephthalhydrazide of oxalyldihydrazide, ODH-T, have been prepared. The tensile properties of fibers from a TDH-T/ODH-T copolymer are given in Table V.

Apparently the great regularity necessary for the attainment of good fiber tensile properties in homopolymers which chain fold is not necessary for the rodlike polymers which probably exist in the chain-extended state. In Tables VII and VIII the tensile properties of selected aramid fibers based on random

TABLE VII

Fibers from Random Copolyterephthalamides
of p-Phenylenediamine Plus Various Diamines[a]

No.	Diamine	Mole %	η_{inh}[b]	$T/E/M_i$[c]
1	$NH_2-C_6H_4-CH_2-C_6H_4-NH_2$	5	3.8	15/4.1/580
2	$NH_2-C_6H_4-CH_2-CH_2-C_6H_4-NH_2$	7.5	3.4	21/3.8/730
3	$NH_2-C_6H_4-O-C_6H_4-NH_2$	5	4.3	24/6.2/520
4	NH_2,NH_2 (1,2-phenylenediamine)	5	4.2	17/5.4/460
5	$NH_2-C_6H_3(Cl)-NH_2$	25	5.7	22/6.9/350
6	$NH_2-C_6H_3(CH_3)-C_6H_3(CH_3)-NH_2$	5	3.4	17/4.9/550

[a] Data taken from Blades (1973); all of the fibers were spun from anisotropic solutions in sulfuric acid by means of the dry jet–wet spinning method. The tensile properties are for *as-spun* fibers.

[b] Determined at 30°C on 0.5 g of fiber dissolved in 100 ml of sulfuric acid.

[c] T = tenacity (g/den); E = elongation-to-break (%); M_i = initial modulus (g/den).

TABLE VIII

Fibers from Random Copolyterephthalamides
of p-Phenylenediamine and Various Diacid Chlorides[a]

No.	Diacid chloride of	Mole %	η_{inh}[b]	$T/E/M_i$[c]
1	—CO—⟨◯⟩—⟨◯⟩—CO—	55	5.3	21/5.5/690
2	—CO—⟨◯⟩—N=N—⟨◯⟩—CO—	5	3.4	19/4.6/500
3	—CO—⟨◯⟩—CO— (Cl)	5	5.5	23/4.2/580
4	—CO ◯ CO— (1,3)	5	3.9	20/4.6/580
5	—CO H \ / C=C / \ H CO—	40	3.3	18/5.7/560
6	(cyclohexane) H—CO / —CO—H	25	4.3	22/4.5/540
7	(cyclohexane) H—CO / —CO—H	50	3.4	18/4.9/490

[a] Data taken from Blades (1973); all of the fibers were spun from anisotropic solutions in sulfuric acid by means of the dry jet–wet spinning method. The tensile properties are for *as-spun* fibers.

[b] Determined at 30°C on 0.5 g of fiber dissolved in 100 ml of sulfuric acid.

[c] T = tenacity (g/den); E = elongation-to-break (%); M_i = initial modulus (g/den).

TABLE IX

Fibers from Aromatic Polyesters[a]

Structure	η_{inh}		$T/E/M_i$[b]	
	Polymer	Final fiber	As-spun	Heat-treated[c]
	3.4	7.1	4.7/2.1/174	11/2.8/249
	2.85	Insol.	5/3.1/189	12/4.7/243
	0.51	1.6	2.6/3.4/142	16/3.3/321

[a] Data taken from Schaefgen et al. (1975).

[b] T = tenacity (g/den); E = elongation-to-break (%); M_i = initial modulus (g/den).

[c] Tensile properties are for fibers that were heated but *not* hot-drawn.

TABLE X

Ultrahigh-Strength/High-Modulus Fibers from Aromatic Polyesters[a]

Structure	Fiber form	$T/E/M_i^{b,c}$
(structure with $-O-CH_2-CH_2-O-$ unit) and (terephthaloyl/chlorohydroquinone units)	Filament	18.0/5.6/423
(biphenylene ester structure) and (terephthaloyl/chlorohydroquinone units)	Yarn	15.0/2.5/528 (20.5/5.3/246)
(naphthalene dicarbonyl structure) and (terephthaloyl/chlorohydroquinone units)	Yarn	30.4/4.7/527

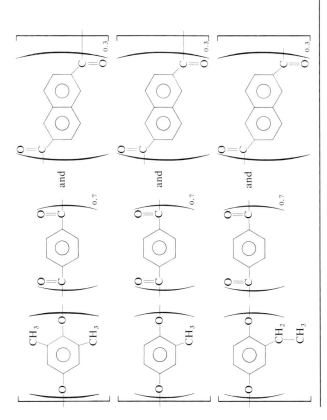

Filament 16.0/3.4/483

Yarn 20.0/4.4/365

Filament 12.0/3.9/355

[a] Data taken from Schaefgen *et al.* (1975).

[b] T = tenacity (g/den); E = elongation-to-break (%); M_i = initial modulus (g/den).

[c] Tensile properties are for fibers which were heated but *not* hot-drawn.

copolyamides may be cited. For copolymers of polyamide–hydrazides and polyhydrazides, the same point may be made and the tensile properties of fibers from such copolymers are found in Tables III and V. These data show that random copolymers yield fibers having tensile properties comparable to homopolymers. In the case of the rodlike aromatic polyesters, certain compositions can be spun from a melt *only* because of the likely random nature of the copolymers, which depresses the melting point below the decomposition temperature. The tensile properties of fibers from some of the random copolyesters are given in Tables IX and X.

As shown by workers at Monsanto Company for fibers from polyamide–hydrazides (Preston *et al.*, 1973b) and by workers at the E. I. du Pont de Nemours and Company for fibers from polyamides (Blades, 1973), the introduction of a few mole percent of rings that are not rigid chain-extending types does not necessarily detract from the tensile properties of these fibers. In fact, on balance, fiber tensile properties may be enhanced considerably.

VI. End Use Applications

The stress–strain curves for the ultrahigh-strength/high-modulus fibers show a strong similarity to those for glass and steel. On a specific basis (i.e., taking into account the lower specific gravity of the fibers from rodlike aromatic polymers in comparison to glass and steel), these fibers are seen to be stronger and stiffer than glass and steel. Taken together, these properties suggest that these fibers should be quite useful in the reinforcement of rigid and flexible composites. Thus, Kevlar fiber has been shown to be useful as a tire cord as a replacement for glass and steel belts in bias-belted and radial-belted tires. Kevlar-49 has been shown to be competitive with the lower modulus types of graphite fibers in rigid composites. Other specific end use applications include conveyor belts; V-belts; ropes and cables; body armor; interior trim, exterior fairings, control surfaces, and structural parts in aircraft; radomes and antenna components; circuit boards; filament-wound vessels; fan blades; sporting goods, e.g., skiis, golf clubs, surfing boards; coated fabrics for use in air-supported shelters. For a particularly useful discussion of the relationship of Kevlar properties to end use applications, the reader is referred to the excellent paper by Wilfong and Zimmerman (1977).

The use of the para-oriented aramid fibers for applications where flame-resistance and dimensional stability are required include: industrial protective clothing, e.g., pants, shirts, coast, and smocks for workers in laboratories, foundries, chemical plants, and petroleum refineries; welder's clothing and protective shields; fire department turnout coats, pants, and shirts; jumpsuits for forest fire fighters; flight suits for military pilots; auto racing drivers' suits.

References

Alfonso, C. C., Bianchi, E., Ciferri, A., Russo S., Salaris, F., and Valenti, B. (1977). *Polym. Prepr., Am. Chem. Soc., Div. Polym. Chem.* **18**(*1*), 179.

Arpin, M., and Strazielle, C. (1976). *Makromol. Chem.* **177**, 581–584.

Bach, H. C., and Hinderer, H. E. (1970). *Polym. Prepr., Am. Chem. Soc., Div. Polym. Chem.* **11**(*1*), 334–338.

Bair, T. I., Morgan, P. W., and Killian, F. L. (1976). *Polym. Prepr., Am. Chem. Soc., Div. Polym. Chem.* **17**(*1*), 59–64.

Ballard, D. G. H. (1963). U.S. Patent No. 3,089,749 (assigned to Courtaulds, Ltd.).

Ballard, D. G. H., and Griffiths, J. D. (1964). U.S. Patent No. 3,121,766 (assigned to Courtaulds, Ltd.).

Ballou, J. W. (1976). *Polym. Prepr., Am. Chem. Soc., Div. Polym. Chem.* **17**(*1*), 75–78.

Black, W. B. (1975). *In* "Wholly Aromatic High-Modulus Fibers" (C. E. H. Bawn, ed.), MTP Int. Rev. Sci., Phys. Chem. Ser. 2, Macromol. Sci. Vol., Ch. 2, pp. 34–122. Butterworth, London,

Blades, H. (1973). U.S. Patent No. 3,767,756 (assigned to du Pont Co.).

Blades, H., and White, J. R. (1963). U.S. Patent No. 3,081,519 (assigned to du Pont Co.).

Burke, J. J. (1973). *J. Macromol. Sci., Chem.* **A7**(*1*), 187–200.

Carter, G. B., and Schenk, V. T. J. (1975). *In* "Ultra-High Modulus Organic Fibers" (I. M. Ward, ed.), Ch. 13. Halsted Press and Wiley, New York.

Ciferri, A. (1975). *Polym. Eng. Sci.* **15**, 191–198.

Dobinson, F., and Pelezo, C. A. (1973). U.S. Patent No. 3,748,298 (assigned to Monsanto Co.).

Federov, A. A., Savinov, V. M., Sokolov, L. B., Zlatogorskii, M. L., and Grechishkin, V. S. (1973). *Vyskomol. Soedin., Ser. B* **15**(1).

Fitzgerald, J. A., and Likhyani, K. K. (1974). U.S. Patent No. 3,849,074 (assigned to du Pont Co.); see also U.S. Patent No. 3,850,888 (1974).

Flory, P. J. (1956). *Proc. R. Soc., Ser. A* **234**, 73–89.

Frazer, A. H. (1972). U.S. Patent No. 3,642,707 (assigned to du Pont Co.).

Hannell, J. W. (1970). *Polym. News* **1**(*1*), 8–13.

Hartzler, J. D. (1976). U.S. Patent No. 3,966,565 (assigned to du Pont Co.).

Hermans, J., Jr. (1962). *J. Colloid Sci.* **17**, 638–648.

Kudryavtsev, G. I., Tokarev, A. V., Avrorva, L. V., and Konstantinov, V. A. (1974). *Khim. Volokna* **6**, 70–71.

Kwolek, S. L. (1972). U.S. Patent No. 3,671,542 (assigned to du Pont Co.).

Kwolek, S. L., Morgan, P. W., Schaefgen, J. R., and Gulrich, L. W. (1976), *Polym. Prepr., Am. Chem. Soc., Div. Polym. Chem.* **17**(*1*), 53–58.

Morgan, H. S. (1968). U.S. Patent No. 3,414,645 (assigned to Monsanto Co.).

Morgan, H. S. (1972). U.S. Patent No. 3,642,706 (assigned to Monsanto Co.).

Morgan, P. W. (1976a). Ger. D. T. 2620-351 (assigned to du Pont Co.).

Morgan, P. W. (1976b). *Polym. Prepr., Am. Chem. Soc., Div. Polym. Chem.* **17**(*1*), 47–52.

Morgan, P. W., and Hartzler, J. D. (1977). *Polym. Prepr., Am. Chem. Soc., Div. Polym. Chem.* **18**(*1*).

Panar, M., and Beste, L. F. (1976). *Polym. Prepr., Am. Chem. Soc., Div. Polym. Chem.* **17**(*1*), 65–68.

Papkov, S. P., Kulichikhin, V. G., Kalmykova, V. D., and Malkin, A. Y. (1974). *J. Polym. Sci., Polym. Phys. Ed.* **12**, 1753–1770.

Preston, J. (1975). *Polym. Eng. Sci.* **15**(*3*), 199–206.

Preston, J., and Black, W. B. (1976). *Witco Award Symp. Honoring P. W. Morgan, Am. Chem. Soc. Meet., New York.*

Preston, J., and Hofferbert, W. L., Jr. (1977). *Polym. Prepr., Am. Chem. Soc., Div. Polym. Chem.* **18**(*1*), 137.

Preston, J., Black, W. B., and Hofferbert, W. L., Jr. (1973a). *J. Macromol. Sci., Chem.* **A7**(*1*), 67–98.

Preston, J., Morgan, H. S., and Black, W. B. (1973b). *J. Macromol. Sci., Chem.* **A7**(*1*), 325–348.

Salaris, F., Valenti, B., Costa, G., and Ciferri, A. (1976). *Makromol. Chem.* **177**, 3073–3076.

Schaefgen, J. R., Pletcher, T. C., and Kleinschuster, J. J. (1975), Belg. Patent No. 828,935; see also Belg. Patent No. 828,936 (1975).

Schaefgen, J. R., Foldi, V. S., Logullo, F. M., Good, V. H., Gulrich, L. W., and Killian, F. L. (1976). *Polym. Prepr., Am. Chem. Soc., Div. Polym. Chem.* **17**(*1*), 69–74.

Sokolova, T. S., Volokhina, A. V., Mil'kova, L. P., Papkov, S. P., Kudryavtsev, G. I., Rogovina, A. A., Fainberg, E. Z, and Khudoshev, I. F. (1974). *Khim. Volokna* **3**, 25–28.

Tohyama, S., Jinda, T., Saito, S., Tsuda, Y., and Shinohara, S. (1976). U.S. Patent No. 3,944,522 (assigned to Toray Ind., Inc.).

Wilfong, R. E., and Zimmerman, J. (1973). *J. Appl. Polym. Sci.* **17**, 2039–2051.

Wilfong, R. E., and Zimmerman, J. (1977). *Appl. Polym. Symp.* **31**, 1.

Yamazaki, N., Matsumoto, M., and Higashi, F. (1975). *J. Polym. Sci. Polym. Chem. Ed.* **13**, 1373–1380.

Zapp, J. A., Jr. (1975). *Science* **190**, 422.

5

Liquid Crystalline Order in Polypeptides†

Edward T. Samulski

Department of Chemistry and Institute of Materials Science
University of Connecticut
Storrs, Connecticut

I. Background

A. Introductory Remarks

Traditionally, research on polymer melts and solutions deals with macromolecules that assume a variety of shapes, i.e., the atoms delineating the polymer backbone chain may adopt any of a large number of conformations consistent with the restrictions of the covalent bonds and valence angles of its primary structure. The secondary structure of such polymers is therefore characterized by a dynamic sequence of rapid changes in the polymer's internal degrees of freedom as the polymer acquiesces to shear

† This work was supported in part by the National Institute of Arthritis Metabolism and Digestive Diseases (AM 17497).

stresses and density fluctuations in its immediate environment. This undulating secondary structure is called the *random coil* conformation. Almost all synthetic polymers exhibit a random coil conformation in solution and in the melt. There are, however, certain biological macromolecules that lie at the other extreme of the conformational spectrum. In proteins and enzymes a combination of covalent and noncovalent forces dictate that certain secondary structures and tertiary structures (three-dimensional spatial arrangements of secondary structure) are energetically favorable, even in solution. And these complex, well-defined, three-dimensional conformations are essential to the highly specific biochemical functions of the proteins and enzymes.

In recent years there has been a growth of interest in designing polymers that demonstrate a preference for a linear, extended conformation in solution and/or in the melt. This interest is predicated on the knowledge that anisotropic molecular shape is a prerequisite for the formation of liquid crystalline aggregates or phases. The attendant anisotropic viscoelasticity and increased fluidity of this phase facilitates the fabrication of highly oriented, high-strength fibers and films.

In this chapter the recent advances in interpreting the behavior of lyotropic polypeptide liquid crystals, i.e., concentrated solutions of α-helical synthetic homopolypeptides, are reviewed. Although the α-helical conformation—or for that matter, the synthetic polypeptides themselves—may be viewed as a rather esoteric phenomenon of limited relevance to the more conventional industrially important polymers, the criteria and principles underlying liquid crystal formation in polypeptide solutions can be used to gain insights into polymeric liquid crystals in general, including thermotropic phases (liquid crystalline polymer melts). The role of the solvent in lyotropics may be grossly equated to that of thermal energy in thermotropics; specific intermolecular forces are attenuated by both. Moreover, the packing constraints imposed on macromolecules with an extreme shape anisotropy are applicable in both types of liquid crystals.

This chapter emphasizes the more recent experimental advances in lyotropic polypeptide liquid crystals, as there have been several earlier reviews of this subject (Robinson, 1966; Tobolsky and Samulski, 1974; DuPré and Samulski, 1978; Iizuka, 1976).

B. The α-Helical Conformation

Synthetic homopolypeptides are represented by the general formula

$$\left(\text{CO}-\overset{\displaystyle \overset{\text{H}}{|}}{\underset{\displaystyle \underset{\text{R}}{|}}{\text{C}^z}}-\text{NH}\right)_n$$

where n is the degree of polymerization and R designates the side chain primary structure. In a covalently linked series of such peptide units, the resulting planar, trans, peptide bond and van der Waals repulsions between neighboring peptide residues cause a marked configurational energy minimum for specific dihedral rotation angles about the C—C^α and C^α—N bonds. In the low energy configuration, the N–H bond of residue i is positioned relative to the C=O of residue $i + 4$ in an orientation that promotes hydrogen bond formation. The configuration of the atoms in the polypeptide backbone is repeated exactly every 18 residues contributing five turns of helix. Stabilized by a network of hydrogen bonds, this conformation is referred to as the *α-helix* (Pauling *et al.*, 1951).

The axial ratio L/d where L is the length of a rigid, extended polymer and d its diameter, is a key parameter in characterizing the potential of the polymer to develop liquid crystalline order. The regular conformation of the α-helix facilitates the calculation of L/d provided n and R are known. Each peptide residue translates a distance $h = 1.5$ Å along the helix axis. Hence for an ideal α-helix, $L = n \times h$ (Å). The appropriate value for d is somewhat ambiguous. The diameter of the helical core of the polypeptide backbone chain is approximately 6 Å. The side chains R extend out radially from this core and mix with solvent molecules. A lower limit for d may be calculated with L using the density of the solid polypeptide. Model building may also give some insight into estimates of d; however, a solvation layer should also be considered.

A discrete, absolute value for L/d cannot be realized because of the uncertainties in d and the fact that in every polymer sample there is polydispersity in the molecular weight, i.e., a distribution of values for n. These uncertainties should be kept in mind when comparing experiment and theory.

A transition from α-helix to random coil (free rotation about the C—C^α and C^α—N bonds) may be induced in dilute solutions of synthetic polypeptides thermally or by addition of denaturing agents (usually strong acids) (Fasman, 1967). However, since a rigid, extended polymer conformation is a prerequisite for liquid crystallinity in polypeptide solutions, we consider only experimental conditions in which the α-helix conformation is obtained. Under such conditions, we may visualize the helical polypeptide as a cylindrical rod supported by an inner core with considerable rigidity sheathed in a soft outer core of flexible side chains mixed with solvent.

Herein we focus primarily on the properties of solutions of a single synthetic homopolypeptide, poly(γ-benzyl-L-glutamate), abbreviated PBLG; $R = -CH_2CH_2COOCH_2C_6H_5$. Solutions of PBLG are probably the most thoroughly characterized lyotropic polymeric liquid crystals; PBLG has been commercially available since about 1950 and has good solubility in a variety of solvents. However, studies using other polypeptides indicate that the characteristics of PBLG liquid crystals are representative of this general class of helical polymers.

II. Preparation of Lyotropic Polypeptide Liquid Crystals

Polypeptide liquid crystals are prepared simply by dissolving the polymer in a solvent that supports the α-helical conformation. The solution becomes liquid crystalline spontaneously when the polymer exceeds a certain limiting concentration in the solution. The homogeneous liquid crystalline phase is preceded by a biphasic regime, isotropic solution in equilibrium with liquid crystalline solution, which is present over a narrow range of polymer concentrations. The polymer concentrations that define the three regimes, isotropic solution, biphasic solution, and liquid crystal, are apparently independent of the solvent; these limiting concentrations depend on the polymer axial ratio L/d (see Section III). There are a variety of organic solvents that have been used to form PBLG liquid crystals. Note that dichloroacetic acid (DCA), a nonhelicogenic solvent (in *dilute* solutions of PBLG in this acid or mixed solvents containing $>75\%$ DCA, the PBLG is a random coil), supports the α-helical conformation of PBLG at the high polymer concentrations used to form the liquid crystal (Robinson, 1961).

The usual methods (heating, agitating, etc.) that expedite the solubilization of polymers are applicable to the polypeptide liquid crystals; a homogeneous phase is obtained only after solubilization, and for concentrated systems this may require equilibration times on the order of days. The addition of trace amounts ($\sim 1\text{--}2\%$) of certain agents (trifluoroacetic acid, formamide, or dimethylformamide) to the solution seems to facilitate the solution process.

The liquid crystalline phase can be readily recognized; it is birefringent and exhibits the optical characteristics of the cholesteric structure (see Section IV). There is also a marked change in the viscosity at the transition from isotropic solution to liquid crystal. Figure 1 shows how the relative viscosity of PBLG solutions in dichloromethane changes with polymer concentration (Hines and Samulski, 1973). In fact, it was the dramatic change in the solution properties that brought about the discovery that synthetic polypeptides form a lyotropic liquid crystal. Elliott and Ambrose (1950) discovered the liquid crystalline phase during the course of evaporating PBLG solutions to form oriented films for IR studies of the conformation of synthetic polypeptides.

The liquid crystalline phase is stable over a considerable temperature range. The low temperature limit is governed by the freezing point of the solution (assuming the polymer does not precipitate) that is dependent on

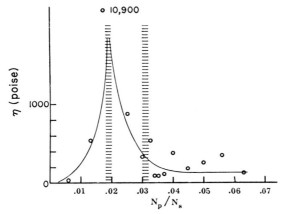

Fig. 1.   The relative flow viscosity η of PBLG solutions in dichloromethane. The concentration of PBLG is expressed in terms of the ratio of the number of peptide units, N_p, to solvent molecules N_s. The vertical dashed lines mark the observed phase boundaries separating isotropic solution from two-phase regime, $N_p/N_s \approx 0.02$, and the latter from the liquid crystalline phase, $N_pN_s \approx 0.03$. (After Hines and Samulski, 1973.)

the solvent used. The upper limit depends on both the concentration and axial ratio of the polypeptide. It is referred to as the clearing point T_c, the temperature at which the liquid crystal "melts" into an isotropic solution. T_c is often in the range 50–100°C (see Section III).

There are no published reports of surface treatments of cell walls to promote specific monodomain, uniform textures in the polypeptide liquid crystals. It does appear, however, that the long rodlike polypeptide molecules align preferentially with the major axes tangent to the container surfaces. Hence, thin cells with a large surface area nucleate a homeotropic texture, i.e., large domains in which the cholesteric axis is normal to the cell wall (Robinson, 1956). Once prepared, these polypeptide lyotropic liquid crystals are stable indefinitely. Aside from the large viscosities characteristic of polymer solutions, the most severe limitation to working with these materials is the need for containers and cells with seals that are impermeable to the volatile and/or chemically reactive solvent component of these liquid crystals.

III.   Theory of Liquid Crystal Formation

Molecular descriptions of the spontaneous transition from an isotropic to an ordered solution of rodlike solute particles focus on the associated changes in the entropy of the system. Straley (1973) has recently reviewed and compared various theoretical approaches to this problem. It appears that the initial work of Onsager (1947) formulating the density dependence of the free energy of a gas (or suspension) of long rods in terms of cluster integrals and excluded volume is still basically correct. The transition from disorder to order is driven by the competition between the "orientational entropy" of the rods (minimized by an isotropic distribution of rods) and their "translational entropy" (minimized by a parallel array of rods). The transition is of the first order, and at the transition point the volume fraction occupied by the rods in the ordered phase (liquid crystal) is

$$\Phi_{lc} = 4.5 \; d/L \tag{1}$$

For the isotropic phase in equilibrium with the liquid crystalline phase the volume fraction of rods is smaller:

$$\Phi_i = 3.3 \; d/L \tag{2}$$

Since the theory is developed for an athermal system of impenetrable rods, Φ_{lc} and Φ_i are independent of temperature.

An approximate expression (Ishihara, 1951) for the angular dependence of the excluded volume makes it possible to calculate the dependence of Φ_i and Φ_{lc} on the axial ratio of the rod (Straley, 1973). The results are shown in Fig. 2a.

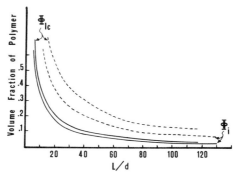

Fig. 2. Calculated values of the polymer volume fractions Φ_i and Φ_{lc} versus polymer axial ration L/d: (———) from the Onsager theory; (– – –) from the Flory theory. (After Straley, 1973.)

Polymer scientists are in general more familiar with lattice models of polymer solutions and may more readily assimilate Flory's (1956) approach to this phase transition. The orientational entropy is specified in terms of the possible number of configurations for arranging rod and solvent particles on a lattice. The approximations used by Flory to minimize the free energy have been criticized (Straley, 1973). Flory's results of the axial ratio dependence of Φ_{lc} and Φ_i are also shown in Fig. 2b.

The critical volume fraction Φ_i can be conveniently determined experimentally; the onset of birefringence (Robinson, 1956) and the abrupt change in viscosity (Hermans, 1962) have been measured as a function of L/D in polypeptide solutions. Measurements of Φ_{1c} are more difficult. Robinson et al. (1958) have estimated concentrations at which polypeptide solutions change from two-phase to liquid crystal by optical microscopy. Hines and Samulski (1973) noted changes in the solvent nuclear magnetic relaxation times at Φ_{1c}.

In Fig. 3 experimental values of Φ_i are compared with calculated values from the Onsager and Flory models. Changes in the solvent do not appreciably affect measured Φ_i suggesting that specific solvent–polypeptide interactions do not play an important role in this phase transition. Figure 3 appears to indicate that the Onsager result agrees with experiment at low values of L/d while the Flory result is in agreement with experiment at high L/d. However, as pointed out earlier, there are significant problems in comparing theory with experiment, in particular:

(a) uncertainties in d,
(b) uncertainties in L (polydisperse molecular weight).

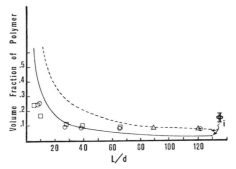

Fig. 3.   The critical volume fraction Φ_i versus calculated values of L/d; $\bigcirc$, various molecular weight fractions of PBLG in dioxane; $\triangle$, PBLG in dimethyl formamide; $\square$, PBLG in dichloromethane, (——) from the Onsager theory; (– – –) from the Flory theory. (After Straley, 1973.)

It is also possible to gain some insights into the driving force for the transition from isotropic to liquid crystalline solution by considering the thermodynamics of these systems. Miller et al. (1974a) have recently reviewed the thermodynamics of PBLG solutions. Experimental findings are discussed in the context of Flory's (1956) lattice theory where the enthalpy term in the free energy of mixing is assumed to be of the Van Laar type, characterized by a parameter X. Typical phase diagrams for $L/d = 50$ and 100

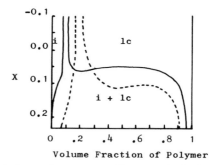

Fig. 4. A plot of the interaction parameter X versus Φ indicating the phase boundaries in polymer solutions; (———) corresponds to $L/d = 100$; (---), $L/d = 50$. (After Okamoto *et al.*, 1974.)

are shown for solutions of rodlike solute particles in Fig. 4. Three regimes are evident:

(i)   isotropic solution, $\Phi < \Phi_i$;
(ii)   a biphasic region with isotropic solution in equilibrium with liquid crystalline solutions;
(iii) liquid crystalline solutions, $\Phi > \Phi_{lc}$.

There is general experimental confirmation of the phase diagram (Wee and Miller, 1971; Miller *et al.*, 1974b) and X has been determined in PBLG solutions (Okamota *et al.*, 1974; Goebel and Miller, 1970). It appears that the transition from isotropic to liquid crystal is endothermic (Miller *et al.*, 1974a). Hence, the enthalpy favors an isotropic solution indicating that the entropic considerations discussed previously dominate the solution thermodynamics. The way in which mixing of solvent with polymer side chains is incorporated into theory—certainly important in determining the solubility of polypeptides—(Flory and Leonard, 1965; Rai and Miller, 1973) and the limitations of lattice theories in general must be considered to obtain a coherent picture of the thermodynamics in these systems.

IV. The Cholesteric Superstructure

In the previous sections we have implied that in lyotropic polypeptide liquid crystals, the rod major axes are aligned, i.e., the rods "close pack" with their major axes parallel in order to facilitate translation. Nearly parallel packing of rods does persist at the microscopic level. On a more macroscopic scale, however, a twisted superstructure is apparent in these liquid crystals. Figure 5 shows a typical photomicrograph of a liquid crys-

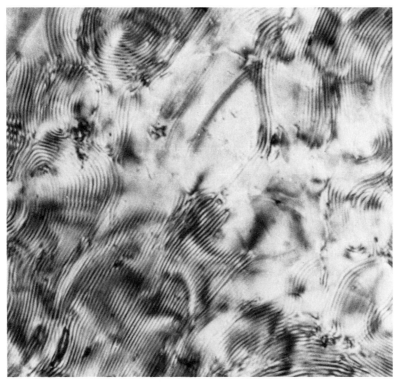

Fig. 5. A photomicrograph of the cholesteric phase under crossed polarizers; 20% PBLG (by weight) in dioxane. The regular spaced retardation lines are characteristic of the cholesteric phase and there separation is one-half the pitch of the cholesteric superstructure (in the photo the separation between retardation lines is $\sim 15 \,\mu$m). The symmetry axis of the twisted superstructure, the cholesteric axis Z is oriented normal to the retardation lines and Z changes direction in a continuous manner throughout the macroscopic sample.

talline PBLG solution; the interpretation of the regularly spaced retardation lines in terms of the relative orientation of the rodlike PBLG molecules was given by Robinson (1956, 1961, 1966). It is summarized in Fig. 6; the relative orientation of two PBLG rods is not a parallel configuration but rather, the equilibrium arrangement is one in which the two rods make a small angle ϕ_0 with respect to one another. This unidirectional twist, when propagated in one direction (along the cholesteric axis Z) results in a periodic superstructure characterized by a pitch $\mathbb{P}$,

$$\mathbb{P} = 2\pi D/|\phi_0| \tag{3}$$

In this expression, D is the distance between two rods (along Z direction) and can be determined from X-ray diffraction studies of polypeptide liquid

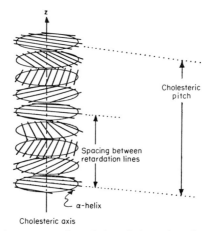

Fig. 6. A schematic representation of the relative orientations of helical polypeptide molecules required for generating the cholesteric superstructure; the parallel lines represent the rodlike polymer molecules.

crystals (Robinson *et al.*, 1958). The pitch $\mathbb{P}$ is twice the distance between the parallel retardation lines (Fig. 5) and is typically on the order of 10 μm. The inter rod spacing D is inversely proportional to the polymer (rod) concentration ($D \propto \Phi^{-2}$) and is typically on the order of 10 Å. Hence, Eq. (3) implies that the requisite twist angle per rod for generating the cholesteric superstructure is on the order of 10^{-2} degree. $\mathbb{P}$ is temperature dependent ($\mathbb{P} \propto T$; DuPré and Duke, 1975) and the cholesteric superstructure has a very high value of the form optical rotation.[†]

Cholesteric superstructures are commonplace in thermotropic liquid crystals composed of a nonracemic mixture of chiral (optically active) molecules. It is of interest to note that the twist is compensated in liquid crystals composed of equal numbers of L and D isomers (a nematic superstructure with a parallel arrangement of rods forms). Conversely, a twisted superstructure can be induced in a nematic phase by adding small amounts of a chiral molecule. Clearly, there is a fundamental connection between molecular chirality and the cholesteric twist, i.e., a nonzero value for ϕ_0.

It is no surprise that the lyotropic polypeptide crystals should exhibit a cholesteric superstructure since the α-helical macromolecules are chiral. They

[†] The form optical rotation should not be confused with the behavior of a chiral or optically active molecule in the vicinity of an absorption band. The latter optical rotation occurs in isotropic dilute solutions that contain optically active molecules and has its origin in the selective absorption of one circularly polarized component of the light. The form optical rotation, which we are discussing, is caused by the supermolecular arrangement of the PBLG rods in the cholesteric structure and originates in the selective reflection of circularly polarized light by this structure.

do, however, manifest a peculiar phenomenon, an anomalous compensation that indicates that the sense of the cholesteric twist is dependent on the achiral solvent in the lyotropic liquid crystal. The sense of the cholesteric twist, i.e., its right- or left-handedness, is directly related to the sign of the form optical rotation. The form optical rotation of PBLG liquid crystals is positive when prepared in dioxane, negative in methylene chloride, while PBLG itself is a righthanded α-helix in both solvents (Robinson, 1961). And in an 8:2 methylene chloride:dioxane mixture, the PBLG liquid crystal is compensated (nematic)! These curious observations seem to violate the empirical rule: "the sense of the molecular chirality [in this case a right-handed α-helix] dictates the sense of the cholesteric twist." This point is discussed further.

As indicated earlier, the magnitude of the mutual twist angle between molecules that is required to produce the observed values of the cholesteric pitch is quite small. A typical PBLG liquid crystal prepared in, for example, the solvent dioxane (20% PBLG by weight) exhibits a cholesteric pitch, $\mathbb{P} = 31$ μm. The intermolecular spacing between helices as determined by X-ray measurements is $D = 30$ Å (Robinson et $al.$, 1958). Since $\mathbb{P} = 2\pi D/|\phi_0|$, one can determine that $|\phi_0| \approx 0.035°$. Recently, Samulski and Samulski (1977) have shown how a comparable value for ϕ_0 results from the Mc-Lauchlan formulation of the van der Waals–Lifshitz forces between chiral dielectric rods embedded in a dielectric medium. Moreover they demonstrated how the McLauchlan analysis of the origin of cholesteric twist articulates the role of the medium and can quantitatively account for solvent-induced anomalous compensation in polypeptide liquid crystals. They derive the following expression for the cholesteric pitch:

$$\mathbb{P} = 2\pi D/\left|\tfrac{1}{2}\tan^{-1}\{(J_{11} - J_{22})/C\}\right| \tag{4}$$

C is an adjustable constant and J_{ij} are proportional to $(\varepsilon_i - \varepsilon_m)(\varepsilon_j - \varepsilon_m)$ where $\varepsilon_{i,j}$ are the principal values of dielectric permittivity normal to the rod major axis, and ε_m is the dielectric permittivity of the medium. It was shown that $\mathbb{P} \to \infty$ (i.e., the cholesteric superstructure is compensated) for a critical value of the medium permittivity, $\varepsilon_m^* = +(\varepsilon_1\varepsilon_2)^{1/2}$ and, for ε_m greater than or less than ε_m^*, ϕ_0 and hence the form optical rotation will change sign.

In Fig. 7 we compare observed values of the pitch with that calculated from Eq. (4) in mixed solvent PBLG liquid crystals. It is clear that the Samulski and Samulski (1977) calculation of the cholesteric superstructure is in good agreement with experiment. The calculation also enables us to understand the origin of anomalous compensation, i.e., the dielectric properties of the medium are an important factor in establishing both the sense and the magnitude of $\mathbb{P}$ in lyotropic cholesterics.

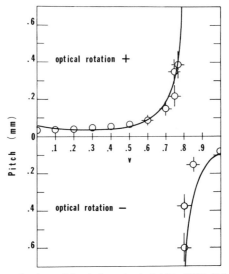

Fig. 7. Experimental values of the cholesteric pitch (○) of PBLG liquid crystals prepared in the mixed solvent system (dioxane + dichloromethane) are shown versus the volume fraction of dichloromethane in the solvent, v. The solid curve is calculated from Eq. (4); compensation of the cholesteric superstructure occurs at the solvent composition 2:8 ratio of dioxane: dichloromethane. The sign of the form optical rotation of the cholesteric superstructure is indicated in the figure. (After Samulski and Samulski, 1977.)

V. Order in the Microstructure

In the preceding section we pointed out that the amount of cholesteric twist per molecule ($\sim 10^{-2}$ degree of arc) does not perturb the local structure in the liquid crystal; on a molecular scale, the polypeptide helices are packed more or less parallel to one another. This microorder can be quantitatively specified in terms of the liquid crystalline order parameter. The order parameter in a system of cylindrical rods is given by

$$S = \tfrac{1}{2}(3\langle \cos^2 \theta \rangle - 1) \qquad (5)$$

θ is the angle that the rod major axis makes with the direction of preferred orientation in the liquid crystal (the nematic director) and $\langle\ \rangle$ indicates an averaging over the orientation of the rods.

The order in lyotropic liquid crystals can be perturbed by two distinct methods: (1) increasing the temperature; and (2) decreasing the polymer concentration. In the former case, increased thermal energy will result in larger librational oscillations of the rod major axes and eventually, at a high enough temperature T_c, the liquid crystal will melt into an isotropic solution $S = 0$. In an analogous manner, an increase in the solvent fraction of the liquid crystal will facilitate librational oscillations of the rods until finally, when the rod volume fraction decreases below Φ_i, the solution will become isotropic.

The degree of order in the microstructure of lyotropic polypeptide liquid crystals can be monitored using X-ray diffraction (Murthy *et al.*, 1976). First, a macroscopically aligned (nematic) liquid crystal must be prepared by removing the cholesteric superstructure with a magnetic field; the diamagnetic susceptibility of the helices is anisotropic (see Section VI). The similarity between magnetically aligned, uniaxial polypeptide liquid crystals and mechanically oriented polymers permits interpretation of the X-ray diffraction data in terms of a formalism commonly used to describe orientation of polymer crystallites in fibers. This procedure is based on an analysis of the *inter*molecular X-ray scattering (Alexander, 1969).

For an oriented fiber with a cylindrical distribution of crystallites about the fiber axis, the degree of orientation of the crystallite c axis (polymer chains are parallel to c) is specified by an orientation function f_c

$$f_c = \tfrac{1}{2}(3\langle \cos^2 \theta_c \rangle - 1) \tag{6}$$

where θ_c is the angle the c axis makes with the fiber axis.

$$\langle \cos^2 \theta_c \rangle = 1 - 2\langle \cos^2 \chi_{hkl} \rangle \tag{7}$$

and

$$\langle \cos^2 \chi_{hkl} \rangle = \int_0^{\pi/2} \frac{I(\chi)\cos^2 \chi \sin \chi \, d\chi}{\int_0^{\pi/2} I(\chi) \sin \chi \, d\chi} \tag{8}$$

χ_{hkl} specifies the orientation of the normal to the diffracting planes hkl and $I(\chi)$ is the azimuthal intensity distribution of the diffracted beam for a given Bragg angle.

In magnetically oriented polypeptide liquid crystals the optic axis of the sample is analogous to the fiber axis. A domain or a "crystallite," with the domain c axis at an angle θ_c relative to the optic axis, is composed of helical polypeptides with their long axes at some angle α relative to the c axis. If α is small, the order parameter of the liquid crystal can be represented by the product

$$S = f_c[\tfrac{1}{2}\langle 3 \cos^2 \alpha - 1 \rangle] \tag{9}$$

The microscopic structural details within a domain suggest that α is indeed very small; the interhelix separation ($D \approx 0{-}10 \,\text{Å}$) and the large helix length ($L = n \times 1.5 \,\text{Å}$) imply that very small angular deviations (minutes of arc) in a parallel array of helices will result in nearest neighbor contacts.

Hence, allowing the c axis of a domain to be specified by a bundle of approximately parallel helices ($\alpha \approx 0$), we have $S = f_c$.

Oriented liquid crystals give diffuse equatorial X-ray diffraction streaks with Bragg spacings of 15–25 Å depending on concentration (Robinson *et al.*, 1958). Measured plots of intensity versus azimuthal angle χ such as that shown in Fig. 8 were used to obtain $I(\chi)$. Equation (8) can be evaluated by numerical integration and S is determined from Eq. (9) with $\alpha = 0$.

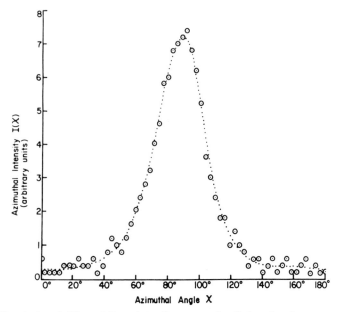

Fig. 8.   A plot of diffracted X-ray intensity versus azimuthal angle χ from a magnetically oriented (uniaxial) PBLG liquid crystal (20% PBLG in dioxane). (After Murthy *et al.*, 1976.)

Figure 9 shows a plot of S versus Φ, the volume fraction of PBLG (axial ratio = 140) in dioxane solutions. At low polymer concentrations, $\Phi < \Phi_i$, $S = 0$; the orientation of the helices is uncorrelated in these isotropic solutions. According to experimental phase diagrams (Robinson, 1966) for PBLG–dioxane liquid crystals with $L/d = 140$, we calculate $\Phi_i = 0.058$ and, when $\Phi_i < \Phi < \Phi_{lc} = 0.083$, a two-phase system viz., isotropic fluid (i) and liquid crystal (lc) exists. As Φ increases between Φ_i and Φ_{lc}, only the relative fraction of the liquid crystalline phase increases. The value of S near the phase boundary Φ_{lc} suggests that the intrinsic order in the liquid crystal at the isotropic nematic phase transition is characterized by a critical order parameter $S_c \approx 0.5$. For $\Phi > \Phi_{lc}$, $S \approx 0.75$ and does not change significantly over the concentration range studied. Also shown in Fig. 9 is the insensitivity

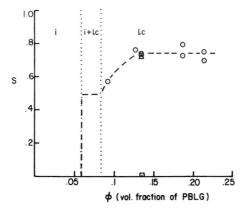

Fig. 9. The measured order parmeter (Eq. (9)) as a function of PBLG volume fraction in dioxane solutions: ($\cdot\cdot\cdot$) experimentally observed phase boundaries separating isotropic solution (i) from liquid crystalline solution (lc) for $L/d = 140$ (◯). S does not change when L/d is decreased from 140 to 70 (△) at constant volume fraction of PBLG; for $L/d = 13$ (▢), the solution is isotropic and $S = 0$. (After Murthy *et al.*, 1976.)

of S to a decrease in L/d from 140 to 70; for the solution with PBLG axial ratio $L/d = 11$, $\Phi = 0.13 < \Phi_i \approx 0.25$ and the solution is isotropic.

The axial ratio of the polypeptide can, in principle, be varied over a considerable range, i.e., polypeptides can be synthesized with a range of molecular weights. This feature, as pointed out in Section III, makes this class of liquid crystals attractive from the standpoint of testing statistical models of liquid crystalline phase transitions. Predicted values of the order parameter at the isotropic–nematic phase transition based on such models range from $S_c = 0.44$ (Maier and Saupe, 1960) to $S_c = 0.84$ (Onsager, 1947). The observed value of $S_c \approx 0.5$ for the PBLG–dioxane liquid crystal is in agreement with the mean field calculation.

As indicated earlier, temperature can disrupt the degree of order in the microstructure. The temperature dependence of $\Delta\chi(T)$, the anisotropy of the liquid crystal diamagnetic susceptibility, can be related to $S(T)$, the temperature dependent order parameter, by

$$\Delta\chi(T) = (\chi_{\parallel} - \chi_{\perp})S(T) \qquad (10)$$

where $\chi_{\parallel}$ and $\chi_{\perp}$ are the molar susceptibilities per residue parallel and normal to the helix major axis, respectively. (Magnetic properties of polypeptide liquid crystals are discussed in the following section.) A recent measurement (Duke *et al.*, 1977) of $\Delta\chi(T)$ in conjunction with $S(T = 25°C) = 0.72$ allows a determination of $S(T)$. In Fig. 10 the temperature dependence of the order parameter is shown as a function of reduced temperature, $T^* = T/T_c$, where $T_c = 65°C$ is the temperature at which the liquid crystal (20% PBLG

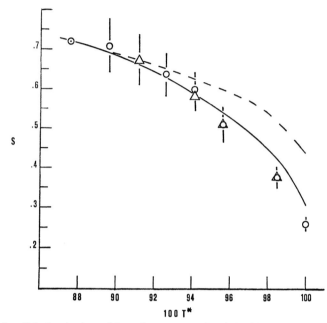

Fig. 10. Calculated values of the order parameter (Eq. (10) with the data in Fig. 13) versus reduced temperature $T^* = T/T_c$ for 20% PBLG in dioxane: (———) form of S versus T^* derived by Humphries *et al.* (1972); (– – –) form of S versus T^* derived by Saupe (1960). (After Duke *et al.*, 1977.)

in dioxane) transforms to an isotropic solution. Figure 10 also displays the form of the universal order parameter predicted by Maier and Saupe (1960) and that derived by Humphries *et al.* (1972).

VI. Response to External Stimuli

The viscoelastic properties of a liquid crystal are characterized by a set of elastic moduli K_{ij} and viscosity coefficients γ_i that are fundamental in the establishment of a homogeneous liquid crystal. These parameters in conjunction with the anisotropic polypeptide magnetic and dielectric susceptibilities $\Delta\chi$ and $\Delta\varepsilon$, respectively, determine the response of the liquid crystal to external perturbations. Both $\Delta\chi$ and $\Delta\varepsilon$ are positive for polypeptide liquid crystals. Hence, in sufficiently strong magnetic (electric) fields, a macroscopically aligned liquid crystal results; the helices are parallel to the field direction. Obviously this alignment process disrupts the cholesteric macrostructure initially present in these phases. In fact, this structural transition from cholesteric to nematic enables one to use a number of techniques to conveniently monitor the transition and to determine the critical value of the field required to induce the transition. The field-induced transition was discovered in these systems using nuclear magnetic resonance (NMR) studies of the solvent molecules (Sobajima, 1967; Samulski and Tobolsky, 1968; Panar and Phillips, 1968). Subsequently, it has been studied using NMR (Orwoll and Vold, 1971), infrared dichroism (Iizuka, 1976), optical measurements (Toth and Tobolsky, 1970), magnetic susceptibility (Guha-Sridhar *et al.*, 1974), and laser light-beating techniques (Duke and DuPré, 1974). One can also monitor this transition by measuring the cholesteric pitch as a function of field strength. In Fig. 11 the normalized cholesteric pitch is shown as a function of the magnetic field strength H. As H increases and approaches H_c, the pitch diverges logarithmically.

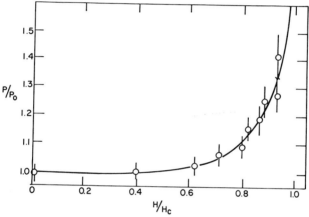

Fig. 11. The reduced pitch $\mathbb{P}/\mathbb{P}_0$ versus reduced magnetic field strength H/H_c for 20% PBLG in dioxane. The pitch of the cholesteric superstructure dilates with increasing field and diverges logarithmically at a critical value of the field $H_c = 5\ kOe$. (After DuPré and Duke, 1975.)

Data of the type shown in Fig. 11 can be used to determine the twist elastic constant K_{22} of polypeptide liquid crystals. Physically, K_{22} describes the force necessary to induce a twist deformation in the cholesteric superstructure, either mechanically, through an external field, or by natural thermal excitations. Meyer (1968) and deGennes (1968) showed that K_{22} could be determined from H_c using

$$H_c = \frac{\pi^2}{2}\left(\frac{K_{22}}{\Delta\chi}\right)^{1/2}\mathbb{P}_0^{-1} \tag{11}$$

$\mathbb{P}_0$ is the cholesteric pitch in the absence of the field. H_c, $\Delta\chi$, and $\mathbb{P}_0$ are of the order of $10^4\ Oe$, $10^{-9}\ emu/cc$, and $10^{-3}\ cm$, respectively, in typical PBLG liquid crystals, giving an order of magnitude for K_{22} of 10^{-7} dynes.

The twist elastic constant is dependent both on short- and long-range orders in the liquid crystal. Polypeptide concentration, molecular weight, and the solvent used in the liquid crystal influence K_{22} (Guha-Sridhar *et al.*, 1975; DuPré and Duke, 1975). In Fig. 12 a plot of K_{22} versus Φ is shown. We arrive at Fig. 12 by assuming that $\Delta\chi$ for the liquid crystal does not contain contributions from the solvent, i.e., $\Delta\chi = \Delta\chi_{\text{rod}}$, where $\Delta\chi_{\text{rod}}$ is the anisotropy per mole of peptide residues. This assumption is justified; NMR studies of the solvent show that the degree of order of the solvent is very low, $S \approx 10^{-3}$ (Samulski and Berendsen, 1972). In contrast, the rods are highly ordered and the intrinsic anisotropy of the rod, $(\chi_{\|} - \chi_{\perp})$ can be determined if the degree of order is known using Eq. (10). For PBLG we find $(\chi_{\|} - \chi_{\perp}) = 8.4 \times 10^{-9}$ emu/mole, where $\chi_{\|}$ is the value along the helix axis; $\chi_{\perp}$ is normal to the helix (Duke *et al.*, 1977).

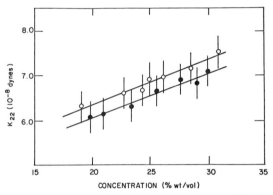

Fig. 12. The dependence of the twist elastic constant K_{22} on PBLG concentration; $\bigcirc$, 20°C; $\bullet$, 25°C. (After DuPré and Duke, 1975.)

Temperature disrupts the degree of order in the liquid crystal and $\Delta\chi$ decreases to zero (discontinuously) at T_c (Fig. 13). $\mathbb{P}_0$ increases linearly with temperature, while H_c is inversely proportional to the temperature (DuPré and Duke, 1975). The combination of these results enables a calculation of K_{22} versus T (Fig. 14). If $K_{22}(T)$ is plotted versus $S(T)^2$, a linear relation is observed and this is predicted by the mean-field theory of liquid crystals (Saupe, 1960).

The response of aligned liquid crystals to a change in the direction of the applied field has been used to calculate the rotational viscosity coefficient γ_1 (Guha-Sridhar *et al.*, 1974; Filas *et al.*, 1974). γ_1 is of the order of $10-10^3$ P; such high values are consistent with the fact that polypeptide liquid crystals are in actuality concentrated polymer solutions.

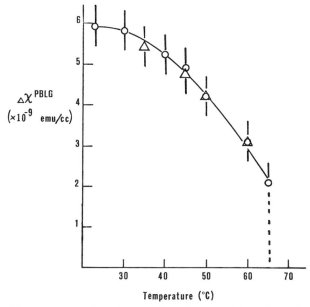

Fig. 13. The temperature dependence of the anisotropy of the volume diamagnetic susceptibility of a PBLG–dioxane liquid crystal (20% PBLG). The anisotropy decreases with increasing temperature reflecting disordering of the microorder (Eq. (10)) until finally at $T_c = 65°C$, the liquid crystal melts into an isotropic solution; $\bigcirc$, increasing T; $\triangle$, decreasing T. (After Duke *et al.*, 1977.)

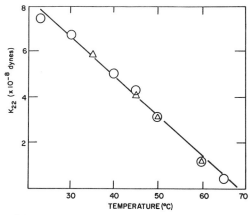

Fig. 14. A plot of the twist elastic constant versus the square of the liquid crystalline order parameter (20% PBLG in dioxane); a linear relation is consistent with the mean field theory of liquid crystals developed for thermotropic systems composed of small organic molecules.

VII. Mesomorphic Structure in the Solid State

In the introduction to this review of polypeptide liquid crystals, it was suggested that the potential for stabilizing mesomorphic structures in the solid state of polymers (and concomitant physical properties of these materials) is a key factor in the renewed interest in the liquid crystalline state. Attempts to retain specific mesomorphic structures in the solid state date from the the 1930s when it was shown that some thermotropic phases could be quenched to low temperatures to form a metastable, brittle, glassy state in which the liquid crystalline structure persisted (Vorlander, 1933). In the case of lyotropic liquid crystals, similar results could be obtained in principle by various techniques, e.g., freeze-drying, slow evaporation of the solvent.

In polypeptide liquid crystals, the equilibrium superstructure—the cholesteric structure—was thought to persist in the solid state of films cast from certain solvents (Samulski and Tobolsky, 1967). Studies of X-ray diffraction patterns, specific volume, and mechanical properties of solid PBLG films indicated that the cholesteric structure was preserved in films cast from chloroform and methylene chloride (McKinnon and Tobolsky, 1966a), while conventionally crystalline films were obtained from dimethyl-formamide (McKinnon and Tobolsky, 1966b). Retention of the cholesteric structure was convincingly demonstrated by casting films of PBLG from binary solvent systems in which one component is nonvolatile (a plasticizer). The characteristic pattern of retardation lines, indicative of the cholesteric superstructure, can be observed optically (Samulski and Tobolsky, 1969). Figure 15 is a photomicrograph of a thin section of a solid PBLG film plasticized with 3,3'-dimethylbiphenyl. This should be compared with the appearance of the fluid liquid crystalline phase (Fig. 5).

The intermolecular spacing D in the solid is considerably smaller than that observed in the liquid crystal. Hence the value of the pitch is correspondingly smaller (of the order of 1 μm). In fact, we have observed the reflection of irridescent colors from some solid polypeptide films indicating that $\mathbb{P}$ may get as small as the wavelength of visible light. Since this cholestric superstructure is locked into the solid state, $\mathbb{P}$ is virtually insensitive to temperature, and change in the reflected color with changes in temperature, characteristic of thermotropic cholesterics, is absent.

The decrease in D in passing from the fluid liquid crystal to the solid films prepared with decreasing amounts of plasticizer is continuous and proportional to the amount of PBLG (Samulski and Tobolsky, 1969). This is additional evidence for the absence of a structural transition in going from the liquid crystalline state to the solid state.

In Section IV we indicated that in the liquid crystal the cholesteric axis Z was preferentially aligned normal to large surface areas. During casting,

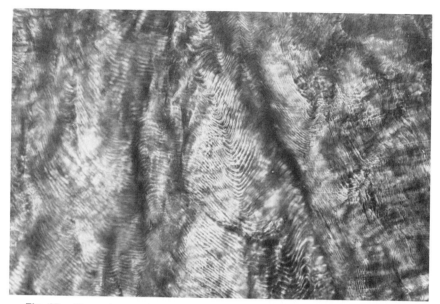

Fig. 15.   Photomicrograph of a 20-μm thick section of a solid PBLG film plasticized with 3,3′-dimethylbiphenyl (30% by weight) taken with crossed polarizers; the distance between retardation lines is about 2 μm. (After Samulski and Tobolsky, 1969.)

the container surface and the solution–air interface present such surface areas. Consequently, the resulting solid films have a uniplanar distribution of the PBLG helices in the plane of the film; Z is normal to the film surface. This is evidenced by the highly anisotropic swelling characteristics of PBLG films (Samulski and Tobolsky, 1968). It is also dramatically illustrated by the fracture pattern of frozen PBLG films (Samulski, 1969). In Fig. 16a, a scanning electron photomicrograph shows the fracture pattern of PBLG films with the residual cholesteric superstructure. Sheathlike layers are apparent; the layers are in the plane of the film (Z is normal to the layers, i.e., vertical in the photomicrograph).

The nematic superstructure can also be retained in the solid state of PBLG. This uniaxial structure can be produced by first aligning the fluid liquid crystal in a magnetic field (larger than H_c), then slowly evaporating the solvent in the presence of the field. A highly oriented, uniaxial solid with the PBLG helices parallel to the initial field direction results. The uniaxial structure is illustrated in the fracture pattern of the film (Fig. 16b); a fibrillar structure is apparent in which the fibril axes are parallel to the initial magnetic field direction (vertical in the photomicrograph). X-ray diffraction patterns from the magnetically oriented films show that the inherent order is comparable to that achieved through mechanical deformation of fibers (Samulski and Tobolsky, 1971).

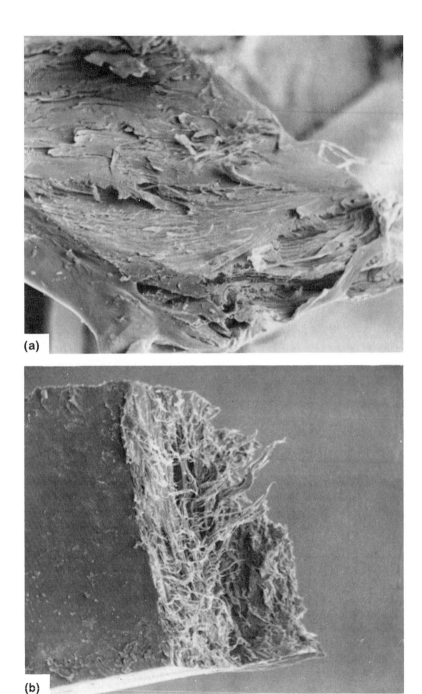

Fig. 16 Scanning electron photomicrographs of fractured solid PBLG film specimens; magnification 300X. (a) Cholesteric supermolecular structure (Z is vertical in photo); (b) magnetically oriented supermolecular structure (nematic axis is parallel to fibrils). (After Samulski, 1969.)

The uniaxial (nematic) superstructure can also be generated by anisotropic shear fields. Drawing fibers of polypeptides from the liquid crystalline phase while flashing off the solvent is a commonly employed technique to prepare highly oriented specimens for X-ray diffraction studies of the α-helical secondary structure.

References

Alexander, L. E. (1969). "X-ray Diffraction Methods in Polymer Science." Wiley (Interscience), New York.
deGennes, P. G. (1968). *Solid State Commun.* **6**, 163.
Duke, R. W., and DuPré, C. B. (1974). *Macromolecules* **7**, 374.
Duke, R. W., DuPré, D. B., and Samulski, E. T. (1977). *J. Chem.* Phys. **66**, 2748.
DuPré, D. B., and Duke, R. W. (1975). *J. Chem. Phys.* **63**, 143.
DuPré, D. B., and Samulski, E. T. (1978). *In* "Liquid Crystals" (F. D. Saeva, ed.). Dekker, New York.
Elliott, A., and Ambrose, E. J. (1950). *Discuss. Faraday Soc.* **9**, 246
Fasman, G. D. (1967). "Poly-γ-amino Acids," Vol. 1. Dekker, New York.
Filas, R. W., Hajdo, L. E., and Eringen, A. C. (1974). *J. Chem. Phys.* **61**, 3037.
Flory, P. J. (1956). *Proc. R. Soc., Ser. A* **234**, 73.
Flory, P. J., and Leonard, W. J. (1965). *J. Am. Chem. Soc.* **87**, 2102.
Goebel, K. D., and Miller, W. G. (1970). *Macromolecules* **3**, 64.
Guha-Sridhar, C., Hines, W. A., and Samulski, E. T. (1974). *J. Chem. Phys.* **61**, 947.
Guha-Sridhar, C., Hines, W. A., and Samulski, E. T. (1975). *J. Phys. (Paris), Colloq.* **36**, 270.
Hermans, J., Jr. (1962). *J. Colloid Sci.* **17**, 638.
Hines, W. A., and Samulski, E. T. (1973). *Macromolecules* **6**, 794.
Humphries, R. L., James, P. G., and Luckhurst, G. R. (1972). *J. Chem. Soc., Faraday Trans. 2* **68**, 1032.
Iizuka, E. (1976). *Adv. Polym. Sci.* **20**, 79.
Ishihara, A. (1951). *J. Chem. Phys.* **19**, 1142
McKinnon, A. J., and Tobolsky, A. V. (1966a). *J. Phys. Chem.* **70**, 1453.
McKinnon, A. J., and Tobolsky, A. V. (1966b). *J. Phys. Chem.* **72**, 1157.
Maier, W., and Saupe, A. (1960). *Z. Naturforsch.* **A15**, 287.
Meyer, R. B. (1968). *Appl. Phys. Lett.* **14**, 208.
Miller, W. G., Wu, C. C., Wee, E. L., Santee, G. L., Rai, J. H., and Goebel, K. G. (1974a). *Pure Appl. Chem.* **38**, 37.
Miller, W. G., Rai, J. H., and Wee, E. L. (1974b). *In* "Liquid Crystals and Ordered Fluids" (R. Porter and J. F. Johnson, eds.), Vol. 2, p. 243. Plenum, New York.
Murthy, N. S., Knox, J. R., and Samulski, E. T. (1976). *J. Chem. Phys.* **65**, 4835
Okamoto, A., Kubo, K., and Ogino, K. (1974). *Bull. Chem. Soc. Jpn.* **47**, 1054.
Onsager, L. (1947). *Ann. N.Y. Acad. Sci.* **51**, 627.
Orwoll, R. D., and Vold, R. L. (1971). *J. Am. Chem. Soc.* **93**, 5335.
Panar, M., and Phillips, W. D. (1968). *J. Am. Chem. Soc.* **90**, 3880.
Pauling, L., Corey, R. B., and Branson, H. R. (1951). *Proc. Natl. Acad. Sci. U.S.A.* **37**, 205.
Rai, J. H., and Miller, W. G. (1973). *Macromolecules* **6**, 257.
Robinson, C. (1956). *Trans. Faraday Soc.* **52**, 571.
Robinson, C. (1961). *Tetrahedron* **13**, 219.
Robinson, C. (1966). *Mol. Cryst.* **1**, 467.
Robinson, C., Ward, J. C., and Beevers, R. B. (1958). *Discuss. Faraday Soc.* **25**, 29.

Samulski, E. T. (1969). Ph.D. Thesis, Princeton Univ., Princeton, New Jersey.

Samulski, E. T., and Berendsen, H. J. C. (1972). *J. Chem. Phys.* **56**, 3921.

Samulski, E. T., and Tobolsky, A. V. (1967). *Nature (London)* **216**, 997.

Samulski, E. T., and Tobolsky, A. V. (1968). *Macromolecules* **1**, 555.

Samulski, E. T., and Tobolsky, A. V. (1969). *Mol. Cryst. Liq. Cryst.* **7**, 433.

Samulski, E. T., and Tobolsky, A. V. (1971). *Biopolymers* **10**, 1013.

Samulski, T. V., and Samulski, E. T. (1977). *J. Chem. Phys.* **67**, 824.

Saupe, A. (1960). *Z. Naturforsch. A* **15**, 810.

Sobajima, S. (1967). *J. Phys. Soc. Jpn.* **23**, 1070.

Straley, J. P. (1973). *Mol. Cryst. Liq. Cryst.* **22**, 333.

Tobolsky, A. V., and Samulski, E. T. (1974). *In* "Liquid Crystals and Plastic Crystals" (G. W. Gray and P. W. Winsor, eds.), Vol. 1, p. 175. Halsted Press, Ellis Horwood, Chichester, England.

Toth, W. J., and Tobolsky, A. V. (1970). *Polym. Lett.* **8**, 537.

Vorlander, E. (1933). *Trans. Faraday Soc.* **29**, 907.

Wee, E. L., and Miller, W. G. (1971). *J. Phys. Chem.* **75**, 1446.

6

Liquid Crystalline Structure of Block Copolymers

Bernard Gallot

Center for Molecular Biophysics
Orléans, France

I. Introduction

Block copolymers exhibit a wide range of original properties. Most of these are related to their ability to form liquid crystalline structures. To understand the formation of such ordered structures, we must recall that the tendency of soaps and, more generally, of lipids to provide mesophases is related to their amphipatic character that is due to the presence in the molecule of two parts with different properties: one part hydrophilic, and the other hydrophobic. In the case of lipids, for example, the hydrophilic moeity is formed by the polar head and the hydrophobic by the hydrocarbon chain (Luzzati et al., 1960).

The simplest block copolymer, an AB copolymer, is also formed of two parts: a homopolymer A resulting from the repetition of a large number of monomer units a, and a homopolymer B resulting from the repetition of a large number of monomer units b, the two homopolymers being linked at one of their ends by a chemical bond to form the copolymer AB.

The A and B blocks are generally incompatible and tend to segregate. The addition to the copolymer of a liquid which is a good solvent for one block, A for instance, and a poor solvent or a precipitant for block B favors the segregation. The copolymer must take a configuration that favors the solvation of the soluble block A and the precipitation of the insoluble block B. Depending on the concentration of the system, different types of organizations are observed (Sadron, 1963).

At very low concentration (less than 0.1% of copolymer), a true solution is obtained. The copolymer AB is molecularly dispersed in the solvent and the number of aggregates is negligible; however, the less soluble sequence B is more tightly coiled than the soluble one. When the concentration increases, the less soluble B sequence precipitates and the copolymer–solvent system becomes a suspension of aggregates formed by a core of B sequences surrounded by A sequences swelled by the solvent. Such aggregates have been observed by electron microscopy (Douy and Gallot, 1971c).

Above a certain limit of concentration (from 30 to 60%, depending upon the nature of the components), the system becomes fully organized with a periodic regular structure, a sort of macro-lattice in which the repeating elements are not atoms or molecules but submicroscopic particles. These particles can be spheres, cylinders, or lamellae (e.g., Sadron, 1963) and their periodic organization generates liquid crystalline structures.

This chapter will be limited to block copolymer–solvent systems that exhibit regular periodic structures of the liquid crystalline type. See Tables I and II for copolymer and block notation.

Dry copolymers are a limiting case ($c = 1$) of liquid crystalline systems. Since they have been the object of two recent reviews (Molau, 1970; Folkes

TABLE I

Notation of Copolymers

Block copolymer	Notation
Polystyrene–polybutadiene	S-B
Polystyrene–polybutadiene–polystyrene	S-B-S
Polybutadiene–polystyrene–polybutadiene	B-S-B
Polystyrene–polyisoprene	S-I
Polystyrene–polyisoprene–polystyrene	S-I-S
Polybutadiene–poly(α-methyl styrene)	B-MS
Polybutadiene–poly(vinyl-2-naphtalene)	B-VN
Polyisoprene–poly(vinyl-2-pyridine)	I-V2P
Polyisoprene–poly(vinyl-4-pyridine)	I-V4P
Polyisoprene–poly(methyl methacrylate)	I-MMA
Polystyrene–poly(vinyl-2-pyridine)	S-V2P
Polystyrene–poly(vinyl-4-pyridine)	S-V4P
Poly(vinyl-2-pyridine)–poly(vinyl-4-pyridine)	V2P-V4P
Polystyrene–polyisoprene–poly(vinyl-2-pyridine)	S-I-V2P
Poly(methyl methacrylate)–poly(hexyl methacrylate)	MMA-HMA
Polystyrene–poly(ethylene oxide)	S-EO
Polybutadiene–poly(ethylene oxide)	B-EO
Poly(ethyl methacrylate)–poly(ethylene oxide)	EMA-EO
Polybutadiene–poly(benzyl-L-glutamate)	B-G
Polystyrene–poly(benzyl-L-glutamate)	S-G
Polybutadiene–poly(N^5-hydroxypropyl glutamine)	B-HG
Polybutadiene–poly(carbobenzoxy-L-lysine)	B-CK
Polystyrene–poly(carbobenzoxy-L-lysine)	S-CK
Polybutadiene–poly-L-lysine	B-K
Polystyrene–poly-L-lysine	S-K
Polysaccharide–poly(benzyl-L-glutamate)	OV-G

and Keller, 1973), we shall only consider a few particular aspects of their behavior that are directly related to mesophases.

Theories predicting the formation, shape, and size of aggregates have also been recently reviewed in detail (Folkes and Keller, 1973) and only some guiding references will be given here (Meier, 1969; Bianchi et al., 1970; Krause, 1970; Kromer et al., 1970; Soen et al., 1972; Krigbaum et al., 1973; Leary and Williams, 1974; Helfand, 1975).

In order to link scattered information dealing with liquid crystalline order in block copolymers, we have preferred to use for this chapter the style of a treatise rather than that of a catalog. We are well aware of the dangers of our choice, especially of the possibility of overlooking some difficulties. Nevertheless, we hope to provide a reasonably representative, if not complete, review of the subject.

TABLE II

Notation of Different Blocks

Block	Notation
Polystyrene	PS
Polybutadiene	PB
Polyisoprene	PI
Poly(vinyl-2-pyridine)	PV2P
Poly(vinyl-4-pyridine)	PV4P
Poly(methyl methacrylate)	PMMA
Poly(ethylmethacrylate)	PEMA
Poly(α-methylstyrene)	PMS
Poly(vinyl-2-naphtalene)	PVN
Poly(ethylene oxide)	PEO
Poly(benzyl-L-glutamate)	PG
Poly(N^5-hydroxypropyl glutamine)	PHG
Poly(carbobenzoxy-L-lysine)	PCK
Poly-L-lysine	PK

II. Methods of Determination of Block Copolymer Structure

The dimensions of liquid crystalline structures (several hundred angstroms) call for the use of electron microscopy and their nature (generally periodic) points to the use of low angle X-ray scattering. Some other techniques have also been fruitfully used: differential scanning calorimetry, dilatometry, polarization microscopy, infrared spectroscopy, and circular dichroism.

We shall briefly describe these techniques and make a few critical comments on their respective advantages and limitations.

A. X-Ray Diffraction

The first technique applied to the study of polymeric liquid crystalline structures was low angle X-ray diffraction (Sadron, 1963, and references herein). This is due to the large amount of work performed in Strasbourg on soap mesophases (Luzatti *et al.*, 1960; Husson *et al.*, 1960) and to the molecular similarity (presence of a hydrophobic and a hydrophilic moiety) between soaps, especially arkopals, and the block copolymers polystyrene–poly(ethylene oxide) that were first studied and to which the structures established for soaps (lamellar, cylindrical, and cubic) were applied.

Due to their periodic structure, block copolymer mesophases exhibit low angle X-ray patterns characterized by a set of sharp lines (a slit collimation beam being generally used) with Bragg spacings in the following ratios:

1, 2, 3, 4, 5 — for an arrangement of equidistant parallel lamellae of infinite lateral extension

$1, \sqrt{3}, \sqrt{4}, \sqrt{7}, \sqrt{9}$ — for a regular hexagonal array of cylinders with infinite length

$1, \sqrt{2}, \sqrt{3}, \sqrt{4}, \sqrt{5}, \sqrt{6}, \sqrt{7}$ — for a set of spheres packed in a body-centered cubic lattice

$1, \sqrt{2}, \sqrt{3}, \sqrt{4}, \sqrt{5}, \sqrt{6}, \sqrt{8}$ — for a set of spheres packed in a simple cubic lattice

$\sqrt{3}, \sqrt{4}, \sqrt{8}, \sqrt{11}$ — for a set of spheres packed in a face-centered cubic lattice.

The spacing sequence of X-ray patterns easily provides the structural type: lamellar, cylindrical, or cubic. It is generally difficult (X-ray patterns of cubic structures being rather poor) to distinguish a centered cubic lattice from a simple cubic lattice that differs only by the presence or absence of the $\sqrt{7}$ reflection.

The Bragg spacing of X-ray patterns also provides the main structural parameters: total thickness d of a sheet for the lamellar structure (Fig. 1), distance D between the axis of two neighboring cylinders for the hexagonal structure (Fig. 2), or between the centers of the spheres for cubic structures (Fig. 3). It gives information neither about the thickness d_A and d_B of the individual layers (Fig. 1) nor about the radius R of the cylinders or spheres (Figs. 2 and 3). In principle, the values of these parameters can be obtained from the relative intensities of X-ray reflections (Luzatti *et al.*, 1960). However, measurement of intensities of low angle reflections is difficult and rather inaccurate, hence this method has rarely been used.

Most authors prefer to calculate the sizes of platelets, cylinders, or spheres by formulas based on simple geometrical considerations using block ratios, solvent content, and specific volumes of different blocks and solvent (Table III). The formulas used involve the hypothesis that specific volumes of

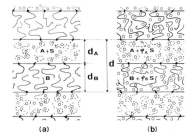

Fig. 1. Schematic representation of the lamellar structure; dots = solvent: (a) case of a selective solvent of the A block; (b) case of a nonselective solvent.

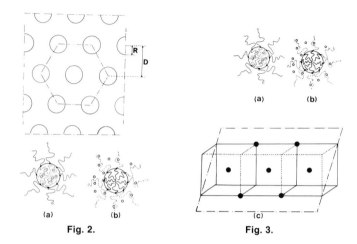

Fig. 2.

Fig. 3.

Fig. 2. Schematic representation of the hexagonal and inverse hexagonal structure: (a) inverse hexagonal structure, the solvent swells the cylinders; (b) hexagonal structure, the solvent swells the matrix.

Fig. 3. Schematic representation of the cubic and inverse cubic structures: (a) inverse cubic structure, the solvent swells the spheres; (b) cubic structure, the solvent swells the matrix; (c) schematic representation of a section of a body-centered cubic structure by a plane(111).

TABLE III

Geometrical Relations between Different Structural Parameters

A. Lamellar structure

$$d = d_A + d_B \tag{1}$$

$$d_A = dZ^{-1} \tag{2}$$

B. Hexagonal structure (cylinders)

$$R^2 = \frac{D^2\sqrt{3}}{2\pi} Z^{-1} \tag{3}$$

C. Simple cubic structure

$$R^3 = \frac{3D}{4\pi} Z^{-1} \tag{4}$$

D. Centered cubic structure

$$R^3 = \frac{3D}{8\pi} Z^{-1} \tag{5}$$

with

$$Z = 1 + \frac{c(1 - X_A)\bar{v}_B + (1 - c)\varphi_B\bar{v}_S}{cX_A\bar{v}_A + (1 - c)\varphi_A\bar{v}_S} \tag{6}$$

where c is the concentration (% weight) of the copolymer; X_A, X_B, percentage (weight) of blocks A and B; $\bar{v}_A$, $\bar{v}_B$, $\bar{v}_S$, specific volumes of the A and B blocks and of the solvent; φ_A, φ_B, partition coefficients of the solvent ($\varphi_A + \varphi_B = 1$).

blocks and of solvent in the microdomains of the liquid crystalline structures are the same as in dry homopolymers and in pure liquids, respectively. They also contain the partition coefficients φ_A and φ_B whose significance and determination will be discussed later (see Section III).

Another parameter that has been calculated in X-ray studies on block copolymer mesophases is the mean cross-sectional area S available for a molecule at the interface. Formulas giving S for all structural types except for the face-centered cubic structure can be found in Grosius *et al.* (1969, 1970a).

It appears from these observations that the determination of structural types and structural parameters is quite easy. This is only true for the lamellar structure where the two layers are equivalent, not for other structures where an internal and an external phase exist. X rays are not able to distinguish between the internal and external blocks in the cylinders or spheres. To resolve this difficulty, Skoulios (Grosius *et al.*, 1969) has proposed to calculate the diameters of cylinders or spheres assuming two possible arrangements for blocks (A inside and A outside) and to study their variation with solvent concentration. The curves obtained generally intersect near the middle of the concentration domain, and it is impossible to choose between the two arrangements. This phenomenon is illustrated in Fig. 4 with S-B copolymers in solution in methyl ethyl ketone.

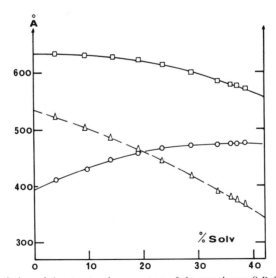

Fig. 4.   Variation of the structural parameters of the copolymer S-B 35 versus solvent concentration for hexagonal and inverse hexagonal structures: $\square$, D = distance between the axes of two neighboring cylinders; $\bigcirc$, $\overline{2R}$ = diameter of the cylinders filled with the PS block swelled by the solvent (inverse hexagonal structure); $\triangle$, $2R$ = diameter of the cylinders filled with the insoluble PB block (hexagonal structure).

B. *Electron Microscopy*

The problem of the localization of different blocks, which in the case of cylindrical and spherical structures can hardly be solved by X rays, is easily overcome by electron microscopy. Before discussing this problem we must examine the special sample requirements of electron microscopy.

The first requirement is the establishment of sufficient contrast between the two types of blocks in order to allow their easy recognition. This can be achieved by staining one of the blocks: a polybutadiene or a polyisoprene block by osmium tetroxide vapor (Kato, 1966), and a poly(vinyl pyridine) block by silver nitrate (Price *et al.*, 1974b). As both these reagents do not stain polystyrene, the polystyrene blocks will appear on micrographs as light sections and the other blocks as dark. The second requirement is the preparation of sufficiently thin samples that are transparent to electrons, i.e., with a thickness between 100 and 2000 Å.

The most popular method of preparing thin samples from solvent free copolymers was film casting from a dilute solution of copolymer (Hendus *et al.*, 1967; Bradford and Vanzo, 1968; Inoue *et al.*, 1969; Matsuo *et al.*, 1969). This method limits the observation to one projection through the film and the structure observed may depend on the conditions of casting (Lewis and Price, 1972; Pedemonte *et al.*, 1975).

A more interesting method is sectioning of massive materials using ultra-microtomy. Sectioning can be performed on dry copolymers whose structure has been improved by annealing (Kampf *et al.*, 1970; Price *et al.*, 1972a; Pedemonte *et al.*, 1973) or extrusion (Dlugosz *et al.*, 1970). It can also be performed on materials containing a diluent after transformation of the gels into organized copolymers. In the latter case, we use as a swelling agent a monomer that is a polymerizable solvent and performs a polymerization of the monomer in order to preserve the structure, to harden the system, and to enable ultramicrotomy (Douy and Gallot, 1971c). Organized co-polymers are generally shaped in the form of disks of one millimeter thickness. The disk surface may serve as a reference plane and the two orthogonal sections (one perpendicular and one parallel to the disk faces) are sufficient to establish the structure. In the case of an extruded S-B-S copolymer, a macrolattice of macroscopic dimensions is present in the material and the direction of extrusion is used as a reference for sectioning of the sample (Dlugosz *et al.*, 1970).

Now we can come back to the problem of the localization of different blocks in cylindrical and spherical structures. Electron microscopy has been applied for the first time to solve such a problem in the case of S-B organized copolymers (Gallot and Douy, 1970a,b) where the diene block was stained with OsO_4 and appeared as black on micrographs. These

authors have shown that, depending on composition, S-B and B-S-B copolymers exhibit two types of cylindrical hexagonal structures: one with
PB cylinders in a PS matrix for copolymers containing between 20 and 35%
of PB (micrographs with black circles hexagonally packed, see Fig. 6) another
with PS cylinders in a matrix of PB for copolymers containing from 65
to 80% of PB (micrographs with white circles hexagonally packed, see
Fig. 7; these structures have been called, respectively, hexagonal and inverse
hexagonal (Douy and Gallot, 1971a,b,c). This method was extended to all
structural types (Gallot, 1974).

The main advantage of electron microscopy is that a visualization of
the structures is obtained even when these are not periodic and regular.
Electron microscopy has allowed the visualization of formation of mesophases from dilute solutions (Douy and Gallot, 1971c).

Another advantage of electron microscopy is the possibility of measuring
all the structural parameters. Not only the lattice parameters d for lamellar
structures and D for hexagonal and cubic structures, but also the thicknesses
d_A and d_B of the different layers of a sheet and the radius R of the cylinders
or the spheres without the complication introduced by a possible presence
of the solvent in both phases (see Section III.B). However, in order to obtain
accurate values of structural parameters, we must use electron micrographs
provided by sections perpendicular to the plane of the lamellae or the
direction of the axis of the cylinders (Gallot and Douy, 1972a) or well-
determined planes of the cubic lattice (Pedemonte *et al.*, 1973). Such micrographs can be easily obtained with a microscope equipped with a goniometer
head.

C. Differential Scanning Calorimetry and Dilatometry

Differential scanning calorimetry (DSC) has been used extensively in the
study of phase diagrams of copolymer–solvent systems (Gervais *et al.*, 1971).

This technique is also useful in the determination of the solvent localization through the study of glass transitions of different blocks (Gallot, 1974).
Another very important application of DSC is the determination of the
degree of crystallinity in copolymers with a block able to crystallize and
the study of the variation of the degree of crystallinity as a function of
parameters, such as crystallization temperature, nature and amount of solvent, molecular weight, and composition of copolymers (Gervais and Gallot,
1973a,b). The advantages of DSC over dilatometry are numerous; DSC is
rapid, requires small samples, and can be applied easily to systems containing
solvent. In contrast to DSC, dilatometry has been more generally used to
follow crystallization kinetics of dry copolymers (Lotz, 1963; Seow *et al.*,
1976a), but even here DSC can still be applied.

D. Polarization Microscopy

Polarization microscopy has been mainly used for detecting the melting of the crystallizable chains by following the disappearance of spherulitic structure as a function of temperature (Gervais et al., 1971; Seow et al., 1976b).

E. Infrared Spectroscopy and Circular Dichroism

Infrared spectroscopy and circular dichroism are of particular interest in the study of copolymers with one or several polypeptide blocks (Perly et al., 1976; Billot et al., 1976, 1977). These methods allow the determination of the conformation of polypeptide chains: α-helix, β-pleated sheet structure, or coiled conformation.

F. Method Suggested for the Determination of Structures

The safest means of structural studies of systems containing solvent is the combined use of both X rays and electron microscopy. The recommended experimental procedure is performed in several steps. Mesomorphic gels are prepared by dissolution of the copolymer in a monomer that is a preferential solvent for one block. The structure of the gel is determined by low angle X-ray diffraction. Polymerization of the solvent is carried out by means of UV light or peroxides, and the polymerization conditions are chosen such that the molecular weight of the polymerized solvent is smaller than the molecular weight of the soluble block; these conditions are easily realized with monomers such as styrene, methyl methacrylate, vinyl acetate, and

TABLE IV

Examples of Structural Parameters Determined by X-Ray
Diffraction (XR) and Electron Microscopy (EM) for
Organized Copolymers[a]

Copolymer	Initial gel % solvent	d (Å)		d_A (Å)		d_B (Å)	
		XR	EM	XR	EM	XR	EM
B-S 41	25 Styr	343	345	236	240	107	105
B-S-B 411	25 Styr	285	290	136	155	149	135
S-B-S 364	35 MMA	358	365	190	185	170	180
B-M-S 31	31 Styr	306	295	183	175	125	120
B-V-N 41	30 MMA	344	335	221	230	123	105
B-G 53	35 diCl Propene	260	240	200	185	60	55

[a] Case of the lamellar structure: d = total thickness of a sheet; d_B = thickness of the polybutadiene layer; d_A = thickness of the nonpolybutadiene layer.

dichloropropene. The structure of the organized solid polymer is determined again by low angle X-ray diffraction to verify that the structure of the mesomorphic gel has not been destroyed by polymerization. The solid sample is then sectioned with an ultramicrotome, one of its blocks is stained, and the structure observed by electron microscopy.

Using this method, we have always obtained in Orléans excellent results. Tables IV–VI illustrate the agreement between values of structural param-

TABLE V

Examples of Structural Parameters
Determined by X-Ray Diffraction (XR) and
Electron Microscopy (EM) for
Organized Copolymers[a]

Copolymer	Initial gel % solvent	D (Å) XR	D (Å) EM	$2R$ (Å) XR	$2R$ (Å) EM
S-B 32	25 Styr	391	370	212	190
B-S-B 374	30 Styr	448	420	236	220
S-B-S 334	25 MMA	480	460	270	240
B-M-S 41	32 Styr	350	365	72	80

[a] Case of the hexagonal structure: cylinders filled with polybutadiene blocks, matrix formed by other blocks swelled by solvent. D = distance between the axis of two neighboring cylinders; $2R$ = diameter of cylinders.

TABLE VI

Examples of Structural Parameters
Determined by X-Ray Diffraction (XR) and
Electron Microscopy (EM) for
Organized Copolymers[a]

Copolymer	Initial gel % solvent	D (Å) XR	D (Å) ME	$2R$ (Å) XR	$2R$ (Å) ME
S-B 36	30 Styr	670	655	460	450
B-S-B 421	26 MMA	423	400	278	290
S-B-S 365	30 Styr	568	580	425	440
B-VN 11	36 MMA	378	360	147	160

[a] Case of the inverse hexagonal structure: matrix or polybutadiene, cylinders filled with other blocks swelled by the solvent. D = distance between the axis of two neighboring cylinders; $2R$ = diameter of cylinders.

eters determined by X rays and electron microscopy in the case of three structures: lamellar, hexagonal, and inverse hexagonal.

III. Copolymers with Amorphous Blocks

Copolymers with amorphous blocks (from S-B to MMA-HMA in Table I) have been probably the most studied. We shall briefly describe their ordered structures, the influence of main parameters governing the formation of such structures, and the value of their geometrical parameters.

A. Structures Observed

Copolymers with two or three amorphous blocks may exhibit five types of structures: lamellar, hexagonal and inverse hexagonal, cubic and inverse cubic. The geometrical parameters d, d_A, and d_B for the lamellar structure (see Fig. 1), D and R for the two types of hexagonal structures (see Fig. 2), and those for the different cubic structures (see Fig. 3) can be directly measured from electron micrographs or calculated from the Bragg spacings of X-ray patterns using the formulas given in Table III. We shall briefly recall the principal features of X-ray patterns and electron micrographs for such copolymers.

1. Lamellar Structure

The lamellar structure provides X-ray patterns exhibiting a set of sharp lines with Bragg spacings in the ratio 1, 2, 3, 4, 5, ...

The lamellar structure appears on electron micrographs as parallel stripes alternatively black (containing the stained blocks) and white (containing the other blocks) (Fig. 5). This striated structure results from the section of the

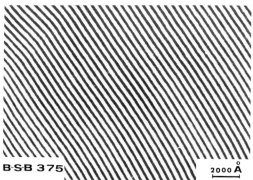

B·S·B 375   2000 Å

Fig. 5. Example of electron micrograph provided by copolymers with a lamellar structure (B-S-B 375 swelled by 25% of polymerized styrene): white stripes, polystyrene layer with thickness d_A; black stripes, polybutadiene layer with thickness d_B stained by osmium.

lamellar phase by a plane perpendicular to the plane of the sheets. The disposition of different blocks and of the solvent in different layers of the lamellar structure is sketched in Fig. 1.

2.   Hexagonal and Inversed Hexagonal Structures

The hexagonal and inverse hexagonal structures exhibit both the same type of X-ray patterns characterized by a set of sharp lines with Bragg spacings in the ratio $1, \sqrt{3}, \sqrt{4}, \sqrt{7}, \sqrt{9}, \ldots$.

However, these two structures can be easily distinguished from one another by electron microscopy (Gallot and Douy, 1970a,b) as it is illustrated by Figs. 6 and 7 representing sections by planes, perpendicular to the axis

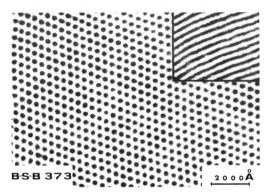

Fig. 6.   Example of electron micrograph provided by copolymers with a hexagonal structure (B-S-B 373 swelled by 28% of polymerized MMA). Main figure: section by a plane perpendicular to the direction of the axis of the PB cylinders; insert: section by a plane parallel to the axis of the cylinders.

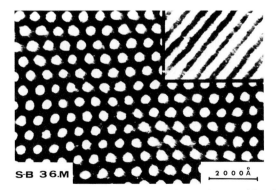

Fig. 7.   Example of electron micrograph provided by copolymers with an inverse hexagonal structure (S-B 36 swelled by 30% of polymerized MMA). Main figure: section by a plane perpendicular to the direction of the axis of the PS cylinders; insert: section by a plane parallel to the axis of the cylinders.

of the cylinders (main figures) and parallel to the axis of the cylinders (inserts), respectively. In the case of copolymers containing a diene block (S-B, S-B-S, B-S-B, S-I, S-I-S, I-VP, I-MMA, B-MS, and B-VN) in solution in a solvent of the other block or in the dry state, the hexagonal structure is characterized by a hexagonal array of black spots on a white background (Fig. 6), and the inverse hexagonal structure by hexagonal array of white spots in a black matrix (Fig. 7). The respective distribution of the blocks and of the solvent are sketched in Fig. 2 for the two types of hexagonal structures.

3. Cubic and Inverse Cubic Structures

Until recently, all structures were characterized by four or less sharp lines with Bragg spacings in the ratio 1, $\sqrt{2}$, $\sqrt{3}$, and $\sqrt{4}$. However, electron microscopy can be used to distinguish a cubic structure from an inverse cubic structure. For copolymers with a diene block (for instance S-B, B-S-B, S-B-S, and S-I-S) placed in a solvent of the other block or in the dry state, the cubic structure is characterized by black spots on a white background (Fig. 8) and the inverse cubic structure by white spots on a black background (Fig. 9). The exact symmetry of the cubic structure has been studied in detail on annealed ultrathin sections of dry copolymers of S-B-S (Pedemonte *et al.*, 1973). The Italian authors have demonstrated that S-B-S copolymers containing 10% of PS exhibit a body-centered cubic lattice of PS spheres. This structure is characterized, in electron microscopy, by hexagonal (111), square (100), and rectangular (110) arrangement of circular spots. They have shown in addition that striated structures with high and low spacings can be generated by proper rotations of the body-centered cubic lattice.

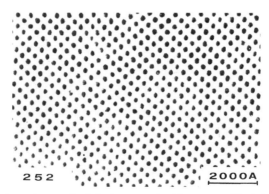

252   2000A

Fig. 8.   Electron micrograph of a body-centered cubic structure. Black dots: polybutadiene spheres.

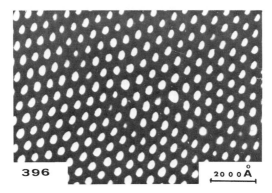

Fig. 9.   Electron micrograph of an inverse body-centered cubic structure. White dots: spheres of polystyrene swelled by the polymerized solvents.

B. *Localization of the Solvent*

We have already mentioned the problem of solvent localization in block copolymers (see Sections II.A and II.B). We shall now examine three different methods that have been used to study the localization of solvent in the liquid crystalline structures of the copolymers.

1.   *Study of the Glass Transitions by DSC*

This method consists in the determination (using a DSC equipped with solvent tight cells) of the T_g of different blocks of the dry copolymers and of mesomorphic gels of different concentrations in different solvents (Gervais *et al.*, 1973). If the T_g of the A block disappears in the mesomorphic gel and the T_g of the B block is the same in the dry copolymer as in the gel, then the solvent does not enter the domain formed by the B blocks. This method gives good results when the T_g of the insoluble or less soluble block is in the vicinity of room temperature [poly(1,2-butadiene) for instance], but cannot be used if the T_g is below or in the vicinity of the melting temperature of the solvent used [poly(1,4-*cis*-isoprene) and toluene for instance).

2.   *Study of the Structure Factor*

The second method is based on the measurement of the intensities of different diffraction orders of X-ray patterns. We then compare the results with the structure factors calculated for different ratios of layer thicknesses in the case of the lamellar structure, or different cylinder and sphere diameters in the case of hexagonal and cubic structures, respectively. Thus the ratio d_A/d is obtained for the lamellar structure (Gervais *et al.*, 1973; Gallot, 1974) and the ratio $D/2R$ for the hexagonal structures (Douy *et al.*, 1976b). The partition coefficients φ_A and φ_B can then be calculated from relations given in Table III.

The accuracy of this method is directly related to the accuracy of the measurement of X-ray intensities, which is particularly difficult in the realm of low angles where the precision rarely exceeds 20%. Nevertheless, it was claimed recently (Ionescu and Skoulios, 1976) that an accuracy better than 1% was obtained by a visual estimation of intensities; however, no X-ray pattern has been published.

3. Study of Micrographs

The third method that has been continuously used in Orléans since 1970, is based on the measurement of all structural parameters from micrographs. The obtained values are introduced in the formulas of Table III giving φ_A and φ_B. For the lamellar structure (Douy and Gallot, 1976a), an accuracy better than 5% is obtained by measuring d, d_A, and d_B with an optic system of high quality (Nikon) on micrographs from ultrathin sections. The sections are examined with an electron microscope equipped with a goniometer head to verify that sections are perpendicular to the plane of lamellae.

It is our opinion that the method using electron microscopy is by far the most accurate; however, it can only be applied with polymerizable solvents.

C. Domain Stability in Ordered Structures

We have to distinguish here between dry copolymers and copolymer–solvent systems.

1. Dry Copolymers

In the absence of diluent, the structure of the compolymers is, as a first approximation, governed by the ratio V_A/V_B of the volumes of the two types of blocks, or more simply by the composition of the copolymer. We shall not enter into the details of the description of the factors governing the morphology of dry copolymers since this problem was recently reviewed by Molau (1970). We shall only recall that the matrix is always formed by the component (A or B) present in larger proportions.

The different lattices observed and their order of succession are given in Table VII. If there is general agreement on the cubic (spheres), hexagonal (long cylinders), and lamellar structures, a controversy exists concerning copolymers with a composition between 15 and 25% of one block. The existence of spheres, cylinders, and a mixture of spheres and cylinders have been claimed by different authors. Perhaps an explanation of such discrepancies is found in the existence of a body-centered cubic lattice of short cylinders of the type discovered by Luzatti *et al.* (1968) for lipids. In such a case, depending upon the direction of the section, we would observe spheres, or cylinders, or even a mixture of the two types of domains.

TABLE VII

Effect of Composition on Domain Morphology of
AB and ABA Type Block Copolymers in Absence of Solvent

Composition	Domains	Matrix	Lattice
100% A			
	B spheres	A	Body-centered cubic
	B short rods	A	Body-centered cubic
	B long cylinders	A	Hexagonal
	A and B lamellae		Lamellar
	A long cylinders	B	Hexagonal
	A short rods	B	Body-centered cubic
	A spheres	B	Body-centered cubic
100% B			

2. Liquid Crystalline Systems

In the case of liquid crystalline systems, we have to divide block copolymers into two categories depending upon the effect of the solvent concentration on the structure (Sadron and Gallot, 1973).

We assign to the first category copolymers whose structural type does not change with the concentration of the solvent, i.e., copolymers in which the structure of the copolymer persists up to solvent concentration of about 40–45%. At these dilutions the liquid crystalline structure disappears. To this category belong S-B, B-S-B, S-B-S, S-I, S-I-S, B-MS, and B-VN. For copolymers in solution in a solvent of the A block, when the B content of the copolymer increases, we observe the following structures: cubic, hexagonal, lamellar, inverse hexagonal, and inverse cubic (Table VIII). If the solvent swells both types of blocks (S-I and S-I-S copolymers in toluene solution, for example), we cannot distinguish between hexagonal and inverse hexagonal structure or between cubic and inverse cubic structures.

TABLE VIII

Effect of Copolymer Composition on Domain
Morphology of AB and ABA Type Block Copolymer[a]

	Domains	Matrix	Lattice
100% A			
	B spheres	A	Body-centered cubic
~18% B			
	B cylinders	A	Hexagonal
~35% B			
	A and B lamellae		Lamellar
~65% B			
	A cylinders	B	Inverse hexagonal
~80% B			
	A spheres	B	Inverse body-centered cubic
100% B			

[a] Copolymers of the first category in solution in a solvent of the A
block.

The second category is characterized by the existence of a polymorphism
depending on solvent concentration. To this category belong copolymers
S-V2P, S-V4P, V2P-V4P (Grosius *et al.*, 1969, 1970a,b) and MMA-HMA
(Ailhaud *et al.*, 1972). Table IX gives all the structural possibilities and the
effect of both copolymer composition and solvent concentration. However,
first all the possibilities have not been experimentally established probably
because of the lack of systematic studies, in particular, no studies were done
on concentrations ranging from the dry copolymer to a solvent concentration
of 15 or 20% because of experimental difficulties. Second, only one structural
change has been observed as a function of solvent concentration, whereas
two changes are theoretically possible: lamellae to cylinders and cylinders
to spheres. Third, the structural change appears for a solvent concentration
of about 40%, i.e., in the concentration range where ordered structures of
systems of the first category disappear.

Copolymers I-V2P, I-V4P, and I-MMA (Rossi and Gallot, 1976, 1977)
seem to belong to the first category, but we cannot definitely reject the possibility of a structural change for copolymers with compositions differing from
that of the studied copolymers.

All the reported observations concerning the effect of the copolymer
composition and the solvent concentration fall into a coherent pattern if we
consider as a basic principle the premise that the structure of a given system
depends only on the ratio of the phase volume: $(V_A + \varphi_A V_S)/(V_B + \varphi_B V_S)$,
where V_A and V_B are volumes of the A and B blocks and $\varphi_A V_S$ and $\varphi_B V_S$ the

TABLE IX

Effect of Copolymer Composition and Solvent Concentration on the Domain Morphology of AB Block Copolymers of the Second Group in Solution in a Solvent of the A Block[a]

Composition	Domains	Lattice
100% A	$0 \xrightarrow{\%\ \text{solvent}} 100$	$0 \xrightarrow{\%\ \text{solvent}} 100$
	B spheres	C
	B cylinders ⎤ B spheres ⎦	H → C
	B cylinders	H
	Lamellae ⎤ B cylinders ⎦	L → H
	Lamellae	L
	A cylinders ⎤ Lamellae ⎦	$\bar{\text{H}}$ → L
	A cylinders	$\bar{\text{H}}$
	A spheres ⎤ A cylinders ⎦	$\bar{\text{C}}$ → $\bar{\text{H}}$
	A spheres	$\bar{\text{C}}$
100% B		

[a] C, cubic; $\bar{\text{C}}$, inverse cubic; H, hexagonal; $\bar{\text{H}}$, inverse hexagonal; L, lamellar.

respective volumes of solvent swelling the A and B blocks (Sadron and Gallot, 1973). Thus the dry copolymers represent the asymptotic case ($c = 1$) of the liquid crystalline systems.

This principle does not explain why copolymers of the first category do not show the existence of structural transitions as a function of concentration. When the dilution increases, the structure changes directly from the periodic solid to a dispersion of aggregates (Douy and Gallot, 1971c). The two types of behavior may reside in different interaction forces between macromolecular chains resulting from the nature of blocks. This explanation is suggested by the fact that polymorphism is observed in the case of sequences with polar or hydrogen bonded groups. Thus for a solvent concentration of about 40%, we observe a collapse of liquid crystalline structure or a structural change, depending upon the magnitude of interactions between polymers and solvents of different nature.

D. Factors Governing Structural Parameters

The principal factors governing the geometrical parameters of liquid crystalline structures are concentration, nature and polymerization of the solvent, temperature, molecular weight of the copolymer, the number of blocks, their relative position, and chemical nature.

1. Influence of the Solvent Concentration

We have divided block copolymers in two categories, depending on the absence of a polymorphism as a function of solvent concentration. Now, we have to consider another type of classification of copolymer–solvent systems: systems where the solvent solvates only one type of blocks, A for instance, and systems where the solvent swells the two types of blocks.

a. Copolymer—Selective Solvent Systems. Copolymers S-B (Douy and Gallot, 1971c), B-S-B (Douy and Gallot, 1972b), S-B-S (Douy and Gallot, 1973), B-MS (Douy *et al.*, 1975), and B-VN (Douy *et al.*, 1976a,b) with poly-(1,2-butadiene) blocks in solution in MEK, MMA, styrene, and α-methyl-styrene, for example, are representative of the first class of systems. Such systems are characterized by a decrease of the lattice parameter (intersheet spacing for the lamellar structure, distance between cylinders or spheres for the hexagonal or cubic structures), with the increase of solvent concentration. If we call A the soluble block, we have $\varphi_A \geq 0.9$, which means that the solvent practically does not enter the domains formed by the B blocks.

The influence of the solvent concentration on various structures is described below.

(i) *Lamellar structure* The influence of solvent concentration on geometrical parameters of the lamellar structure is illustrated by Fig. 10, in which the structural parameters of the copolymer B-S-B 375 are plotted versus the swelling ratio of the polystyrene block.

We can see that as the amount of solvent increases, the intersheet spacing d and the thickness d_B of the insoluble polybutadiene layer decrease, and the thickness d_A of the soluble polystyrene layer and the average surface S available to a molecule increase.

(ii) *Hexagonal and inverse hexagonal structures* The influence of the solvent concentration on geometrical parameters of the hexagonal and inverse hexagonal structures are illustrated by Figs. 11 and 12, where structural parameters of the copolymers B-S-B 373 and B-S-B 421 are plotted versus the swelling ratio of the polystyrene block.

We can see that for both hexagonal and inverse hexagonal structures, the distance D between the axis of two neighboring cylinders decreases and the specific surface S increases as the amount of solvent increases. The variation of the diameter of cylinders is different for the two structures. The diameter $2R$ of the cylinders filled with the insoluble polybutadiene blocks in the case of the hexagonal structure decreases (Fig. 11) while the diameter $2R$ of the cylinders filled with the polystyrene blocks swelled by the solvent in the inverse hexagonal structure increases (Fig. 12).

(iii)   Cubic and inverse cubic structures   For the cubic structure, the diameter of spheres filled by the insoluble block decreases as the amount of solvent increases, and for the inverse cubic structure, the diameter of spheres filled by the soluble block swelled by the solvent increases.

For S-V2P and S-V4P copolymers in solution in octanol or toluene (Grosius et al., 1969, 1970a) a behavior similar to that of S-B copolymers is observed, with the exception of a polymorphism that could occur depending upon solvent concentration.

The decrease of the characteristic parameter of the insoluble domains with the increase in solvent concentration is due to the requirement that these domains keep a constant volume (Sadron and Gallot, 1973).

b. Copolymer—Nonselective Solvent Systems.   In these systems an important partition of solvent between the two types of blocks $(0.4 < \varphi_A < 0.6)$ takes place. Such systems are generally characterized by an increase of the lattice parameter with solvent concentration. Copolymers S-I and S-I-S, with a poly (1,4-*cis*-isoprene) block, in solution in toluene or styrene are typical of such systems. To illustrate the behavior of this second category of systems, we have plotted in Fig. 13 the variation of the geometrical parameters of the copolymer S-I 921 versus toluene concentration. The structure is lamellar and the partition coefficient $\varphi_A = \varphi_{PS} = 0.6$ (Gervais et al., 1973). We can see that the structural parameters (the intersheet spacing d, the thickness d_A of the polystyrene layer, and the thickness d_B of the polyisoprene layer) all increase with toluene concentration. We have also plotted (in dotted lines) the variation of d_A and d_B calculated assuming $\varphi_A = 1$ (Douy et al., 1969) in order to show the influence of the partition coefficient on the values of d_A and d_B.

It is evident that such systems cannot exhibit polymorphism with solvent concentration since the ratio of the volumes of the A and B phases changes only slightly with solvent concentration.

2.   Influence of the Nature of the Solvent

The nature of the solvent has no effect on the characteristic parameter of the insoluble block as long as it remains a selective solvent of the same block (Douy and Gallot, 1972b, 1973). But the geometrical parameters of the A and B domains change with the value of the partition coefficient of the solvent as illustrated in Fig. 13 for d_A and d_B in the case of a lamellar structure.

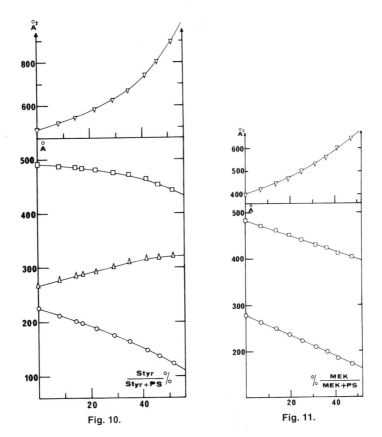

Fig. 10. Evolution of parameters of the lamellar structure with the swelling ratio of the soluble block. Copolymer system B-S-B 375/styrene. d, □ ; d_A, △ ; d_B, ○ ; S, ▽ .

Fig. 11. Evolution of parameters of the hexagonal structure with the swelling ratio of the soluble block. Copolymer system B-S-B 373/MEK. D, □ ; $2R$, ○ ; S, ▽ .

3. Influence of the Polymerization of the Solvent

If liquid crystalline structures can be obtained by dissolution of the co-polymer in a monomer (styrene, vinyl acetate, methyl methacrylate), they can be converted into solids by polymerization of the monomer by means of UV light or by a peroxide (e.g., Douy and Gallot, 1971c). Low angle X-ray diffraction patterns show that the structural type before and after polymerization of the solvent remains unchanged. The question is what is the effect of the solvent polymerization on the geometrical parameters? Figures 14 (B-S-B 343/MMA), 15 (B-S-B 374/styrene), and 16 (S-B-S 365/MMA) illus-

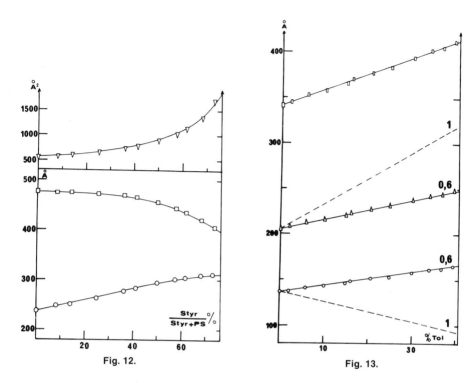

Fig. 12.   Evolution of parameters of the inverse hexagonal structure with the swelling ratio of the soluble block. Copolymer system B-S-B 421/styrene. D, □ ; $2R$, ○ ; S, ▽.

Fig. 13.   Evolution of parameters of the lamellar structure with solvent concentration in the case of a nonselective solvent. Copolymer system S-I 921/toluene. d, □ ; d_A, △ ; d_B, ○ ; $\varphi_A = 1$ (−−−); $\varphi_A = 0.6$ (——).

trate the answer. Examination of these figures shows that the polymerization of the monomer entails a contraction of the characteristic parameter of the soluble polystyrene block: d_A for the lamellar structure (Fig. 14), $D - 2R$ for the hexagonal structure (Fig. 15), and $2R$ for the inverse hexagonal structure (Fig. 16). When the partition of solvent between different blocks is important, the ordered structure is generally destroyed by polymerization.

4.   Influence of the Temperature

Study by DSC equipped with solvent-tight cells and by low angle X-ray diffraction of copolymer–solvent systems of different concentrations as a function of temperature has shown that S-B, B-S-B, S-B-S, B-MS, and B-VN block copolymers exhibit only one type of liquid crystalline structure. The structure is body-centered cubic, hexagonal, lamellar, inverse hexagonal, or

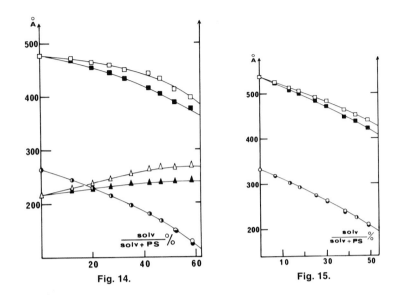

Fig. 14.

Fig. 15.

Fig. 14. Variation of parameters of the lamellar structure during the polymerization of solvent. B-S-B 343/MMA: d, □ ; d_A, △ ; d_B, ○ . B-S-B 343/pMMA: d, ■ ; d_A, ▲ ; d_B, ● .

Fig. 15. Variation of parameters of the hexagonal structure during the polymerization of solvent B-S-B 374/styrene: D, □ ; $2R$, ○ . B-S-B 374/polystyrene: D, ■ ; $2R$, ● .

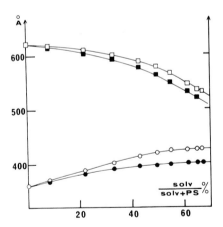

Fig. 16. Variation of parameters of the inverse hexagonal structure during the polymerization of solvent. S-B-S 365/MMA: D, □ ; $2R$, ○. S-B-S 365/pMMA: D, ■ ; $2R$, ●.

inverse body-centered cubic depending on copolymer composition. Figure 17 gives an example of such a phase diagram in the case of the copolymer B-S-B 372.

For copolymer S-B, B-S-B, and S-B-S the structural parameters decrease when the temperature increases.

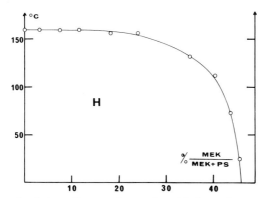

Fig. 17.   Example of phase diagram of amorphous block copolymers. Case of the hexagonal structure of the copolymer B-S-B 372 in solution in methyl ethyl ketone.

5.   Influence of the Molecular Weight of Copolymer

The molecular weight of the copolymer determines the scale of the microphase structure. Inside the domain of stability of a structural type, lamellar for example, the lattice parameter monotonically increases with the molecular weight of the copolymer (Douy and Gallot, 1972b). Nevertheless, if the molecular weight is very small the boundaries of the domain of stability of different structures can be shifted as a function of the copolymer composition (Sadron and Gallot, 1973).

6.   Influence of the Number and the Relative Position of the Blocks

a. Comparison of Copolymers with Two and Three Blocks.   Copolymers B-S and symmetric B-S-B obtained from B-S copolymers by addition of a B block have been studied by low angle X-ray diffraction and electron microscopy (Douy and Gallot, 1976a). It has been demonstrated that the addition of a third block to the diblock copolymer achieves an important conformational change of the macromolecular chains. It consists in a stretching of both the polybutadiene and the polystyrene chains. This stretching is observed in the lamellar dry copolymer obtained by evaporation of the solvent from mesophase as well as in the lamellar liquid crystalline structure itself. The stretching of the macromolecular chains in B-S-B copolymers is certainly related to the large domain of stability of the lamellar structure in B-S-B copolymers (the lamellar structure exists for PB compositions between 33 and 70%).

b. Comparison of ABA and BAB Copolymers.   The main difference between ABA and BAB copolymers is in the corresponding influence of the molecular weight of the soluble and insoluble blocks upon the geometrical parameters of their ordered structures.

For instance, in the case of the lamellar structure of S-B-S and B-S-B copolymers in solution in a solvent of the PS block, the following behavior has been observed: for B-S-B copolymers, the thickness d_B of the insoluble PB layer is independent of the molecular weight M_A of the soluble PS blocks and increases linearly with the molecular weight M_B of the PB blocks, while the thickness d_A of the PS layer depends upon M_B (Douy and Gallot, 1972b). On the contrary, for S-B-S copolymers there is a mutual interaction between the soluble and insoluble blocks (Douy and Gallot, 1973).

c. Comparison of AB and Star Copolymers. The study of films of linear S-I copolymers and star copolymers $(S-I)_nY$ with $n = 2, 3, 4$ has been performed in England. For copolymers containing 25% of PS, a hexagonal structure with PS cylinders has been observed, and it has been found that the values of structural parameters were independent of the geometry of the molecule (Price *et al.*, 1972b).

d. Graft Copolymers. For a copolymer containing 26% of PS obtained by grafting polyisoprene chains onto a polystryrene backbone, a phase separation has been observed by electron microscopy. The morphology of such films has been interpreted as due to polystyrene spheres in a polyisoprene matrix (Price *et al.*, 1974a).

7. Influence of the Nature of the Blocks

We have already seen (Section III.C.2) that the nature of blocks governs the existence of a polymorphism as a function of solvent concentration. The nature of blocks also influences the thermal behavior of copolymers and the boundaries of the stability domains of the mesophases; however, in the later case, the influence is minimal.

IV. Copolymers with an Amorphous Block and a Crystallizable Block

Three types of copolymers have been studied: polystyrene–poly(ethylene oxide) (S-EO), polybutadiene–poly(ethylene oxide) (B-EO), and poly(ethyl methacrylate)–poly(ethylene oxide) (EMA-EO). Copolymers S-EO and B-EO have been studied in the dry state and in solution in preferential solvents for each type of block; copolymers EMA-EO have only been studied in the dry state.

A. *Studies of Systems Containing a Solvent of the Amorphous Block*

Copolymers S-EO have been studied in solution in toluene, xylene, and diethyl phtalate, which are preferred solvents of the polystryrene blocks. Since diethyl phtalate is a poor solvent of polybutadiene, copolymers B-EO were studied only in xylene and toluene.

Since both types of copolymers give similar results, we shall describe the behavior of copolymers S-EO and use copolymers B-EO in order to show the effect of the nature of the amorphous block.

1. Phase Diagrams

The study of systems copolymer S-EO–diethyl phtalate of different concentrations as a function of temperature by DSC, polarization microscopy, and X-ray diffraction (Gervais *et al.*, 1971) has led to phase diagrams such as represented in Figure 18.

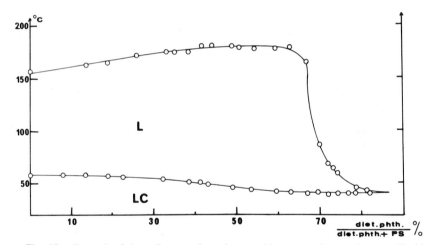

Fig. 18. Example of phase diagram of copolymers with an amorphous and a crystallizable block in solution in a preferential solvent of the amorphous block. System S-EO 3/diethyl phtalate. LC = lamellar structure with crystallized PEO chains; L = lamellar structure with melted PEO chains (Gervais and Gallot, 1973a).

All these diagrams exhibit two mesophases as a function of temperature (Gervais and Gallot, 1973a). The LC phase, found at a lower temperature has a lamellar structure with folded and crystallized poly(ethylene oxide) chains. The phase found at a higher temperature depending on copolymer composition exhibits hexagonal, lamellar, or inverse hexagonal structures.

For these three structures the poly(ethylene oxide) chains are melted. The melting temperature of poly(ethylene oxide) chains decreases when the solvent concentration increases (Fig. 18). Study of copolymers of different composition has shown that the melting temperature of the poly(ethylene oxide) chains increases with the poly(ethylene oxide) content and the molecular weight of the copolymer (Gervais *et al.*, 1972).

2. Observed Structures

Structures of phases observed at temperatures higher than the melting temperature of the PEO chains are similar to the structures of copolymers with two amorphous blocks; hence, we shall describe here only the structure of the LC phase.

In the LC phase, PEO chains are crystallized as shown by DSC, polarization microscopy, and wide angle X-ray diffraction patterns (Gervais *et al.*, 1971). Furthermore, the structure of the LC phase is lamellar as shown by low angle X-ray diffraction and consists of plane, parallel equidistant sheets.

We have sketched the LC structure in Fig. 19. Each sheet results from the superposition of two layers: a layer of thickness d_A formed by the solvated polystyrene blocks and a layer of thickness d_B formed by the crystallized and folded PEO chains (Gervais and Gallot, 1973a). In the LC structure the PEO chains are folded, therefore, an interesting parameter of this structure is the number v of folds of the PEO chains; v is given by a formula based on simple geometrical considerations:

$$v + 1 = 2\bar{M}_n^{PEO}/m d_B \tag{7}$$

where m is the weight per unit of length of a crystallizable PEO chain ($m = 15.95 \, \text{g Å}^{-1}$ from Tadokoro *et al.*, 1964); d_B is the thickness of the PEO layer; $\bar{M}_n$, the molecular weight of the PEO blocks; the factor 2 is due to the model in which PEO chains are crystallized in two layers (Fig. 19) as suggested by electron microscopy of copolymer monocrystals (Lotz *et al.*, 1966) and also shown by the study of polymers in a preferential solvent of the crystallizable block (Gervais and Gallot 1973b). A one-layer folding has been recently postulated for copolymers EMA-EO (Seow *et al.*, 1976b).

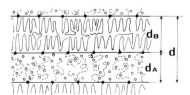

Fig. 19. Schematic representation of the lamellar crystalline structure LC: (——), crystallized PEO chains; (----), amorphous chains; dots, solvent.

We think that this postulate is incorrect since it was based on a study carried out in the dry state. Had it been made in a preferential solvent of the PEO block, it would have shown a two-layer folding.

3.   Factors Governing the Folding of the Crystallizable Chains

The principal factors governing the number of folds of the PEO chains are the concentration and the nature of the solvent, the crystallization temperature, the molecular weight of the two blocks, and the nature of the amorphous block. We shall examine them successively.

a. Effect of the Solvent Concentration.   The effect of the solvent concentration on the folding of the PEO chains is clearly illustrated by the copolymer S-EO 4 whose molecular characteristics are given in Table X. As shown in Fig. 20, the variation of its structural parameters with the swelling ratio c_1 of the PS block

$$c_1 = \frac{\text{weight of solvent}}{\text{weight of (PS + solvent)}} \tag{8}$$

shows two discontinuities, for $c_1 = 37\%$ and $c_1 = 49\%$. For $0 < c_1 < 37\%$, the thickness d_B of the PEO layer and the specific area S are constant and the same as in the dry copolymer ($d_B = 85\,\text{Å}$ and $S = 175\,\text{Å}^2$). For $c_1 = 37\%$, d_B decreases suddenly and S increases suddenly; for $37\% < c_1 < 49\%$, d_B and S remain constant ($d_B = 75\,\text{Å}$ and $S = 200\,\text{Å}^2$). Then d_B and S vary continuously with c_1. Calculation of the folding parameter v for different swelling ratios show that $v = 7$ for $c_1 < 37\%$, $v = 8$ for $37\% < c_1 < 49\%$, and then v increases continuously with c_1. Thus v at first increases stepwise, and then continuously with the swelling of the amorphous block. For

TABLE X

Effect of Molecular Weight of the Amorphous
PS Block on the Number of Folds of the
PEO Chains

S-EO	$\bar{M}_n$ PS[a]	$\bar{M}_n$[a] PEO	% PS	v_0[b]	v_{39}[b]
5	8.800	5.500	61.5	13	16
4	3.500	5.500	39	7	8
3	14.500	20.400	41	17	22
8	22.100	20.400	52	23	27

[a] $\bar{M}_n$ = average number molecular weight.
[b] v_0 and v_{39} = number of folds of the PEO chains for $c_1 = 0$ and 39%.

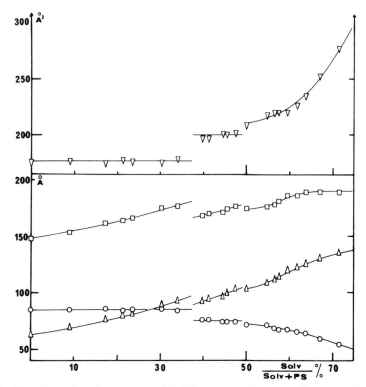

Fig. 20. Evolution of parameters of theLC structure with the swelling ratio of the PS block. Copolymer S-EO 4; solvent = diethyl phtalate; $T = 25°C$; d, $\square$; d_A, $\triangle$; d_B, $\bigcirc$; S, $\triangledown$ (Gervais and Gallot, 1973a).

copolymers where $v > 10$ in the dry state, we observe a continuous increase of v (Gervais and Gallot, 1973a).

  b. Effect of the Nature of Solvent.   Three solvents of the amorphous block have been used: toluene, xylene, and diethyl phtalate. Copolymers S-EO and B-EO exhibit the same behavior in toluene and xylene. On the contrary, polystyrene chains are differently swelled by xylene and diethylphthalate; the PS chains are less stretched and occupy a higher surface in diethyl phthalate than in xylene, consequently PEO chains are more folded in diethyl phthalate than in xylene (Gervais and Gallot, 1977b).

  c. Effect of the Crystallization Temperature.   Copolymer systems of S-EO–diethyl phtalate of different concentrations have been studied by low angle X-ray diffraction as a function of the crystallization temperature. A crystallization temperature–concentration diagram has been established (Gervais and Gallot, 1973a). Figure 21 gives an example of such a diagram

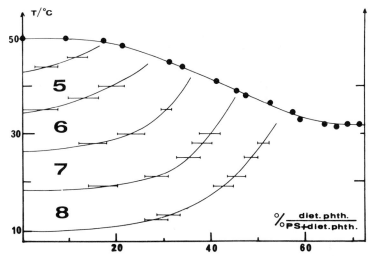

Fig. 21.   Crystallization temperature/PS swelling ratio diagram for the system S-EO 4/ diethyl phtalate. 5, 6, 7, 8 = number of folds of the PEO chains (Gervais and Gallot, 1973a).

for the copolymer S-EO 4 that displays four domains in which the folding parameter remains constant and successively equal to 8, 7, 6, and 5 when the crystallization temperature increases. The PEO chains unfold in successive jumps and, for example, for a swelling ratio of the PS block of 21%, the stepwise increase of v takes place at 19°, 29°, and 41°C.

It has also been shown (Gervais and Gallot, 1973a) that the crystallinity of the PEO block increases with the increase of both crystallization temperature and solvent concentration, and that for copolymers with a small content of PEO, a fractionation during crystallization can occur when the swelling of the amorphous block reaches about 20%.

d. Effect of the Molecular Weight and Composition of the Copolymer. The influence of the molecular weight and the composition of the copolymer on the folding of the crystallizable chains has been studied in detail for copolymers S-EO (Gervais and Gallot, 1973a).

(i)   Molecular weight of the amorphous block   Comparison of copolymers 4 and 5 on one hand and 3 and 8 on the other hand, with the same molecular weight of the PEO block but different content of PS (Table X) show that the number of folds v increases with the PS content of the copolymer.

(ii)   Molecular weight of the crystallizable block   Copolymers 7, 8, and 9 (Table XI) have the same molecular weight for the PS block (22 100) but they contain, respectively 32, 48, and 70.5% of PEO; their folding parameters

TABLE XI

Effect of Molecular Weight of the PEO Block on
the Number v of Folds of the PEO Chains

S-EO	$\bar{M}_n{}^a$ PS	$\bar{M}_n{}^a$ PEO	% PEO	$v_0{}^b$	$v_{20}{}^b$
7	22.100	10.400	32	23	25
8	22.100	20.400	48	23	25
9	22.100	52.900	70.5	28	29

[a] $\bar{M}_n$ = average number molecular weight.
[b] v_0 and v_{20} = number of folds of the PEO chains for $c_1 =$
0 and 20%.

v are 23, 23, and 28 in the dry state, 25, 25, and 29 for a swelling ratio of the PS block of 20%. Thus for copolymers containing less than 50% of PEO, the folding parameter v is governed by the molecular weight of the PS block; for copolymers with higher PEO content, there is an equilibrium between the effect of the two blocks.

(iii) Molecular weight of the copolymer The comparison of copolymers 3 and 4 (Table X), which have nearly the same composition but very different molecular weights, shows that the number of folds increases with the molecular weight of the copolymer.

e. Effect of the Nature of the Amorphous Block. The effect of the amorphous block clearly appears if one compares block copolymers S-EO (Gervais and Gallot, 1973a) and B-EO (Gervais *et al.*, 1976; Gervais and Gallot, 1977a) with a similar composition ~60% of PEO and the same molecular weight for their PEO blocks (12.200). As shown in Table XII, the surface S available for a PEO molecule at the interface of the two layers is larger when it is fixed to PS (S-EO 1) than when it is fixed to PB (B-EO 1). In order to fill this surface, the PEO chains must fold more in S-EO copolymers ($v = 12$) than in BEO copolymers ($v = 7$). This result is confirmed by the fact that in order to obtain a given number of folds (7, for example) we

TABLE XII

Effect of Nature of the Amorphous Block on the
Number v of folds of the PEO Chains

Copolymers	% PEO	$\bar{M}_n$ 1st block	$\bar{M}_n$ PEO	S (Å)	v
S-EO 4	61	3.500	5.500	176	7
B-EO 1	59	8.500	12.200	176	7
S-EO 1	60	8.500	12.200	290	12

need PEO blocks of molecular weight of 12.200 in B-EO copolymers and only of 5.500 in S-EO copolymers.

It should be interesting to compare the folding of PEO chains in S-EO and EMA-EO copolymers. The polydispersity of EMA-EO copolymers is rather high (Seow *et al.*, 1975), hence in these copolymers the thickness of the PEO layer is not known exactly and some uncertainty on the calculated value of v remains, even though v is of the same order of magnitude as for S-EO copolymers of similar molecular characteristics. Nevertheless, it can be concluded that the conformation of the amorphous block has a big influence on the surface available for a PEO chain and, therefore, on the number of folds of the crystallizable block.

B. Studies of Systems Containing a Solvent of the Crystallizable Block

Copolymers S-EO have been studied in solution in acetic acid and in nitromethane, and copolymers B-EO in solution in acetic and acrylic acids.

1. Phase Diagrams

Phase diagrams of copolymers B-EO and S-EO in solution in a preferential solvent of the PEO block have been obtained by using the same techniques as in the case of a solvent of the amorphous block, namely, *DSC*, polarization microscopy, and X-ray diffraction (Gervais *et al.*, 1971). S-EO and B-EO copolymers exhibit two mesophases and Fig. 22 gives an example of such

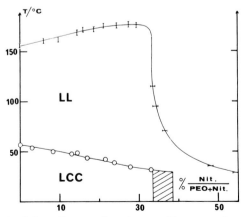

Fig. 22.   Example of phase diagram of copolymers with an amorphous and a crystallizable block in a preferential solvent of the crystallizable block. System S-EO 3/nitromethane. LCC = lamellar structure with crystallized PEO chains; LL = lamellar structure with dissolved PEO chains (Gervais and Gallot, 1973b).

diagrams for copolymer S-EO 3 in solution in nitromethane. The LCC phase found at the lower temperatures and small solvent concentrations is always characterized by a lamellar structure with crystallized and folded PEO chains. The phase found at higher temperatures and higher solvent concentrations exhibits a hexagonal, lamellar, or inverse hexagonal structures depending upon the composition of the copolymer and contains solvated PEO chains.

The effect of solvent concentration and of molecular characteristics on the melting of the PEO chains and on the stability of different phases has been described in detail by Gervais and Gallot (1973b, 1977a).

2. Observed Structures

By using low and wide angle X-ray diffraction, polarization microscopy, and DSC, it has been shown (Gervais and Gallot, 1973b) that the LCC structure is lamellar and characterized by crystallized and folded PEO chains. In this structure (see Fig. 23), the layer of thickness d_B is formed by the insoluble amorphous block (PS or PB); the layer of thickness d_A contains the PEO chains and the solvent; this lamella of thickness d_A has a complex structure and results from the superposition of three layers: the two layers formed by crystallized PEO chains are separated from one another by a layer of solvent.

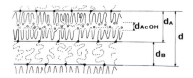

Fig. 23. Schematic representation of the lamellar crystalline structure LCC.

The study of the variation of structural parameters of the LCC phase and of the degree of crystallinity as a function of solvent concentration has shown that the solvent solvates only the amorphous part of the PEO chains located between the two layers formed by the crystallized and folded PEO chains (Gervais and Gallot, 1973b).

3. Factors Governing the Folding of the Crystallizable Chains

The PEO chains are folded in two layers separated by a layer of solvent. The number of folds can be calculated by formula (7) as in the case of a solvent of the amorphous block.

a. Effect of the Solvent Concentration. Copolymer S-EO 9 whose molecular characteristics are given in Table XI is representative of the behavior of both copolymers S-EO and B-EO. This copolymer exhibits two structures as a function of solvent concentration: at first a lamellar LCC structure and then, after a small domain of separation, a hexagonal structure with solvated PEO chains (Fig. 24). Here we are only interested in the LCC structure. All the parameters of the LCC structure present a discontinuity for $c_2 = 20\%$:

$$c_2 = \frac{\text{weight of solvent}}{\text{weight of (PEO + solvent)}} \tag{9}$$

On each side of the discontinuity, the values of the thickness d_B of the PS layer and the specific area S are constant. These values of d_B and S are the results of crystallinity of the PEO chains. Furthermore, for $c_2 < 20\%$, the values of d_B and S are the same as in the dry copolymer.

Inside the layer of thickness d_A, the thickness of the PEO (d_{PEO}) and the thickness of the acetic acid (d_{HOAc}) can be calculated using the formula

$$d_{PEO} = d_A \left(1 + \frac{\bar{v}}{\bar{v}_B} \frac{1 - c_3 X_{PEO}}{c_3 X_{PEO}} \right)^{-1} \tag{10}$$

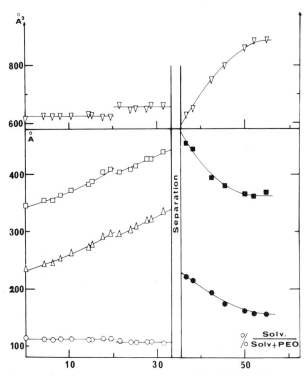

Fig. 24. Evolution of the structural parameters of the LCC and H structures of the co-polymer S-EO 9. Solvent = acetic acid; T = 25°C; d, $\square$; d_A, $\triangle$; d_B, $\bigcirc$; D, $\blacksquare$; $2R$, $\bullet$; S, $\triangledown$ (Gervais and Gallot, 1973b).

where

$$c_3 = 1 - c_2 = \frac{\text{weight of PEO}}{\text{weight of (PEO + solvent)}} \qquad (11)$$

X_{PEO} is the PEO content of the layer of thickness d_A, $\bar{v}$ the specific volume of the solvent, and $\bar{v}_B$ the specific volume of the PEO.

The variation of d_{PEO} and d_{HOAc} are plotted in Fig. 25 as a function of c_4:

$$c_4 = \frac{\text{weight of solvent}}{\text{weight of PEO}} 100\% \qquad (12)$$

We can see that when $c_4 < 26\%$, the thickness d_{PEO} of the layer of crystallized PEO chains remains constant and equal to their thickness in the dry co-copolymer (230 Å); the thickness of the acetic acid layer increases linearly with c_4.

When c_4 reaches 26%, the thickness d_{PEO} decreases suddenly and then remains constant and equal to 220 Å until the disappearance of the LCC structure; the evolution of the thickness of the acetic acid layer shows a small discontinuity, then increases again linearly with c_4 but with a smaller slope.

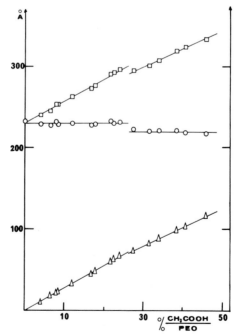

Fig. 25. Evolution of the geometrical parameters of the lamellae containing the PEO chains and the acetic acid. d_A, □; d_{PEO}, ○; d_{HOAc}, △.

The discontinuity in the variation of d_{PEO} corresponds to an increase in the number of folds of one unit for the PEO chains and is accompanied by a decrease of the degree of crystallinity of the PEO chains.

Thus when the solvent concentration increases, the solvent progressively solvates the amorphous PEO located between the two layers of crystallized PEO and the thickness of the solvent layer increases. When the amorphous PEO is saturated by the solvent, two possibilities occur: the LCC structure disappears (this is the case if thickness of the PEO layer is in the vicinity of 50 Å) for higher values of d_{PEO}, the PEO chains crystallize on a smaller thickness and with a higher number of folds. The supplementary folding is accompanied by a decrease of the degree of crystallinity generating amorphous PEO that can be solvated until a new folding of the PEO chains or the disappearance of the LCC structure occurs (Gervais and Gallot, 1973b).

b. Effect of the Nature of the Solvent. When acetic acid is replaced by nitromethane, the number of discontinuities in the variation of the thickness of the PEO layer with solvent concentration (i.e., the number of changes in the folding of the PEO chains), increases by one or more units (Gervais and Gallot, 1973b).

c. Effect of Other Parameters. The effect of the molecular weight of the two blocks and of the nature of the amorphous block is similar to that described in the case of a solvent of the amorphous block.

V. Copolymers with a Polypeptide and a Nonpolypeptide Block

Copolymers with a polypeptide and a nonpolypeptide block are a new class of copolymers that have been synthesized and studied in Orléans since 1973 in order to obtain models for biological systems. Depending upon the nature of the blocks, they can be divided into three families.

A. Copolymers with a Polyvinyl Block and a Hydrophobic Polypeptide Block

1. Copolymers Studied

The following copolymers have been studied: polybutadiene–poly(γ-benzyl-L-glutamate) (B-G), polystyrene–poly(γ-benzyl-L-glutamate) (S-G), polybutadiene–poly(carbobenzoxy-L-lysine) (B-CK), polystyrene–poly(carbobenzoxy-L-lysine) (S-CK), and polybutadiene–poly(N^5-hydroxypropyl-glutamine) (B-HG). These copolymers have been studied in the dry state and in solution in different solvents by X-ray diffraction, electron microscopy, infrared spectroscopy, and circular dichroism. In B-HG copolymers, the poly(N^5-hydroxypropylglutamine) chains are soluble in both polar and nonpolar solvents, but, even in solution in acrylic acid, their conformation is similar to that of hydrophobic polypeptides. The properties of B-HG copolymers will be described below.

2. Observed Structures

In the dry state and in concentrated solution (less than 60% of solvent) for copolymers B-G, S-G, B-CK, and S-CK mesophases have been observed in solvents such as dioxane, 1,2-dichloroethane, and 2,3-dichloropropene. Copolymers B-HG in the same conditions gave a mesophase in acrylic acid.

The structure of mesophases is lamellar for any copolymer and can be described by lamellae of thickness d resulting from the superposition of two layers: one of thickness d_A formed by chains of the polyvinyl block in a more or less random coiled conformation; the other of thickness d_B formed by polypeptide chains in an α-helix conformation, arranged in an hexagonal array and generally folded.

The lamellar character of the structure is demonstrated by low angle X-ray diffraction (presence in the central region of X-ray patterns of a set of 3 to 6 sharp lines with Bragg spacings in the ratio 1, 2, 3, 4, 5, . . .) and by electron microscopy. Figure 26 gives an example of electron micrographs obtained with B-G and B-CK copolymers. On this micrograph, we can see parallel stripes alternately as white and black; the white stripes contain polypeptide blocks and the black stripes, polybutadiene blocks stained with osmium.

The hexagonal packing of the polypeptide chains is established by X-ray results (Perly *et al.*, 1974, 1976). The α-helix conformation is demonstrated by X-ray diffraction, infrared spectroscopy, and circular dichroism (Perly

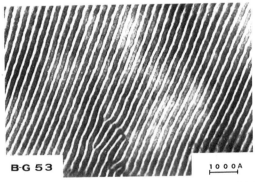

Fig. 26. Electron micrograph of the copolymer B-G 53 swelled with 30% of polymerized 2,3-dichloro-1-propene and strained with OsO_4. Polybutadiene layers appear in dark and poly-(benzyl glutamate) layers in light.

TABLE XIII

Characteristics of Some Copolymers B-G and S-CK[a]

Copolymer	$\bar{M}_n$ PV[b]	% PP	$\bar{M}_n$ PP[c]	$\bar{P}_n$ PP[d]	L (Å)	d_B (Å)	v
B-G 537	25.600	21	6.800	31	46.5	48	0
B-G 532	25.600	48	23.700	108	162	90	1
S-CK 15	37.000	36	20.800	80	120	66	1
S-CK 16	37.000	57	49.000	187	280.5	101	2

[a] This illustrates the influence of the nature of the blocks, the composition, and the molecular weight of the copolymers on the number v of folds of the polypeptide chains.
[b] $\bar{M}_n$ PV = number molecular weight of the polyvinyl block.
[c] $\bar{M}_n$ PP = number molecular weight of the polypeptide block.
[d] $\bar{P}_n$ PP = number polymerization degree of the polypeptide block.

et al., 1976; Billot *et al.*, 1977). The folding of the polypeptide chains (Table XIII) is deduced from the comparison of the thickness d_B of the polypeptide layer with the average length $\bar{L}$ of polypeptide chain calculated by the formula

$$\bar{L} = \bar{h}\bar{P}_n = 1.5\bar{P}_n \qquad (13)$$

where $\bar{h} = 1.5$ Å is the projection of the distance between two peptide residue on the helix axis, and $\bar{P}_n$ the number average degree of polymerization of the polypeptide block.

The number of folds is given by

$$v + 1 = L/d_B \qquad (14)$$

for the polypeptide chains crossing the entire thickness d_B before folding (Fig. 27a). We obtain a number of folds $v = 0, 1$, or 2 depending upon the nature of blocks, the molecular weight, and the composition of the co-polymers (Table XIII). The folding of the polypeptide chains after crossing half of the thickness of their layer is not very probable (Perly *et al.*, 1976).

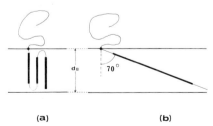

(a)   (b)

Fig. 27. Schematic representation of a macromolecular chain in the lamellar structure of the copolymer S-CK 16: (a) folded polypeptide chain ($v = 2$); (b) tilted polypeptide chain ($\theta = 70°$).

Another possibility of polypeptide chains packing within the lamellae would be a tilting of chains such as shown in Fig. 27b. But, in that case, we obtain for two very similar copolymers S-CK 15 and S-CK 16 different tilt angles θ equal to 57 and 70°, respectively. No justification for such differences in θ exists. Furthermore, the tilt of the polypeptide chains is forbidden by both the hexagonal packing and the polydispersity of the polypeptide chains.

3. Influence of the Solvent Concentration

Figure 28 illustrates the variation of the geometrical parameters of the lamellar structure for copolymers S-CK 15 and S-CK 16, with solvent concentration. These are representative of copolymers containing less than 50% of polypeptide (S-CK 15) and more than 50% of polypeptide (S-CK 16).

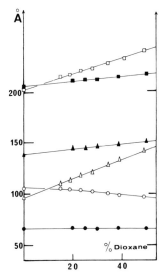

Fig. 28.   Evolution with dioxane concentration of the parameters of the lamellar structure of the copolymers S CK 15 (d, ■ ; d_A, ▲ ; d_B, ●) and S-CK 16 (d, □ ; d_A, △ ; d_B, ○).

When the solvent concentration increases, the intersheet spacing d increases, the thickness d_A of the solvated polyvinyl layer increases (the partition coefficient of the solvent φ_A in favor of the polyvinyl block varies from 0.7 to 0.8 depending on the nature of blocks), while the thickness d_B of the polypeptide layer remains constant (S-CK 15) or nearly constant (S-CK 16). This thickness invariance of polypeptide layer as demonstrated by X rays is due to an expansion of the hexagonal lattice of polypeptide chains on swelling.

B. Copolymers with a Polyvinyl Block and a Hydrophilic Polypeptide Block

In order to describe the structural behavior of block copolymers with a polyvinyl block and a hydrophilic polypeptide block, we shall take the example of copolymers polystyrene—poly(L-lysine) (S-K). Copolymers polybutadiene–poly(L-lysine) (B-K) exhibit a similar behavior (Billot *et al.*, 1976). The study of copolymers with other hydrophilic polypeptide blocks is in progress in our laboratory.

The study of S-K copolymers by X-ray diffraction and infrared spectroscopy (Billot *et al.*, 1976) has shown that S-K copolymers exhibit a lamellar structure in the dry state and in water solution (for solvent concentration smaller than about 50%). As in all lamellar structures, the lamellar structure of copolymers S-K results from the superposition of plane, parallel equidistant sheets; each sheet contains two layers, one formed by the insoluble polystyrene chains, the other by the poly(L-lysine) chains swelled by water. The originality of this structure consists in the state of organization of the polylysine chains that are roughly 15% in a β-chain conformation, 35% in an α-helix conformation, and 50% in a coiled conformation. This structure shows no periodic arrangement.

Copolymers S-K exhibit a solvent concentration dependence similar to that of many copolymers with two amorphous blocks. When the swelling ratio of the poly(L-lysine) block increases (Fig. 29), the intersheet spacing

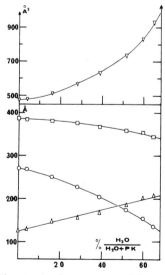

Fig. 29. Evolution of the structural parameters of the copolymer S-K 16 versus the swelling ratio of the polylysine block. d, □; d_A, △; d_B, ○; S, ▽ (Billot *et al.*, 1976).

d and the thickness d_B of the insoluble polystyrene layer decreases, while the thickness d_A of the polylysine layer and the average surface S per molecule increases.

C. Copolymers with a Polysaccharide Block and a Polypeptide Block

1. Copolymers Studied

The last class of amphipatic block copolymers that have been synthesized and studied is formed by copolymers with a hydrophilic polysaccharide block and a hydrophobic polypeptide block. In such copolymers, the polysaccharide block is the carbohydrate fraction α or β extracted from ovo-mucoide by enzymatic degradation of the polypeptide chain of this glyco-protein, followed by column chromatography fractionation and purification.

The α fraction has a molecular weight of 1850 and contains 1 residue of sialic acid in the terminal position, 3 of mannose, and 7 of N-acetyl-glucosamine (Bayard, 1974). The β fraction has a molecular weight of 3200 and contains 1 residue of galactose in terminal position, 5 of mannose, and 10 of N-acetylglucosamine (Bayard, 1974). The two oligosaccharides are terminated by an asparagine residue that has allowed the synthesis of block copolymers with a hydrophobic polypeptide as a second block (Douy and Gallot, 1977).

2. Structures Observed

When the polypeptide is a poly(benzylglutamate) block, polysaccharide–polypeptide copolymers exhibit mesophases in dimethyl sulfoxide. This occurs for DMSO concentrations ranging from zero to a limiting value that depends upon the composition of the copolymer and the nature of the carbohydrate block (Douy and Gallot, 1977).

Low angle X-ray diffraction has shown that the structure of the meso-phases is lamellar and presents many analogies with the lamellar structure of copolymers with a polyvinyl block and a hydrophobic polypeptide block (see Section V.A).

The lamellar structure results from the superposition of plane, parallel equidistant sheets. The total thickness of a sheet varies between about 60 and 150 Å, depending upon the nature of the carbohydrate block, the com-position of the copolymer, and the amount of solvent. Each sheet contains two superposed layers: one formed by the carbohydrate block; the other formed by the polypeptide chains in their α-helix conformation, arranged in a hexagonal bidimensional array for copolymers with a β-carbohydrate block and probably in a centered rectangular array for copolymers with an α-carbohydrate block (Douy and Gallot, 1977).

VI. Concluding Remarks

We have seen that copolymers with 2 or 3 amorphous blocks may exhibit five types of structures: cubic, hexagonal, lamellar, inverse hexagonal, and inverse cubic. In contrast, copolymers with a crystallizable block or a polypeptide block exhibit only lamellar structures; the crystallization of the chains or their helical conformation prevent the formation of cubic and hexagonal structures, possibly because curved interfaces are incompatible with a well-developed organization of the macromolecular chains. Another similarity between these copolymers is the ability of folding of both the crystallizable and the polypeptide chains. The conformation of the chains thus largely determines the type of liquid crystalline structure adopted by the copolymer.

Copolymers with a polypeptide and a nonpolypeptide block are of particular interest for biologists because they form simplified models of proteins, especially membrane proteins. Recently Singer has proposed a fluid mosaic model for membranes (Singer and Nicolson, 1972). In this model, the membrane matrix is formed by the lipid bilayer and the integral proteins are globular molecules, partly embedded in and partly protruding from the membrane. According to this model, the protein molecules are amphipatic, their nonpolar residues are to a large extent embedded in the lipid bilayer and their ionic residues protrude from the surfaces. Thus block copolymers with a hydrophobic polyvinyl and hydrophilic polypeptide blocks may be considered as models of amphipatic proteins, and copolymers with a polysaccharide and hydrophobic polypeptide blocks as models of glycoproteins. This latter type of copolymer is probably the most interesting one from a biological point of view, because the thickness of its lamellae is of the same order of magnitude as the thickness of the membrane bilayers and the interactions of such copolymers with lectines and with immunological probes could simulate biological interactions, and, perhaps, help to throw some light on problems as important as cell recognition and cell contact inhibition.

Acknowledgments

The author wishes to acknowledge the considerable contribution made by Dr. Douy and Dr. Gervais in the field of liquid crystalline structure of block copolymers, work which made this review possible.

Thanks are expressed to Mr. Breton and Mr. Labbe for their help in the preparation of the manuscript.

References

Ailhaud, H., Gallot, Y., and Skoulios, A. (1972). *Makromol. Chem.* **151**, 1.
Arridge, R. G. C., and Folkes, M. J. (1972). *J. Phys. D* **5**, 344.
Bayard, B. (1974). Thèse Doctorat, Univ. de Lille, Lille.

Beecher, J. F., Marker, L., Bradford, R. D., and Aggarwal, S. L. (1969). *J. Polym. Sci., Part C* **26**, 117.

Bianchi, U., Pedemonte, E., and Turturro, A. (1970). *Polymer* **11**, 268.

Billot, J. P., Douy, A., and Gallot, B. (1976). *Makromol. Chem.* **177**, 1889; Hüthig and Wepf Verlag, Basel.

Billot, J. P., Douy, A., and Gallot, B. (1977). *Makromol. Chem.* **178**, 1641.

Bradford, E. B., and Vanzo, E. (1968). *J. Polym. Sci., Part A-1* **6**, 1661.

Dlugosz, J., Keller, A., and Pedemonte, E. (1970). *Kolloid Z. Z. Polym.* **242**, 1125.

Dlugosz, J., Folkes, M., and Keller, A. (1973). *J. Polym. Sci., Part A-2* **11**, 929.

Douy, A., and Gallot, B. (1969). *C.R. Acad. Sci., Ser. C* **268**, 1218.

Douy, A., and Gallot, B. (1971a). *C.R. Acad. Sci., Ser. C* **272**, 440.

Douy, A., and Gallot, B. (1971b). *C.R. Acad. Sci., Ser. C* **272**, 1478.

Douy, A., and Gallot, B. (1971c). *Mol. Cryst. Liq. Cryst.* **14**, 191.

Douy, A., and Gallot, B. (1972a). *C.R. Acad. Sci., Ser. C* **274**, 498.

Douy, A., and Gallot, B. (1972b). *Makromol. Chem.* **156**, 81.

Douy, A., and Gallot, B. (1973). *Makromol. Chem.* **165**, 297.

Douy, A., and Gallot, B. (1976). *C. R. Acad. Sci., Ser. C* **282**, 895.

Douy, A., and Gallot, B. (1977). *Makromol. Chem.* **178**, 1595.

Douy, A., Mayer, R., Rossi, J., and Gallot, B. (1969). *Mol. Cryst. Liq. Cryst.* **7**, 108.

Douy, A., Jouan, G., and Gallot, B. (1975). *C.R. Acad. Sci., Ser. C* **281**, 355.

Douy, A., Jouan, G., and Gallot, B. (1976a). *C.R. Acad. Sci., Ser. C* **282**, 221.

Douy, A., Jouan, G., and Gallot, B. (1976b). *Makromol. Chem.* **177**, 2945.

Finaz, G., Skoulios, A., and Sadron, C. (1961). *C.R. Acad. Sci.* **253**, 265.

Folkes, M., and Keller, A. (1973). *In* "Physics of Glassy Polymers" (R. N. Haward, ed.), p. 548. Appl. Sci. Publ., London.

Folkes, M., Keller, A., and Scalisi, F. (1973). *Kolloid Z. Z. Polym.* **251**, 1.

Gallot, B. (1971). *Colloq. Int. Methodes Anal. Rayonnent X, 2nd*, Toulouse, p. 53.

Gallot, B. (1974). *Pure Appl. Chem.* **38**, 1.

Gallot, B., and Douy, A. (1970a). *Int. Liq. Cryst. Conf., 3rd, Berlin*, p. 80.

Gallot, B., and Douy, A. (1970b). *IUPAC Symp. Leiden* **1**, 99.

Gallot, B., and Douy, A. (1972a). *In* "Quelques aspects de l'état solide organique," Vol. 5, p. 13. Masson, Paris.

Gallot, B., and Douy, A. (1972b). *IUPAC Symp. Helsinki* **3**, 125.

Gallot, B., and Gervais, M. (1974). *IUPAC Symp. Rio de Janeiro* **B.7**, 128.

Gallot, B., and Sadron, C. (1971). *Macromolecules* **4**, 514.

Gallot, B., Mayer, R., and Sadron, C. (1966). *C.R. Acad. Sci., Ser. C* **263**, 42.

Gallot, B., Mayer, R., and Sadron, C. (1967). *Rubber Chem. Technol.* **40**, 932.

Gallot, B., Perly, B., and Douy, A. (1974). *IUPAC Symp. Rio de Janeiro* **D.6**, 250.

Gervais, M., and Gallot, B. (1973a). *Makromol. Chem.* **171**, 157; Hüthig and Wepf Verlag, Basel.

Gervais, M., and Gallot, B. (1973b). *Makromol. Chem.* **174**, 193; Hüthig and Wepf Verlag, Basel.

Gervais, M., and Gallot, B. (1977a). *Makromol. Chem.* **178**, 1577.

Gervais, M., and Gallot, B. (1977b). *Makromol. Chem.* **178**, 2071.

Gervais, M., Douy, A., and Gallot, B. (1971). *Mol. Cryst. Liq. Cryst.* **13**, 289.

Gervais, M., Jouan, G., and Gallot, B. (1972). *C.R. Acad. Sci., Ser. C* **275**, 1243.

Gervais, M., Douy, A., and Gallot, B. (1973). *C.R. Acad. Sci., Ser. C* **276**, 391.

Gervais, M., Jouan, G., and Gallot, B. (1976). *C.R. Acad. Sci., Ser. C* **282**, 919.

Grosius, P., Gallot, Y., and Skoulios, A. (1969). *Makromol. Chem.* **127**, 94.

Grosius, P., Gallot, Y., and Skoulios, A. (1970a). *Makromol. Chem.* **132**, 35.

Grosius, P., Gallot, Y., and Skoulios, A. (1970b). *Makromol. Chem.* **136**, 191.

Gulik, T., Rivas, E., and Luzzati, V. (1967). *J. Mol. Biol.* **27**, 303.
Helfand, E. (1975). *Macromolecules* **8**, 552.
Hendus, H., Illers, K., and Ropte, E. (1967). *Kolloid Z. Z. Polym.* **216/217**, 110.
Hoffmann, M., Kampf, G., Kromer, H., and Pampus, G. (1971). *Adv. Chem. Ser.* **99**, 351.
Husson, F., Mustacchi, H., and Luzzati, V. (1960). *Acta Crystallogr.* **13**, 668.
Inoue, T., Soen, T., Hashimoto, T., and Kawai, H. (1969). *J. Polym. Sci., Part A-2* **7**, 1283.
Inoue, T., Soen, T., Hashimoto, T., and Kawai, H. (1970). *Macromolecules* **3**, 87.
Ionescu, M. L., and Skoulios, A. (1976). *Makromol. Chem.* **177**, 257.
Kampf, G., Hoffmann, M., and Kromer, H. (1970). *Ber. Bunsenges. Phys. Chem.* **74**, 851.
Kampf, G., Kromer, H., and Hoffmann M. (1972). *J. Macromol. Sci., Phys.* **6**, 167.
Kato, K. (1966). *J. Polym. Sci., Part, B* **4**, 35.
Keller, A., Pedemonte, E., and Willmouth, F. M. (1970). *Kolloid Z. Z. Polym.* **238**, 385.
Krause, S. (1970). *Macromolecules* **3**, 84.
Krigbaum, W., Yazgan, S., and Tolbert, W. (1973). *J. Polym. Sci., Part A-2* **11**, 511.
Kromer, H., Hoffmann, M., and Kampf, G. (1970). *Ber. Bunsenges. Phys. Chem.* **74**, 859.
Leary, D. F., and Williams, M. C. (1973). *J. Polym. Sci., Part A-2* **11**, 345.
Leary, D. F., and Williams, M. C. (1974). *J. Polym. Sci., Part A-2* **12**, 265.
Lewis, P. R., and Price, G. (1972). *Polymer* **13**, 20.
Lotz, B. (1963). Thèse 3ème Cycle, Univ. de Strasbourg, Strasbourg.
Lotz, B., Kovacs, A., Bassett, G. A., and Keller, A. (1966). *Kolloid Z. Z. Polym.* **209**, 115.
Luzzati, V., Mustacchi, H., Skoulios, A., and Husson, F. (1960). *Acta Crystallogr.* **13**, 660.
Luzzati, V., Gulik, T., and Tardieu, A. (1968). *Nature (London)* **218**, 1031.
Matsuo, M., Sagae, S., and Asai, H. (1969). *Polymer* **10**, 79.
Meier, D. J. (1969). *J. Polym. Sci. Part C* **26**, 81.
Molau, G. E. (1970). *In* "Block Copolymers" (S. L. Aggarwal, ed.), p. 102. Plenum, New York.
Oster, G., and Riley, D. P. (1952). *Acta Crystallogr.* **5**, 272.
Pedemonte, E., Turturro, A., Bianchi, U., and Devetta, P. (1973). *Polymer* **14**, 145.
Pedemonte, E., Dondero, G., Alfonso, G., and de Candia, F. (1975). *Polymer* **16**, 531.
Perly, B., Douy, A., and Gallot, B. (1974). *C.R. Acad. Sci., Ser. C* **279**, 1109.
Perly, B., Douy, A., and Gallot, B. (1976). *Makromol. Chem.* **177**, 2569.
Price, C., Watson, A., and Chow, M. T. (1972a). *Polymer* **13**, 333.
Price, C., Lally, T. P., Watson, A., Woods, D., and Chow, M. T. (1972b). *Br. Polym. J.* **4**, 413.
Price, C., Singleton, R., and Woods, D. (1974a). *Polymer* **15**, 117.
Price, C., Lally, T. P., and Stubbersfield, R. (1974b). *Polymer* **15**, 541.
Rossi, J., and Gallot, B. (1976). *Makromol. Chem.* **177**, 2801.
Rossi, J., and Gallot, B. (1977). *Makromol. Chem.*, in press.
Sadron, C. (1962). *Pure Appl. Chem.* **4**, 347.
Sadron, C. (1963). *Angew. Chem., Int. Ed. Engl.* **2**, 248.
Sadron, C. (1965). *Rev. Gen. Caoutch. Plast., Ed. Plast.* **2**, 112.
Sadron, C., and Gallot, B. (1973). *Makromol. Chem.* **164**, 301.
Seow, P. K., Gallot, Y., and Skoulios, A. (1975). *Makromol. Chem.* **176**, 3153.
Seow, P. K., Gallot, Y., and Skoulios, A. (1976a). *Makromol. Chem.* **177**, 177.
Seow, P. K., Gallot, Y., and Skoulios, A. (1976b). *Makromol. Chem.* **177**, 199.
Singer, S., and Nicolson, G. (1972). *Science* **175**, 720.
Skoulios, A., and Finaz, G. (1961). *C.R. Acad. Sci.* **252**, 3467.
Soen, T., Inoue, T., Myoshi, K., and Kawai, H. (1972). *J. Polym. Sci., Part A-2* **10**, 1757.
Tadokoro, H., Chatani, Y., Yoshihara, T., Tamara, S., and Murahashi, S. (1964). *Makromol. Chem.* **73**, 109.
Uchida, T., Soen, T., Inoue, T., and Kawai, H. (1972). *J. Polym. Sci., Part A-2* **10**, 101.

7

Rheology of Polymers with Liquid Crystalline Order

Donald G. Baird

Monsanto Textiles Company
Pensacola, Florida

I. Introduction

A. Significance of Rheology of Polymers with Liquid Crystalline Order

Liquid crystals are substances that possess mechanical properties resembling those of fluids yet are capable of transmitting polarized light under static conditions, and in some cases may show Bragg reflections characteristic of a well-defined molecular spacing. They are sometimes referred to as anisotropic fluids or mesophases, and we shall use these terms interchangeably in this chapter.

Organic compounds that exhibit liquid crystalline behavior have been studied for nearly a century. However, it has been only in the last thirty years that liquid crystalline order has been recognized in polymer systems. Some of the earliest reports of polymers with liquid crystalline order include the studies by Oster (1950) on aqueous solutions of tobacco mosaic virus (TMV) and by Robinson (1956) on solutions of poly(γ-benzyl-L-glutamate) (PBLG) in various solvents. More recently a number of synthetic macromolecules have been reported to exhibit liquid crystalline behavior (Blades, 1972; Kwolek, 1972; Kwolek et al., 1976).

The significance of polymer systems with liquid crystalline order arises from their presence in living systems and their use in the production of ultrahigh strength synthetic fibers. According to Mishra (1975) many of the features of the living state such as specificity, asymmetry, dynamic transformation, rhythmicity, control, and communication in molecular domains and evolution can be attributed to the liquid crystalline state. Fluctuations in structure of these systems are a consequence of the input of mass, momentum, or electromagnetic fields. Thus, a knowledge of the rheology of biological systems with liquid crystalline order could be useful in understanding some of the processes that occur within the human body.

Recently a number of patents (Blades, 1972; Kwolek, 1972) have appeared in the literature claiming the production of fibers from polymer solutions existing in the liquid crystalline state. These fibers have been reported to have tenacities of the order of 20–30 g/den in textile units (or ~ 1.76–2.64×10^{10} dyne/cm^2) and moduli as high as 800–1200 g/den (7.05–12.58×10^{11} dyne/cm^2), which are greater in magnitude than those reported for steel on a weight basis (tenacity = 3.0–4.5 g/den; modulus = 280–300 g/den). Reports of fibers spun from polyester melts in the liquid crystalline state having tenacities greater than 30 g/den (3.8×10^{10} dyne/cm^2) have also appeared in the patent literature (Schaefgen et al., 1976). Processing conditions described in these patents indicate that these materials exhibit unique rheological properties of which little is known.

In addition, a number of block copolymer systems of the general structure A–B–A, where A is a thermoplastic block polymer and B is an elastomeric block polymer, have been found to have mechanical properties similar to those of vulcanized rubbers but flow like thermoplastics above the glass transition temperature of the thermoplastic block (Holden et al., 1969). At these temperatures, liquid crystalline order has been observed (Skoulios, 1975; Skoulios et al., 1960).

B. Types of Liquid Crystalline Order in Polymer Systems

Before discussing the rheological properties of polymer systems with liquid crystalline order, it is essential to review briefly the types of polymers that exhibit this order and the types of order found in these systems. We first review the various types of mesomorphic arrangements (i.e., liquid crystalline types) found in organic liquid crystalline systems. As we shall see, polymer systems, for the most part, can be classified into these categories.

A schematic representation of the three types of mesophases is given in Fig. 1. For the nematic phase, the centers of gravity of the molecules have no long-range order. There is however order in the direction of the molecules, and they tend to be parallel to a common axis labeled by a unit vector **n**. The molecules are free to slide over one another but maintain their relative orientations. The cholesteric phase, being very similar to the nematic phase, is described as a twisted nematic. The centers of gravity again have no long-range order, and the molecular orientation shows a preferred direction given by **n**. However, **n** is not constant in space but varies in a periodic fashion. Because of the periodicity of the planes, these substances exhibit Bragg scattering of light beams. The term smectic has been given to those materials in which the molecules lie in planes with well-defined interlayer spacing that can be detected by X-ray diffraction. A somewhat more detailed description of the different mesophases is given in Chapter 1. For more advanced treatment we note in particular the work of de Gennes (1974) and Saupe (1969).

NEMATIC   SMECTIC   CHOLESTERIC

Fig. 1. Schematic representation of the molecular structure of the three basic types of liquid crystals.

Molecules that are rodlike in solution or in the undiluted state apparently form either the cholesteric or nematic mesophase. According to Robinson (1956) the optical properties of solutions of PBLG in several solvents were reminiscent of those exhibited by cholesteric liquid crystals. The aromatic polyamides such as poly(1,4-benzamide) (PBA) or poly(1,4-phenylene tere-phthalamide) (PPPT) in strong acids or diakylamides form the nematic mesophase (Kwolek, 1972). These systems agree at least qualitatively with the predictions of Flory (1956) for the separation of solutions of rodlike molecules into ordered and disordered phases (Hermans, 1962). The important variables affecting the onset of anisotropy are molecular weight, polymer concentration, temperature, and solvent type.

Block copolymers form the smectic mesophase and in many respects resemble soap molecules (Richards and Szwarc, 1959; Skoulios et al., 1960;

see also Chapter 6). For a block copolymer of the A–B type, the blocks are incompatible and segregate into organized domains. These microdomains distribute periodically in space producing the same type of liquid crystalline structure as found in soap systems. There are basically three types of organization: (i) lamellar, shown schematically in Fig. 2; (ii) cylindrical; and (iii) spherical (Skoulios, 1975). The lamellar structure is the simplest and most frequently encountered and represents a perfect example of a smectic liquid crystal. Depending upon their chemical structure, blocks can either crystallize, be in the glassy state, or act as a fluid. These systems may exist as undiluted polymers (i.e., melts) or in solutions in which one of the blocks is insoluble in the given solvent. The variables affecting the formation of a given structure are the molecular weight of the blocks, the temperature, and in the case of solutions, the solvent. A theory that gives criteria for the formation of domains and their size in terms of molecular and thermodynamic variables for block copolymers of the A–B type has been offered by Meier (1969).

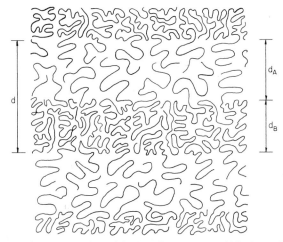

Fig. 2.   Schematic representation of the lamellar structure of block copolymer systems in the smectic mesophase. (Courtesy of Skoulios, 1975.)

Finally, in the last few years a number of polymers have been produced from monomers which themselves form liquid crystals. Typical examples are the polymeric esters of alkoxybenzoic acids. These systems differ from those mentioned in that liquid crystalline order is a result of interaction between side groups along the chain rather than interaction between backbone segments. These types of polymers have been termed comblike polymers with liquid crystalline order. All three types of mesophases have been observed in these systems (Bouligand *et al.*, 1974).

C. Scope of the Review

The most recent general review of the rheology of liquid crystals was given by Porter and Johnson (1967). At that time, the only polymeric systems recognized to have liquid crystalline order were solutions of biological molecules and the synthetic polypeptides. We shall begin our review with these systems for the sake of completeness and because of the similarity of their rheological behavior to that of solutions of aromatic polyamides.

The review will be organized according to polymer structure. In particular we shall study first the systems of rodlike molecules followed by the block copolymer systems, and then the comblike polymers with liquid crystalline order. The major emphasis will be on the rodlike molecules primarily because more data on these systems exist in the literature.

We shall conclude this review with a summary of various approaches taken for developing constitutive equations for fluids with liquid crystalline order. Major emphasis will be on the continuum theory of Ericksen and Leslie and where the opportunity arises, we shall compare experimental data with the theory.

II. Rodlike Systems

A. Synthetic Polypeptides and Related Molecules

A number of polypeptides and biological molecules such as ribonucleic acid (RNA), deoxyribonucleic acid (DNA), collagen, and TMV exist in various solvents as helical coils. The aspect ratio (i.e., length-to-width ratio) of these molecules is large, and for hydrodynamic purposes they can be considered as rodlike. When a critical concentration c^* is achieved, an ordered phase spontaneously occurs as a result of the energetically favorable packing of these rods in the solvent. The theory of separation into ordered and disordered phases for systems of rods seems well established (Flory, 1956).

Hermans (1962) undertook one of the first rheological studies of anisotropic solutions. He investigated the viscosity (η) of concentrated solutions of PBLG of molecular weight (MW) from 85,000 to 340,000 in *m*-cresol using a capillary and a couette-type viscometer. The most characteristic behavior of these solutions is the concentration dependence of the low shear rate viscosity. In Fig. 3, it is seen that the viscosity first increases with increasing concentration but then passes through a maximum and decreases with further increases of concentration. The concentration c^* at which η goes through a maximum is associated with the onset of liquid crystalline order. Hermans found that c^* depended on molecular weight in a manner that was in agreement with the predictions of Flory.

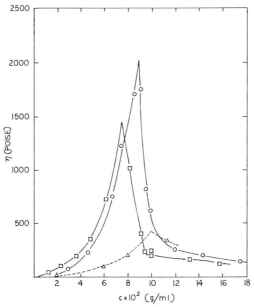

Fig. 3. Viscosity at low shear stress ($\tau < 100$ dyne/cm^2) versus concentration for fractions of PBLG in m-cresol with molecular weights of 342,000 ($\square$); 270,000 ($\bigcirc$); and 220,000 ($\triangle$). (From Hermans, 1962, used with permission.)

Hermans (1962) also studied the effect of shear stress (τ) on the reduced viscosity (η_{sp}/c). In Fig. 4, values of η_{sp}/c versus c are presented for various τ levels. These data point out two interesting aspects. First, the onset of anisotropy occurs at lower values of c with increasing τ, and second, for values of τ greater than 5×10^3 dyne/cm^2, there is no distinction between the viscosities of the two phases. In fact, Hermans claimed that the distinction between the anisotropic and isotropic phases disappeared at high τ levels and the ordering produced by shear was much greater than that produced by the energetically favorable packing of the molecules in the solution.

Miller and co-workers (1974) studied the viscosity dependence of concentrated solutions of PBLG in dimethylformamide (DMF) on temperature and polymer concentration. The critical concentration for the onset of

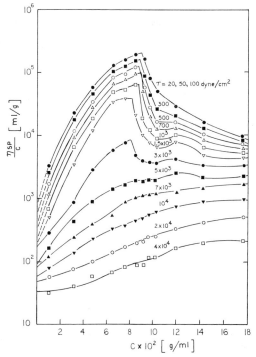

Fig. 4. Reduced viscosity (η_{sp}/c) versus concentration (c) at various shear stress levels for solutions of PBLG in *m*-cresol. (From Hermans, 1962, used with permission.)

anisotropy was found to increase with increasing temperature. They attributed this to an increase in the flexibility of the molecules that necessitated higher concentrations before anisotropy could be observed. The viscosity of the anisotropic phase was found to depend on temperature in an unusual but characteristic way, as shown in Fig. 5. The viscosity is seen to decrease rapidly with increasing temperature passing through a minimum followed by another rise with further increases in temperature. This behavior was associated with the transition from the anisotropic to isotropic phase.

The flow curves (i.e., log τ versus log $\dot{\gamma}$) of PBLG in both DMF and *m*-cresol for both the isotropic and anisotropic states indicate that the solutions are Newtonian at low shear rates ($\dot{\gamma}$). For $\dot{\gamma} > 100\,\text{sec}^{-1}$, the solutions are non-Newtonian, exhibiting power law behavior (i.e., $\tau \propto \dot{\gamma}^n$) (Hermans, 1962; Miller *et al.*, 1974).

At present there have been very few measurements of material parameters other than viscosity reported in the literature for these systems. Iizuka (1974) has recently measured, in addition to viscosity, normal stresses in steady shear flow and viscoelastic properties under oscillatory flow for concentrated

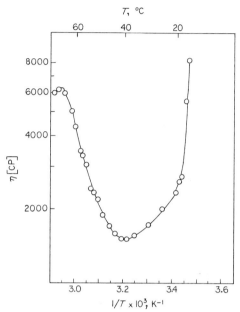

Fig. 5. Temperature dependence of viscosity at $\dot{\gamma} = 19.4\,\mathrm{sec}^{-1}$ for a PBLG(MW = 310,000) DMF solution. (Courtesy of Miller *et al.*, 1974.)

solutions of poly(γ-ethyl-L-glutamate), its equimolar mixture with the D-isomer, and poly(γ-benzyl-L-glutamate). PBLG was found to exist in the nematic mesophase in CH_2Br_2 and in the cholesteric mesophase in 1,4-dioxane. There was no distinguishable difference in the rheological properties of these two systems suggesting that the rheological properties for these two mesophases may be identical for polymer systems. The loss modulus (G'') and the storage modulus (G') depended on concentration in a manner similar to that observed for viscosity.

Several mechanisms have been offered to explain the flow behavior of the polypeptide solutions. Hermans (1962) attributed the lower viscosity of the anisotropic phase at low shear rates to the orientation of the rodlike molecules with the flow direction. In the anisotropic phase there is a correlation in the orientation of neighboring molecules. At a small value of $\dot{\gamma}$, the molecules become completely oriented in the direction of flow, resulting in a reduction of the viscosity. On the other hand, the molecules are still randomly oriented at low $\dot{\gamma}$ for the isotropic solutions. Miller *et al.* (1974) proposed that in the liquid crystalline state the correlation of rod motion leads to the ordering of the rods into a mean parallel arrangement and thereby promotes considerably easier bulk flow. Iizuka (1974) believed that rodlike clusters in which the molecules are in a nearly parallel orientation are produced in the liquid crystalline solutions upon the application of a shear stress. He attributed the non-Newtonian flow behavior and the existence of normal stress differences to these clusters. Although these explanations are reasonable, they do not seem satisfactory for explaining the distinct drop in viscosity in the anisotropic phase in the very low $\dot{\gamma}$ range (further discussion of this subject will be given in the next section).

Several rheological studies have been reported on isotropic solutions of PBLG. We find them of significance for they suggest rheological behavior that may be indicative of molecules that can form anisotropic solutions. Yang (1958) investigated the dependence of η on τ for solutions of PBLG in m-cresol, a helicogenic solvent, and dichloracetic acid, in which PBLG is a random coil. A distinct difference in the flow curves of the two systems was noted. The viscosity of the rodlike system was highly non-Newtonian, whereas the system of random coils showed very little shear dependence of viscosity. The viscoelastic behavior of dilute solution of PBLG in m-methoxyphenol, another helicogenic solvent, was studied by Tschoegel and Ferry (1964). The frequency dependence of G' and G'' was intermediate between the predictions of the Kirkwood–Auer theory for rigid rods and the Zimm theory for flexible coils with dominant hydrodynamic interaction.

A number of other biological molecules that also form helical coils have been reported to form liquid crystals. A review of the rheology of isotropic solutions of DNA was given by Robins (1966). Three aspects of flow were covered in this review: non-Newtonian flow, viscosity measurement methods, and degradation. Attempts were also made to utilize the sharpness of fall in viscosity with $\dot{\gamma}$ to deduce information about the rigidity of the molecule. Eisenberg (1957) concluded from his studies on the shear dependence of viscosity that the DNA molecule is best described as a Gaussian chain with internal viscosity.

Katz and Ferry (1953) studied the viscosity and dynamic rigidity (G') of solutions of sodium deoxyribonucleic acid (SDNA) in aqueous solutions containing 10% NaCl and 1% glycerine at concentrations less than 0.04 g/ml, the concentration at which the solutions became anisotropic. From the concentration dependence of viscosity and the low activation energies of flow, they concluded that the high viscosities are primarily due to hydrodynamic and steric interference. The relaxation time (α) was of the order of 1 sec, yet the rigidity (G') was between 10 and 100 dyne/cm^2. In synthetic polymers, G' is approximately 10^3 and α is of the order of 10^{-3} sec. From these results they further concluded that the viscosity and rigidity involve orientation of rigid particles but no interconnecting structure through bonds or local attractive forces.

Similar salt-free DSNA solutions in a lower concentration range (0.19–18.3 $\times$ 10^{-3} g/ml) were observed to be highly non-Newtonian with a power law slope of -0.9 (Helders and Ferry, 1956). The magnitude of this value is considerably larger than that found for dilute solutions of many flexible polymer molecules and may be an indication of the rodlike nature of SDNA. At concentrations less than 0.38×10^{-3} g/ml, Helders and Ferry (1956) reported that the viscosity data reduced well using η/η_0 and $\eta_0\dot{\gamma}$, where η_0 is the zero shear viscosity, which is the form predicted by Saito for rigid ellipsoids. For concentrations greater than 0.75 g/ml, the data reduced well using η/η_0 versus $\eta_0\dot{\gamma}/c$, where η_0 is the zero shear viscosity, which is similar to the form predicted by the theory of Bueche for free-draining flexible coils. They interpreted their results to mean that for the more concentrated solutions, flow is accompanied by intramolecular changes of configuration as a result of some degree of flexibility. In more dilute solutions, the inherent stiffness of the coil becomes more important and flow is consistent with that of a rigid rod.

Collagen is also a molecule of helical structure consisting of three polypeptide strands and is believed to be a long rigid rod in some solvents (Boedtker and Doty, 1956). Fukada and Date (1963) measured the dynamic rigidity and viscosity as well as the steady flow viscosity of collagen solutions in dilute hydrochloric acid for concentrations from 0.5 to 2.0 g/ml. The strong concentration dependence of the rigidity was attributed to the gellike nature of the solutions that occurred as a result of the association of the rodlike molecules into bundles in which crystalline order may have existed. These bundles served as cross-links giving the solutions a networklike structure.

Aqueous solutions of wood and cotton cellulose are thought to form liquid crystals under the action of shearing (Marchessault et al., 1959). Hermans (1963) found that these solutions exhibited unique rheological behavior. Solutions in the range of concentrations from 10.8 to 17.6% demonstrated yield stresses, and for $\dot{\gamma}$ from about 70 to 1000 sec^{-1}, τ was independent of $\dot{\gamma}$. Solutions of cotton cellulose exhibited two values of yield stress depending on the rate at which they were sheared just prior to the yield stress measurement. Above $\dot{\gamma} = 75$ sec^{-1}, the higher yield stress value was observed. The higher yield stress was attributed to the formation of liquid crystalline regions due to the alignment of the molecules into a side-by-side arrangement (Marchessault et al., 1959). Marchessault and co-workers claimed that the action of the shear field was critical to producing a system of particles with sufficient degree of parallel alignment so that intermolecular forces could maintain the liquid crystalline phase.

B. Solutions of Extended Chain Polymers

Poly(1,4-benzamide) (PBA) and poly(1,4-phenylene terephthalamide) (PPPT) were two of the first nonpeptide synthetic condensation polymers reported to form liquid crystalline solutions (Kwolek *et al.*, 1976). These polymers are thought to form liquid crystalline solutions because of the inherently extended rigid structure (Schaefgen *et al.*, 1976; Papkov *et al.*, 1974; Baird, 1977) as opposed to the helical conformation of the polypeptides.

For the most part, the rheological studies of solutions of these polymers reported in the literature have been concerned with the concentration dependence of viscosity that is similar to the behavior exhibited by solutions of PBLG (Blades, 1972; Kwolek, 1972; Kwolek *et al.*, 1976; Papkov *et al.*, 1974; Sokolova *et al.*, 1973; Kulichikhin *et al.*, 1974). Representative data selected from the work of Papkov and co-workers (1974) for solutions of PBA in dimethylacetamide (DMAC) are presented in Fig. 6. The viscosity of these anisotropic solutions was also found to vary with temperature in a characteristic way. We note also that Papkov's data agreed well with the predictions of Flory's (1956) theory for rodlike molecules.

An exceptional study of the viscous properties of solutions of PBA in DMAC was reported by Papkov and co-workers (1974). The viscosity versus

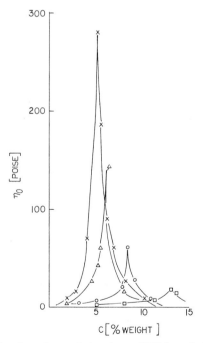

Fig. 6.   Concentration dependence of viscosity at 20°C for solutions of PBA of various molecular weights in *N,N*-dimethylacetamide/LiCl: (□), 10,900; (○) 16,000; (△) 23,800; (×) 69,000. (Adapted from Papkov *et al.*, 1974.)

τ is shown in Fig. 7 for solutions of varying concentration (for $c \geq 5\%$, the fluids are anisotropic). For the isotropic solutions, a Newtonian viscosity region is observed followed by a region of non-Newtonian flow. However, in the anisotropic state, the solutions exhibit a yield stress followed by a region of quasi-Newtonian flow. The anisotropic fluids also exhibit a region of non-Newtonian flow. In addition, they noted that the viscosity measured at high shear rates no longer shows the characteristic maximum at c^* but increases monotonically with increasing concentration.

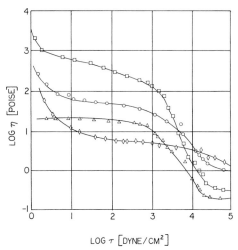

LOG τ [DYNE/CM^2]

Fig. 7. Dependence of viscosity on shear stress of solutions of PBA in DMAC/LiCl: $\bar{M}_w$ = 69,000; ($\triangle$) 3%, ($\square$) 5%, ($\bigcirc$) 7%, ($\diamondsuit$) 9.5%. (Adapted from Papkov et al., 1974, with permission.)

Two important observations can be made from the viscosity data obtained by Papkov et al. (1974) for the isotropic solutions. First, the flow curves could be plotted in reduced form using η/η_0 versus $\eta_0\dot{\gamma}$, which again is the form predicted by Saito for rigid ellipsoids. The viscosity of solutions of flexible polymer molecules, on the other hand, can be plotted in reduced form using η/η_0 versus $\eta_0\dot{\gamma}/c$. Second, η_0 was found to be proportional to $\bar{M}_w{}^8$, where $\bar{M}_w$ is the weight average molecular weight, whereas for flexible polymer coils η_0 is proportional to $\bar{M}_w^{3.5}$ (Graessley, 1974). These may represent two ways of predicting whether a given polymer system has the potential to form a liquid crystalline phase.

At the time of writing, the only rheological measurements of quantities other than viscosity are reported by Baird (1977). In addition to viscosity, values of the primary normal stress difference N_1 and the dynamical mechanical properties (G', G'', and η^*) were measured for solutions of poly(1,4-phenylene terephthalamide) in H_2SO_4. As is expected, N_1, G', and G'' depend on concentration in a manner similar to that of η. The zero shear viscosity obtained from isotropic solutions was found to depend on $\overline{M}_w{}^{6.4}$, which is similar to the result reported earlier for solutions of PBA.

Papkov et al. (1974) explained the flow behavior of the solutions of PBA using the concept of a fluctuating entanglement network.[†] In the isotropic state (i.e., $c < c^*$) the flow properties are governed by an entanglement network, which arises as a result of the intermolecular interaction of randomly oriented molecules. With the onset of anisotropy, highly ordered regions are formed (Kalmykova et al., 1971) connected by disordered regions of entanglements. As the concentration increases, the size of the ordered regions increases at the expense of the disordered regions. The disappearance of entangled regions and thus the number of entanglements plus the ease of orientation of the ordered regions with the flow direction could account for the less pronounced non-Newtonian flow of the anisotropic solutions.

It may be possible to extend the above mechanism to explain the anomalous concentration dependence of viscosity at low $\dot{\gamma}$. With the onset of anisotropy, there is in addition to a lower number of entanglements, a change in the flow unit. In the anisotropic state, X-ray diffraction studies have shown that aggregates in which the molecules are arranged in nearly parallel order exist (Kalmykova et al., 1971). At low $\dot{\gamma}$, these aggregates or domains may act as a rigid flow unit that dissipates less energy than would be dissipated by each individual molecule in the aggregate and thereby account for the lower solution viscosities at low $\dot{\gamma}$.

C. Polymer Melts

There is at present very little known about the rheology of anisotropic polymer melts. The only information available in the literature at the time of this writing was given by Jackson and Kuhfuss (1976) for copolyesters of poly-(ethylene terephthalate) (PET) modified with p-hydroxybenzoic acid (PHB). They reported that η at a given $\dot{\gamma}$ first increased with increasing PHB content up to 30 mole % and then decreased as the PHB content was increased further to 60 mole %. In addition, they found that the viscosity became $\dot{\gamma}$ dependent at lower $\dot{\gamma}$ with increasing PHB content. This behavior was attributed to the presence of liquid crystalline order.

[†] The concept of an entanglement network, which in recent years has been used to explain many of the rheological properties of polymer melts and solutions, is described in detail in a recent review by Graessley (1974). For the case of a rigid rod, it is difficult to imagine the form of an entanglement, but it probably arises as a result of a frictional interaction between two adjacent molecules.

III. The Rheology of Other Polymer Systems

A. Block Copolymers

The studies of the rheological properties of block copolymer systems with mesomorphic order are by far less numerous than those reported on the molecular structure of these systems. A number of reports of the dilute solution properties have appeared in the literature (Gallot *et al.*, 1962; Krause, 1964; Kotaka *et al.*, 1972; Ohnuma *et al.*, 1970; Demin *et al.*, 1974) that were concerned primarily with configurational changes with the addition of nonsolvent. Even in the case of dilute solutions, there is evidence of micelle formation but no indication of the formation of ordered domains distributed periodically in the solvent (Krause, 1964).

Relatively few papers on the flow of melts or solutions of block copolymers have been published, and most of these have pertained to the butadiene–styrene copolymers (Kraus and Gruver, 1967; Arnold and Meier, 1970; Paul *et al.*, 1970; Kraus *et al.*, 1971; Kotaka and White, 1973). These studies have indicated striking behavior for solutions of styrene–butadiene copolymers of the–ABA–type. The viscosity exhibited major increases as the solvent quality for the styrene segments decreased. This behavior was associated with the separation of the block copolymer into separate phases. Similar behavior has been observed for other copolymer systems (Ohnuma *et al.*, 1970; Kotaka *et al.*, 1972; Tanaka *et al.*, 1972).

Kotaka and White (1973) measured, in addition to the viscosity, the primary normal stress difference (N_1) for solutions of butadiene–styrene copolymers in decalin, decane, and their mixtures. The striking behavior of solutions of SBS block copolymer is illustrated in Fig. 8. Values of η_0, the compliance J_e defined as $J_e = N_1/2\tau^2$, and the relaxation time defined as $\alpha = N_1/\tau\dot{\gamma}$ are plotted as a function of the solvent ratio (note: decane is a nonsolvent for polystyrene). It is observed that both η_0 and N_1 increase drastically with increasing decane content, but that J_e remains nearly unaffected. The flow behavior of these solutions is attributed to the formation of a three-dimensional network of the type envisioned by Holden and co-workers (1969) and Meier (1969). At moderate levels of decane, the styrene blocks become immiscible and serve as cross-links of very high functionality giving the solution a networklike character. Solutions of the simple block copolymer (SB) behave in a similar manner but exhibit a yield stress and are thixotropic. In this case, however, there is no formation of a three-dimensional network, but rather a micelle structure in which the polystyrene segments form a rigid core with the polybutadiene solubilizing the micelles.

Mesomorphic order has also been observed in undiluted (or melt) SBS block copolymers. Arnold and Meier (1970) reported that the SBS copolymer

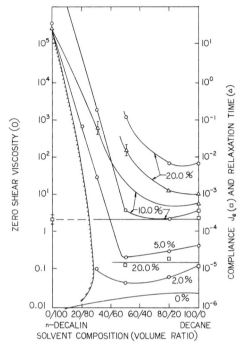

Fig. 8. Zero shear viscosity (η_0), compliance (J_e), and relaxation time (α) as a function of solvent ratio for a SBS block copolymer. (Courtesy of Kotaka and White, 1973.)

exhibited no Newtonian flow region at low $\dot{\gamma}$ whereas most polymer melts do. However, at intermediate shear rates a power law region was observed in the flow curve that is similar to most polymer melts. Arnold and Meier proposed that the SBS copolymer could exist in three states depending on $\dot{\gamma}$. At low $\dot{\gamma}$, a three-dimensional network is intact. At intermediate $\dot{\gamma}$, this network is disrupted and the domains of polystyrene would be partially separated from the polybutadiene regions. At high $\dot{\gamma}$, the system would behave as an assemblage of individual nonaggregated molecules.

Erhardt and co-workers (1970) studied the rheological properties of styrene–ethylene oxide block copolymers. The melt viscosity and elasticity were found to be greater than values for either homopolymer. This again was attributed to the phase separation of the two blocks resulting in a three-dimensional network.

B. Comblike Polymers

Although the synthesis of comblike polymers with liquid crystalline order has received considerable attention recently, there are very few reports on the rheology of these systems. Several papers have appeared in the literature that have been concerned with the conformation of these molecules in dilute solutions (Tsvetkov *et al.*, 1969, 1971, 1973). Platè and Shibaev (1974) studied the flow properties of polyacrylates and polymethacrylates that are crystalline in the solid state. These polymers began to flow below the melting temperature and exhibited a strong anomaly in the flow curves. It was not certain whether this behavior was associated with the formation of some type of mesomorphic state. Vinogradov and co-workers (1974) have reported some rheological data for poly(alkyl acrylate) and poly(alkyl methacrylate) melts with liquid crystalline order. These melts were found to exhibit a yield stress.

IV. Rheological Theories for Anisotropic Fluids

The early attempts at formulating continuum theories for liquid crystals can be traced to the work of Oseen (1933), Anzelius (1931), and Frank (1958). More recently, Ericksen (1960, 1961, 1962a,b, 1966a,b, 1967, 1969a,b) and Leslie (1966, 1968a,b, 1969, 1971) have developed a complete set of equations based on continuum mechanics to describe the various conservation laws (mass, momentum, angular momentum, and energy) for nematic and cholesteric liquid crystals. In addition a number of alternative presentations for nematic liquid crystals have appeared in the literature, of which we mention those of Davison (1967, 1969), Davison and Amos (1969), Martin *et al.* (1970), Aero and Bulygin (1970), (1971a,b, 1972), Helfrich (1969a,b, 1970, 1972), Forster and co-workers (1970), Stephen (1970), Huang (1971), Lee and Eringen (1971a,b,c), Eringen and Lee (1973), and Jähnig and Schmidt (1972). According to Shahinpoor (1976), these theories with the exception of that of Lee and Eringen are in essential agreement with that of Ericksen and Leslie. As shown recently by Shahinpoor (1976), the first version of the Lee and Eringen theory is inconsistent, and the second appears to be similar to the Ericksen–Leslie theory. Because of the similarities in the predictions of the other approaches with those of the Ericksen–Leslie theory, we shall primarily consider the Ericksen–Leslie theory in this review.

Ericksen (1959, 1960) first formulated a simple theory for describing the flow properties of anisotropic fluids. However, the theory seemed inadequate for describing the behavior of liquid crystals. First, it did not reduce to the static theory of Oseen (1933) and Frank (1958), and second, there was no way to incorporate the influence of solid boundaries on preferred orientations within the fluid. Motivated by his hydrostatic theory of liquid crystals, Ericksen (1961) proposed general conservation laws for the dynamical behavior of liquid crystals. Adopting these laws, Leslie (1966) formulated constitutive equations that essentially were similar to the simple theory of Ericksen (1960). Subsequently, Leslie (1968a) employed a generalization of the entropy inequality due to Müller (1967) and reformulated his theory. However, for the case of rigid molecular regions, the simple theory of Ericksen is presumably recovered. For the case of a simple shear flow, the more complicated theory of Leslie leads to results that are similar to those predicted by the simple theory of Ericksen for regions of flow sufficiently far from the walls (Leslie, 1968b).

It is a formidable task to evaluate the components of stress even for a simple flow using the more complicated theory of Leslie (1968b). It would seem sensible then to attempt to understand the concepts involved in the simpler theory of Ericksen and then evaluate its ability to describe experimental behavior for simple well-defined flows. We shall then compare the predictions of Ericksen's simple theory with those of a more complicated one based on the class of theories considered by Leslie (1968a).

Ericksen considered a liquid crystalline fluid to consist of tiny cylindrically symmetric packets of rodlike molecules. Within each packet or domain the molecules were free to move parallel to each other so that the length of the packet was variable. The direction of the packet was described by the vector **n** that has cartesian components n_i ($i = 1$–3). The orientation and motion of this packet was essentially controlled by the fluid motion although electric and magnetic fields could have an effect. He idealized this moving packet as a line segment of variable length and direction.

He next assumed that, to within an arbitrary hydrostatic pressure p, the stress at any oriented particle was given in cartesian components as (the Einstein summation convention is used here)

$$p_{ij} = -p\,\delta_{ij} + f_{ij}(n_k; v_{m,n}) \tag{1}$$

where δ_{ij} is the Kronecker delta and $v_{m,n} = \delta v_m/\delta x_n$ is the velocity gradient. He further assumed that

$$Dn_i/Dt = g_i(n_k, v_{m,n}) \tag{2}$$

where D/Dt is the material time derivative. With the requirement that the functions f_{ij} and g_i be invariant under simple rigid body motions, Ericksen expanded f_{ij} and g_i as polynomial functions of the rate of deformation tensor $\dot{\gamma}$ which is defined in cartesian components as

$$\dot{\gamma}_{km} = \tfrac{1}{2}(v_{k,m} + v_{m,k}) \tag{3}$$

The linearization of f_{ij} and g_i with respect to $\dot{\gamma}_{km}$ plus a number of mathematical simplifications yielded the following expression for the components of stress:

$$p_{ij} = -p\delta_{ij} + 2\mu\dot{\gamma}_{ij} + (\mu_1 + \mu_2\dot{\gamma}_{km}n_k n_m)n_i n_j + 2\mu_3(\dot{\gamma}_{jk}n_k n_i + \dot{\gamma}_{ik}n_k n_j) \tag{4}$$

With the restriction that $\sum_{i=1}^{3} \sum_{i=1}^{3} n_i n_i = 1$, which is the case for rigid particles, the vector $\mathbf{n}$ is related to $\dot{\gamma}_{ij}$ by

$$Dn_i/Dt = \lambda(\dot{\gamma}_{ij}n_j - \dot{\gamma}_{km}n_k n_m n_i) + \omega_{ij}n_j \qquad (5)$$

where $\omega_{ij} = \frac{1}{2}(v_{i,j} - v_{j,i})$ and λ is an undefined constant.

For a simple shear flow for which the velocity field is given by $v_1 = \dot{\gamma}x_2$ and $v_2 = v_3 = 0$, and $\dot{\gamma}$ is constant, the three independent quantities of stress are (for steady-state conditions)

$$N_1 = p_{11} - p_{22} = \mu_1(n_1^2 - n_2^2) + \mu_2(n_1^2 - n_2^2)n_1 n_2 |\dot{\gamma}| \qquad (6)$$

$$N_2 = p_{22} - p_{33} = \mu_1 n_2^2 + (\mu_2 n_2^2 + 2\mu_3)n_1 n_2 |\dot{\gamma}| \qquad (7)$$

$$p_{21} = p_{12} = \mu_1 n_1 n_2 + [\mu + \mu_2 n_1^2 n_2^2 + \mu_3(n_1^2 + n_2^2)]\dot{\gamma} \qquad (8)$$

where for $\lambda > 1$ and $n_3 = 0$

$$n_1 = \pm[(\lambda + 1)/2\lambda]^{1/2} \qquad (9)$$

$$n_2 = \pm[(\lambda - 1)/2\lambda]^{1/2} \qquad (10)$$

As is readily observable, there are four material constants μ, μ_1, μ_2, μ_3, and one undefined constant λ and only three independent equations of stress to solve for these unknowns. Thus, it is impossible to evaluate quantitatively the theory.

On the other hand, it is interesting to consider the qualitative predictions and compare them with existing experimental data. Yield stress values are predicted for N_1, N_2, and p_{21}. We have seen in Section II. B that for some anisotropic solutions yield stress values are observed, but that this may not in general be characteristic of all anisotropic fluids. Equation (8) predicts that the viscosity defined as $p_{21}/\dot{\gamma}$ should be independent of $\dot{\gamma}$. Although Papkov et al. (1974) (see Fig. 7) showed one set of data for which η is quasi-Newtonian, in general, anisotropic fluids are highly non-Newtonian. N_1 and N_2 are shown to depend on $\dot{\gamma}$ rather than $\dot{\gamma}^2$, which has been found to be the case for many polymer solutions and melts at low $\dot{\gamma}$. In Fig. 9, values of N_1 and $2G'$ obtained by Baird (1977), for solutions of PPPT in H_2SO_4 are plotted versus $\dot{\gamma}$ and ω, respectively. For this data, N_1(or G') is not proportional to $\dot{\gamma}$ (or ω) but to a power less than one.

In the more complicated theory of Leslie (1968a) and Ericksen (1969a), the effects of body and surface couples were included. However, in order to evaluate the components of stress even for the case of a simple shear flow, there appears to be no way to include the effects of orientation at the wall. Thus, the more complicated theory must be evaluated assuming high $\dot{\gamma}$ and wide gap widths where the flow dominates. For a simple shear flow the com-

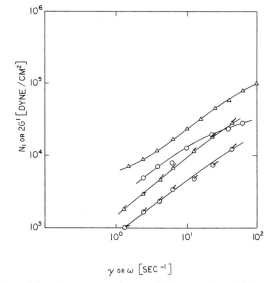

Fig. 9. Values of the primary normal stress difference (N_1) and $2G'$ versus shear rate or frequency for anisotropic solutions of poly(1,4-phenylene terephthalamide) in 100% H_2SO_4 at 60°C: $\bar{M}_W = 40,100$, $c = 16.0\%$, ($\triangle$) N_1, ($\triangle$) $2G'$; $\bar{M}_W = 23,700$, $c = 15.0\%$, ($\bigcirc$) N_1, ($\bigcirc$) $2G'$. (Courtesy of Baird, 1977.)

ponents of stress are

$$p_{13} = p_{31} = p_{23} = 0 \tag{11}$$

$$p_{21} = [2\mu_1 n_1{}^2 n_2{}^2 + \mu_4 + (\mu_5 - \mu_2)n_1{}^2 + (\mu_6 + \mu_3)n_2{}^2]\dot{\gamma} \tag{12}$$

$$p_{12} = [2\mu_1 n_1{}^2 n_2{}^2 + \mu_4 + (\mu_5 - \mu_2)n_1{}^2 + (\mu_6 - \mu_3)n_2{}^2]\dot{\gamma} \tag{13}$$

$$N_1 = p_{11} - p_{22} = \tfrac{1}{2}n_1 n_2[2\mu_1(n_1{}^2 - n_2{}^2) - 2(\mu_2 + \mu_3)]|\dot{\gamma}| \tag{14}$$

$$N_2 = p_{22} - p_{33} = \tfrac{1}{2}n_1 n_2[2\mu_1 n_2{}^2 + \mu_2 + \mu_3 + \mu_5 + \mu_6]|\dot{\gamma}| \tag{15}$$

Equations (12)–(15) differ from Eqs. (6)–(8) in that they predict $p_{21} \neq p_{12}$ and no yield stress values.

At high $\dot{\gamma}$, for which $n_i = (1, 0, 0)$, the following results are predicted from Eqs. (12)–(14):

$$N_1 = N_2 = 0 \tag{16}$$

$$p_{21} = p_{12} = \mu\dot{\gamma} \tag{17}$$

where $\mu = \mu_4 + \mu_6 + \mu_3 = \mu_4 + \mu_5 + \mu_2$. Thus, at high $\dot{\gamma}$, these equations predict that the fluid behaves as a Newtonian fluid.

The more complicated theory suffers from the same problems as the simple theory. There are too many material constants to be determined from flow experiments alone. We note, however, that Ericksen (1969a) has proposed that by measurements of viscosity and normal stresses for specified orientations of $\mathbf{n}$ imposed by strong static magnetic fields one could in principle measure all six μs. For example, for $n_i = (0, 0, 1)$, $2\mu = \mu_4$. Other imposed orientations lead to similar relations among the μs.

We mention two approaches for presenting a more tractable theory of anisotropic fluids. Gordon and Schowalter (1972) presented a theory that was a combination of molecular and continuum approaches. However, it may only be applicable to dilute polymer solutions. Hand (1962) proposed a theory in which the stress tensor was a function of $\dot{\gamma}$ and a symmetric tensor describing the microscopic structure of the fluid. This theory predicted expressions for the stresses that are more readily evaluated. In addition, Ericksen's simple theory is contained in Hand's theory as a special case.

We find that very little theoretical consideration has been given to the flow behavior of the other two mesomorphic states. The fundamental equations for cholesterics have again been discussed by Leslie (1969). For the case of very weak twist, incompressibility, and uniform temperature, the equations are identical to those for nematics (de Gennes, 1974). There are apparently no complete theories for describing the flow of smectic polymer systems.

V.   Conclusions

This chapter covers the literature up to the present (December 1977), and has pointed out that there are very few rheological studies of properties other than viscosity for polymers with liquid crystalline order. Further measurements of normal stresses, dynamic properties such as the storage and loss moduli, and transient stresses are needed. A complete rheological characterization of these materials would not only be useful in understanding the structure of polymer liquid crystals but in processing of these fluids. Studies of the elongational viscosity, die swell, and flow instabilities would be of great practical interest.

The development of rheological theories for each mesophase type are incomplete. This is especially the case for the smectic and cholesteric mesophase. Our understanding of the nematic mesophase is on a somewhat sounder foundation. However, the theory of Leslie and Ericksen contains too many material constants to be of practical value even for simple flows. Finally, there are not enough data available to readily evaluate the qualitative or quantitative predications of the Leslie–Ericksen theory.

Acknowledgments

I would like to express my gratitude to the Monsanto Company for their assistance in the preparation of this chapter.

References

Aero, E. L. (1970). *Izv. Akad. Nauk SSSR, MZLG* **3**.

Aero, E. L., and Bulygin, A. N. (1971a). *Soviet Phys.—Solid State* **13**, 1701.

Aero, E. L., and Bulygin, A. N. (1971b). *PPM* **35**, 879.

Aero, E. L., and Bulygin, A. N. (1971c). *All-Union Inst. Tech. Inf.* No. 290, 6.

Aero, E. L., and Bulygin, A. N. (1972). *Zh. Tekh. Fiz.* **42**, 880.

Anzelius, A. (1931). *Arsskr. Mat. Acb. Nat.* p. 1

Arnold, K. R., and Meier, D. J. (1970). *J. Appl. Polym. Sci.* **14**, 427.

Baird, D. G. (1977). *J. Polym. Sci. Polym. Chem. Ed.*, in press.

Blades, H. (1972). U.S. Patent No. 3,767,756.

Boedtker, H., and Doty, P. (1956). *J. Am. Chem. Soc.* **78**, 4267.

Bouligand, Y., Cladis, P. E., Liébert, L., and Strzelecki, L. (1974). *Mol. Cryst. Liq. Cryst.* **25**, 233.

Davison, L. (1967). *Phys. Fluid* **10**, 2333.

Davison, L. (1969). *Phys. Rev.* **180**, 232.

Davison, L., and Amos, D. E. (1969). *Phys. Rev.* **183**, 288.

de Gennes, P. G. (1974). "The Physics of Liquid Crystals." Oxford Univ. Press (Clarendon), London and New York.

Demin, A. A., Rudkovskii, G. D., Dmitrenko, L. A., Ovsyannikova, T. A., Sokolova, T. A., Samsonov, G. V., and Nikonova, I. N. (1974). *Vysokomol. Soedin., Ser. A* **16**, 2206.

Eisenberg, H. (1957). *J. Polym. Sci.* **25**, 257.

Erhardt, P. F., O'Malley, J. J., and Crystal, R. G. (1970). *In* "Block Copoloymers" (S. L. Aggarival, ed.), p. 195. Plenum, New York.

Ericksen, J. L. (1959). *Arch. Ration. Mech. Anal.* **4**, 231.

Ericksen, J. L. (1960). *Kolloid-Z.* **173**, 117.

Ericksen, J. L. (1961). *Trans. Soc. Rheol.* **5**, 23.

Ericksen, J. L. (1962a). *Arch. Ration. Mech. Anal.* **9**, 371.

Ericksen, J. L. (1962b). *Arch. Ration. Mech. Anal.* **10**, 89.

Ericksen, J. L. (1966a). *Phys. Fluids* **9**, 1205.

Ericksen, J. L. (1966b). *Arch. Ration. Mech. Anal.* **23**, 266.

Ericksen, J. L. (1967). *Trans. Soc. Rheol.* **11**, 5.

Ericksen, J. L. (1969a). *Trans. Soc. Rheol.* **13**, 1, 9.

Ericksen, J. L. (1969b). *Mol. Cryst. Liq. Cryst.* **7**, 153.

Ericksen, J. L. (1970). *Liq. Cryst. Ordered Fluids*, pp. 181–193, New York–London.

Eringen, A. C., and Lee, J. D. (1973). *Liq. Cryst. Ordered Fluids* **2**, 315.

Flory, P. J. (1956). *Proc. Roy. Soc. London* A**234**, 60.

Forster, D., Lubensky, T. C., Martin, P. C., Swift, J., and Pershan, P. S. (1970). *Phys. Rev. Lett.* **26**, 1016.

Frank, F. C. (1958). *Discuss. Faraday Soc.* **25**, 19.

Fukada, E., and Date, M. (1963). *Biorheology* **1**, 101.

Gallot, Y., Leng, M., Benoit, H., and Rempp, R. (1962). *J. Chim. Phys. Phys.-Chim. Biol.* **59**, 1093.

Gordon, R. J., and Schowalter, W. R. (1972). *Trans. Soc. Rheol.* **16**, 79.

Graessley, W. W. (1974). *Adv. Polym. Sci.* **16**. 49.

Hand, G. L. (1962). *J. Fluid Mech.* **13**, 33.

Helders, F. E., and Ferry, J. D. (1956). *J. Phys. Chem.* **60**, 1536.

Helfrich, W. (1969a). *J. Chem. Phys.* **50**, 100.

Helfrich, W. (1969b). *J. Chem. Phys.* **51**, 4092.

Helfrich, W. (1970). *J. Chem. Phys.* **53**, 2267.

Helfrich. W. (1972). *J. Chem. Phys.* **56**, 3187.

Hermans, J., Jr. (1962). *J. Colloid. Sci.* **17**, 638.

Hermans, J., Jr. (1963). *J. Polym. Sci., Part C* **2**, 129.

Holden, G., Bishop, E., and Legge, N. (1969). *J. Polym. Sci., Part C* **26**, 99.

Huang, H. W. (1971). *Phys. Rev. Lett.* **26**, 1525.

Iizuka, E. (1974). *Mol. Cryst. Liq. Cryst.* **25**, 287.

Jackson, W. J., Jr., and Kuhfuss, H. F. (1976). *J. Polym. Sci., Polym. Chem. Ed.* **14**, 2043.

Jähnig, F., and Schmidt, H. (1972). *Ann. Phys. (U.S.A.)* **71**, 129.

Kalmykova, V. D., Kudriavtev, G. I., Papkov, S. P., Volokhina, A. V., Iovleva, M. M., Milkova, L. P., Kulichikhin, V. G., and Bandurian, S. I. (1971). *Vysokomol. Soedin., Ser. B* **13**, 707.

Katz, S., and Ferry, J. D. (1953). *J. Am. Chem. Soc.* **75**, 1589.

Kotaka. T., and White, J. L. (1973). *Trans. Soc. Rheol.* **17**, 587.

Kotaka, T., Tanaka, T., and Inagaki, H. (1972). *Polym. J.* **31**, 327.

Kraus, G., and Gruver, J. T. (1967). *J. Appl. Polym. Sci.* **11**, 2121.

Kraus, G., Naylor, F. F., and Rollman, K. W. (1971). *J. Polym. Sci., Part A-2* **9**, 1839.

Krause, S. (1964). *J. Phys. Chem.* **68**(7), 1948.

Kulichikhin, V. G., Malkin, A. Y., Papkov, S. P., Korolkova, O. N., Kalmikova, V. D., Volokhuma, A. V., and Semenov, O. B. (1974). *Vysokomol. Soedin., Ser. A* **16**, 169.

Kwolek, S. L. (1972). U.S. Patent No. 3,671,542.

Kwolek, S. L., Morgan, P. W., Schaefgen, J. R., and Gulrich, L. W. (1976). *Polym. Prepr., Am. Chem. Soc., Div. Polym. Chem.* **17**, 53.

Lee, J. D., and Eringen, A. C. (1971a). *J. Chem. Phys.* **54**, 5027.

Lee, J. D., and Eringen, A. C. (1971b). *J. Chem. Phys.* **55**, 4504.

Lee, J. D., and Eringen, A. C. (1971c). *J. Chem. Phys.* **55**, 4509.

Leslie, F. M. (1966). *Q. J. Mech. Appl. Math.* **19**, 357.

Leslie, F. M. (1968a). *Arch. Ration. Mech. Anal.* **28**, 265.

Leslie, F. M. (1968b). *Proc. Roy. Soc., Ser. A* **307**, 359.

Leslie, F. M. (1969). *Mol. Cryst. Liq. Cryst.* **7**, 407.

Leslie, F. M. (1971). *Rheol. Acta* **10**, 91.

Marchessault, R. H., Morehead, F. F., and Walters, N. M. (1959). *Nature (London)* **184**, 632.

Martin, P. C., Pershan, D. S., and Swift, J. (1970). *Phys. Rev.* **25**, 844.

Meier, D. J. (1969). *J. Polym. Sci., Part C* **26**, 81.

Miller, W. G., Wu, C. C., Wee, E. L., Santee, G. L., Rai, J. H., and Goebel, K. G. (1974). *Pure Appl. Chem.* **38**, 25.

Mishra, R. K. (1975). *Mol. Cryst. Liq. Cryst.* **29**, 201.

Müller, I. (1967). *Arch. Rational. Mech. Anal.* **26**, 118–141.

Ohnuma, H., Kotaka, T., and Inagaki, H. (1970). *Polym. J.* **1**, 716.

Oseen, C. W. (1933). *Trans. Faraday Soc.* **29**, 883

Oster, G. (1950). *J. Gen. Physiology* **33**, 445.

Papkov, S. P., Kulichikhin, V. G., Kalymykova, V. D., and Malkin, A. Y. (1974). *J. Polym. Sci., Polym. Chem. Ed.* **12**, 1953.

Paul, D. R., St. Lawrence, J. E., and Troell, J. H. (1970). *Polym. Eng. Sci.*, **10**, 270.

Platė, N. A. and Shibaev, V. P. (1974). *J. Poly. Sci., Macromol. Rev.* **8**, 117.

Platė, N. A., Malkin, A. Y., Shibayev, V. P., Poolak, T. K., and Vinogradov, G. V. (1974). *Polym. Sci. USSR* **16**, 510.

Porter, R. S., and Johnson, J. F. (1967). *In* "Rheology" (F. R. Eirich, ed.), Vol. 4, p. 317. Academic Press, New York.

Richards, D. H., and Szwarc, M. (1959). *Trans. Faraday Soc.* **55**, 1644.

Robins, A. B. (1966). *Biorheology* **3**, 153.

Robinson, C. (1956). *Trans. Faraday Soc.* **52**, 571.

Saupe, A. (1969). *In* "Liquid Crystal" (G. H. Brown, ed.), Vol. 2, p. 59. Gordon & Breach, New York.

Schaefgen, J. R. (1975). Belg. Patent No. 828,935.

Schaefgen, J. R., Foldi, V. S., Logullo, F. M., Good, V. H., Gulrich, L. W., and Killian, F. L. (1976). *Polym. Prepr., Am. Chem. Soc., Div. Polym. Chem.* **17**, 69.

Shahinpoor, M. (1976). *Rheol. Acta* **15**, 99.

Skoulios, A. (1975). *Adv. Liq. Cryst.* **1**, 169.

Skoulios, A., Finoz, G., and Farrod, J. (1960). *C. R. Acad. Sci.*, **251**, 739.

Sokolova, T. S., Yefimova, S. G., Volokhina, A. V., Kudryavtsev, G. I., and Papkov, S. P. (1973). *Vysokomol. Soedin., Ser. A* **15**, 2832.

Stephen, M. J. (1970). *Phys. Rev. A* **2**, 1558.

Tanaka, T., Kotaka, T., and Inagaki, H. (1972). *Polym. J.* **3**, 338.

Tschoegel, N. W., and Ferry, J. D. (1964). *J. Am. Chem. Soc.* **86**, 1474.

Tsvetkov, V. N., Shtennikova, I. N., Ryumtsev, Y. I., Kolbina, G. F., Konstantinov, I. I., Amerik, Y. B., and Krentsel, B. A. (1969). *Polym. Sci. USSR* **11**, 2874.

Tsvetkov, V. N., Ryumtsev, Y. I., and Shtennikova, I. M. (1971). *Polym. Sci. USSR* **13**, 579.

Tsvetkov, V. N., Kimmtsev, E. I., Shtennikova, I. N., Korneeva, E. V., Krentsel, B. A., and Amerik, Y. B. (1973). *Eur. Polym. J.* **9**, 481.

Vinogradov, G. V., Platè, N. A., Mulkin, Y. A., Shibaev, V. P., and Poolak, T. H. (1974). *Vysokomol. Soedin., Ser. A* **16**, 437.

Yang, J. T. (1958). *J. Am. Chem. Soc.* **80**, 1783.

Yang, J. T. (1959). *J. Am. Chem. Soc.* **81**, 3902.

8

Liquid Crystalline Order in Biological Materials

Y. Bouligand

E.P.H.E., Histophysique et Cytophysique
C.N.R.S., Centre de Cytologie Expérimentale
Ivry-sur-Seine, France

I. Introduction

Many biological molecules are elongated. The examples considered in this chapter will be mainly biological polymers and specifically, certain polysaccharides, many proteins, and the nucleic acids. Smaller molecules
can also have elongated forms. For geometrical reasons, rod-shaped molecules can align themselves when they are sufficiently concentrated (see de
Gennes, 1974); the solution is then birefringent.

Many of these biological molecules do not give true solutions but rather
form colloids. The solute molecules tend to form groups called micelles that
are sufficiently large to scatter light strongly. A sol is a liquid colloid in which
the micelles are independent. In a gel, on the contrary, the micelles are linked
here and there and the liquid character is abolished. The problem of colloidal
liquid crystals has been studied by Ostwald (1931). This question is particularly interesting in the biological domain and the earliest work is reviewed
in the classic book of Frey-Wyssling (1953).

Birefringent biological materials are often colloidal; some are true sols
and deserve the name of liquid crystals. Most of the birefringent biological
colloids however are gels and, in a strict sence, are not liquid crystals. Nevertheless, their structure is often in close analogy to various nematic and smectic
states.

Gels that are liquid crystal analogues can be more or less dense or con-
solidated. They are sometimes very hard and some of them are included in
skeletal tissues (Bouligand, 1972a). We shall call these systems that are
liquid crystal analogues but are not liquid, "pseudomorphoses" (Bouligand,
1969).

Each type of liquid crystal has its own geometric and optical properties.
On the molecular level, an order which can be defined by a group of symmetries is present (Friedel and Kléman, 1970). Most of the birefringent biological systems show structures whose symmetries are those of the various
well-known mesomorphic phases (Bouligand, 1978a). The different types of
mesomorphic orders are thus widespread in nature. We should not ignore
the fact that true crystals with three-dimensional arrays are also present
(see Oncley, 1959). The importance of mesomorphs and their colloidal
counterparts is shown by their presence in the membranes of cells and
cellular organelles, in the cell nucleus, and chromosomes of certain
microorganisms, in the myelin surrounding the axon of certain nerve cells
(particularly abundant in the white matter of the vertebrate brain), and in

muscles and skeletal tissues (see Nageotte, 1936; Needham, 1942; Frey-Wyssling, 1953; Stewart, 1966, 1969, 1974; Bouligand, 1978a).

The birefringent materials that are genuine liquids will be reviewed first. Biological membranes will be studied in Section III; they are fluid, but are extremely thin and do not allow a direct observation of their birefringent character. However, the birefringence can be measured in lamellar systems resulting from parallel accumulation of membranes. The following sections deal with the nonfluid analogues of liquid crystals.

II. Liquid Crystalline Biopolymers

A. Fatty Acids and Related Molecules

Let us first consider biological polymers that form true liquid mesophases. Fatty acids are small polymers by virtue of their paraffinic chains. They are components of some of the most important biological liquid crystals (see Chapman, 1965, 1966, 1970, 1974). In particular, cell membranes and their principal derivatives contain phospholipids forming a bilayer similar to a smectic structure reduced to only two layers. The amphiphilic character of these molecules and the relatively constant length of the paraffinic chains are responsible for the smectic order. The principal formulas of these compounds are well known and can be found in Luzzati (1968).

Many liquid crystalline inclusions have been observed in endocrine tissues such as the *adrenal cortex* and the *corpus luteum* (Stewart, 1966). These secretions are in the form of nematic droplets with or without cholesteric twist and they contain either free cholesterol or its esters. We also find considerable amounts of triglycerides and phospholipids. The *adrenal cortex* produces steroid hormones involved in many aspects of metabolism. The *corpus luteum* of the human ovary produces progesterone, another steroid hormone involved in the implantation of the placenta and in the retention of the embryo by the uterus.

Mesomorphic droplets with similar compositions have been seen in the walls of arteriosclerotic blood vessels (Stewart, 1966).

Lison (1936) was one of the first to note the existence of mesomorphic inclusions in tissues, but he remarked with reason that it should be ascertained if these droplets are still true mesophases at the body temperature of the organism.

B. Proteins

Various synthetic polypeptides are known to form nematic and cholesteric liquid crystals in solution. Research in this area is reviewed in Chapter 6. These polypeptides are not at all soluble in water and are not natural substances. However, some protein secretions (Kenchington and Flower, 1969; Neville and Luke, 1971) resemble certain cholesteric droplets formed by synthetic polypeptides (Robinson *et al.*, 1958). Among these are the secretion of the oothecal glands of the praying mantis (Kenchington and Flower, 1969). It seems also that certain silks are cholesteric when they are in the fluid state, as can be seen from a micrograph (Rudall, 1965, Fig. 6).

C. Nucleic Acids

Nucleic acids are mesomorphic in concentrated aqueous solutions. Robinson (1961) was the first to describe desoxyribonucleic acid (DNA) in its cholesteric form. The X-ray fiber diagrams that established the double helix configuration (Watson and Crick, 1953) were done on wet DNA drawn out into a thread, i.e., a nematic gel. The mesomorphic nature of DNA under similar conditions has also been studied by Luzzati (1963).

Lerman studied cholesteric phases that separated from a mixture with polyoxyethylene. The orientations of the polymer were studied with the polarizing microscope, analyzing either the birefringence or the polarized fluorescence obtained with acridine orange or quinacrine mustard (Lerman, 1973). The molecules of these dyes intercalate themselves between pairs of successive bases in the double helix, and thus serve as good indicators of polymeric orientations.

Magnificent cholesteric spherulites were formed from concentrated solutions of ribonucleic acid (RNA) forming double helices. This was at first thought to be transfer RNA (Spencer *et al.*, 1962), but turned out to be fragmented RNA (Spencer, 1963) probably of ribosomal origin (Spencer and Poole, 1965).

It is probable that the helical structures of polypeptides, proteins, and nucleic acids are responsible for the cholesteric twist of the mesophases.

D. Supramolecular Rods or Filaments

Hemoglobin S of red blood cells in the disease known as sickle cell anemia has been described as liquid crystalline (Perutz *et al.*, 1951). This molecule differs from normal hemoglobin A by a single amino acid substitution in the sequence of the β chain of the proteins. Upon deoxygenation, this small difference is sufficient to permit the self-assembly into microtubules of the hemoglobin, which is normally a globular molecule. The mesogenic behavior of these microtubules leads to a cholesteric geometry (Bouligand, 1972a). Electron microscopic sections show the presence of rows of parallel arcs (Döbler and Bertles, 1968), which, as we shall see, correspond to a cholesteric arrangement.

Birefringent fluid phases have long been known to occur in viruses forming large inclusions in host cells and, in particular, in the rod-shaped viruses such as tobacco mosaic virus. This virus is cylindrical in form, with definite length and diameter. Bernal and Fankuchen (1937) indicated either smectic or nematic characters by means of polarizing light microscope observations. More recent electron microscopic studies (Warmke and Christie, 1967; Milicic *et al.*, 1968) confirm the existence of a smectic arrangement. Gourret (1975) has taken a micrograph of a nematic arrangement for the same virus (Fig. 29). Recent electron microscopic preparations with freeze-etching show layers as in a smectic and figures of hexagonal packing in each layer (Willison, 1976). The virus particles of adjacent layers are tilted so as to produce a zigzag or a herringbone appearance. It is clear that these figures are very close to those given by a true three-dimensional crystal. The mutual arrangement of these cylindrical viruses presents a remarkable crystalline or mesomorphic polymorphism.

We should also note that some inclusions linked to the presence of viruses may have a cholesteric structure (Shikata and Galvez, 1969; Hull, 1970; Wilson and Tollin, 1970; Rassel, 1972; Horne, 1971).

True mesomorphic structures occurring in living organisms are formed either of relatively small molecules or of edifices made by a great number of macromolecules, as is the case for the rod-shaped viruses. Biological mesophases are often made of polymers that are functionally among the most important (nucleic acids, proteins). Some of these phases are likely to form gels to a certain extent.

III. Membranes and Lamellar Systems

Let us now consider biological systems that are very similar to liquid crystals but which differ in certain respects. We shall begin with biological membranes where the smectic stacking is limited to only two layers.

Most of the research on biological liquid crystals concerns cell membranes and water–lipid systems (natural or artificial). Water–lipid systems are interesting models in the study of several cellular mechanisms (ion permeability, water flow, electric resistance and capacity, etc.). Research is also stimulated by the importance of the amphiphilic lipids in several industrial fields such as soaps, cosmetics, pharmacy, and food manufacture. Books and review articles are numerous and contain detailed informations on chemistry, phase diagrams, differential thermal analysis, infrared spectroscopy, magnetic resonance, and X-ray diffraction patterns analysis. We find it necessary to discuss here the concept of liquid crystallinity applied to the cell membranes and its physiology.

The cell membrane is a very thin film enclosing each cell and has remarkable properties. It allows deformations of the cytoplasm to take place; it does not tear. Even more, it is capable of pinching off certain fragments of itself in the shape of closed surfaces to form internal vesicles that can play a digestive role, or it makes external vesicles used in secretion processes.

A. Principal Materials Studied

The first studies on membrane biochemistry were undertaken using red blood cells. Membranes are easily separated and have been shown to contain phospholipids (phosphatidyl-ethanolamine, -choline, -serine, and -inositol; mono- and diphosphatidylglycerol; cholesterol; a great variety of proteins; and small amounts of polysaccharides). Gorter and Grendel (1925) studied the liquid films formed on water, by the phospholipidic molecules extracted from red blood cells. They found that each molecule occupies an area that is identical in the red cell membrane and in the reformed film. This area is very small and this shows that the molecules are normal to the membrane. From this work and other observations, such as measurements of the thickness of superimposed bilayers by interferential methods (see Langmuir, 1917; Perrin, 1918; later, see Waugh and Schmitt, 1940) and X-ray diffraction pattern analysis, Danielli and Dawson (1935) derived a model of the cell membrane, which consists of a phospholipidic bilayer sandwiched between two protein layers.

Another approach to the properties and the composition of the cell membrane was made possible when it was shown by means of electron microscopy that myelin of nerve fibers was formed by an extensive accumulation of superimposed membranes (Geren, 1954). The fluid character of the myelin was known by many ancient authors (see Frey-Wyssling, 1953) and this characteristic probably applies to the cell membrane itself.

B. Planar Fluidity

We have pointed out in recent papers (Bouligand, 1978a) that the cell membrane deformations are explained by the fluidity in its own plane and not by the concept of elasticity. A rubber sheet is elastic and not liquid. Two neighboring molecules of polymer A and B remain very close even in the case of a strong deformation. The thickness of the rubber sheet depends on the tension forces and a hole can exist in a rubber sheet and will deform according to the distribution of tensions. On the contrary, a constant diffusion occurs in a liquid film. The density of molecules remains constant, but the neighborhood of a specific molecule is rapidly changed. The existence of diffusion in water–lipid systems and biological membranes is also shown by techniques of magnetic resonance (see Chapman, 1970, 1974; see also Johansson and Lindman, 1974). The thickness of the plasma membrane is constant when observed in thin section by electron microscopy and does not depend on its various shapes (see Robertson, 1969). The fluid character of the membrane of the red blood cells is necessary to interpret the passage from a biconcave disk to an echinoid form. Such morphological changes are inconceivable by pure elasticity, the thickness of the membrane being constant during these transformations. Another reversible transition between normal biconvave shape and spherical shape in the red blood cells has been studied by Helfrich and Deuling (1975). These authors assume that the membrane behaves as a two-dimensional liquid; the only elastic constant is that of splay. They have shown the existence of a spontaneous curvature opposite to the curvature of the spheric shape. The proteins and, possibly, the phospholipids are different on the two opposite sides of the red cell membrane. A liquid film of soap and water cannot subsist if a hole with sufficient dimensions appears. Such holes do not exist in plasma membranes due to their liquid character. A free edge is thus unstable and the plasma membrane forms only closed surfaces.

The liquid character of biological membranes has also to be considered to explain the behavior of cells when they are studied with the help of a micromanipulator (Chambers, 1930; Chambers and Chambers, 1961). Fluid streams and diffusion in the plane of the plasma membrane are directly visible in the microscope. The fluidity also appears necessary in the formation of free vesicles by pinching off fragments of the cell membrane. Such processes occur when external materials are taken up within invaginations of cell membrane, which separate to give a vacuole (endocytosis or pinocytosis, Stockem and Wohlfarth-Bottermann, 1969). Bennett (1956) has confirmed the existence of fluid streams in the interpretation of his electron micrographs of pinocytosis.

Cellular materials can be released from cells within membrane-limited vesicles (exocytosis). The liquid character of the cell membrane is also involved in the fusion of vesicles with the plasma membrane as it occurs in numerous secretory processes. It has been impossible to observe the successive steps of vesicle separation or fusion, which need a brief rearrangement of the phospholipidic bilayer.

The cell contains a nucleus bounded by a double membrane that has a chemical composition and physical properties very similar to those of the external membrane of the cell. In the cytoplasm surrounding the nucleus, there are also organelles made of membranes showing the fluidity and a chemical structure very close to those of the limiting cell membrane. These organelles are principally mitochondria (involved in the energy production of the cell), ergastoplasm (protein synthesis), and Golgi apparatus (packaging and export of synthetized materials). All these organelles can be seen with the light microscope and show in the living cell very strong deformations that undoubtedly require the fluidity of their membranes.

C. Model of the Membrane Structure

A synthetic model of the structure of cell membranes is presented in Fig. 1. Phospholipids form a bilayer with external hydrophilic extremities. The paraffinic chains have a highly disordered conformation (α type according to a notation introduced by Luzzati, 1968). Molecules of cholesterol are present in general and prevent the transition of the paraffinic chains to a stiff configuration with parallel alignment, hexagonal packing, and rotational disorder (β type in the Luzzati notation). We must note that cholesterol is absent in the membranes of bacteria and mycoplasms and in the inner membrane of the mitochondria. In the case of mycoplasms, the α–β transition has been observed. Most proteins float on the polar ends of the phospholipids. Certain proteins are more or less deeply embedded in the paraffinic layer and can even span the entire thickness of the bilayer (Singer and Nicholson, 1972). The position of these proteins with respect to the bilayer is due mainly

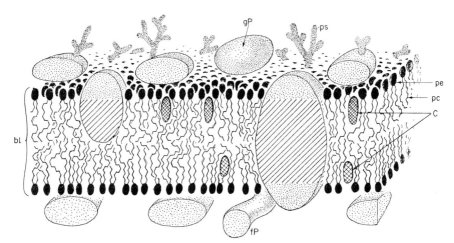

Fig. 1. Model of the cell membrane. Proteins are more or less embedded in the phospholipidic bilayer. The stippled zones represent the polar residues of proteins and the hatched ones the nonpolar amino acids. This model draws its inspiration from Singer and Nicholson (1972), but differs by the absence of hexagonal packing of phospholipids (type α) and by the presence of surface proteins as in models of Danielli and Dawson (1935) and Robertson (1959). bl, Phospholipidic bilayer; C, cholesterol; fP, filament protein or microtubule; gP, globular protein; pc, paraffinic chain; pe, polar extremity of phospholipids; and ps, polysaccharides.

to the distribution of the polar and nonpolar residues. Protein filaments and microtubules are frequent in the vicinity of the cytoplasmic side of the membrane. Branched polysaccharides are more or less linked to either proteins or phospholipids on the extracellular side and form a cell coat or glycocalyx. Proteins and polysaccharides can change the fluidity or destroy it in certain regions of the cell membrane.

Biological membranes can form lamellar systems that are either tightly packed (myelin surrounding axons, myelinic forms corresponding to concentric accumulations of closed membranes, rods and cones in visual cells, etc.) or are loosely spaced with more or less regularity (ergastoplasm, Golgi apparatus, chloroplasts, etc.). The morphology of these organelles is well illustrated in Porter and Bonneville (1964). The tightly packed lamellar systems are birefringent. This does not clearly appear in the case of less ordered systems with a loose packing. Lamellar arrangements are also obtained with water–lipid models. The birefringence depends on the amount of added water and is first positive, indicating the average orientation of phospholipids lying normally to the lamellae. The birefringence progressively decreases when more water is added and becomes negative. The water separates the lamellae and introduces a negative form birefringence. The mean distance between successive bilayers can be estimated from X-ray diffraction patterns.

The study of model systems has a strong bearing on the question of morphogenesis of cell organelles and the cell itself. Biological membranes form tubes or vesicles that can be more or less hexagonally packed. It is worth remembering that certain water–lipid models also give hexagonal phases made of parallel tubes. The transitions from lamellae to rods or spheres observed in models cannot be directly compared to the various morphologies of membranes occurring in cells during their differentiation, but nevertheless suggest an interesting approach, that must not be ignored.

Membranes and lamellar systems are smectic analogues whose birefringence is often difficult to study but show numerous signs of fluidity. We shall now consider the biological materials that are birefringent analogues of liquid crystals but are not fluid. Examples are found in muscles, skeletal tissues, and certain cell nuclei.

IV. Liquid Crystalline Superphases in Muscles

A. *Fine Structure of the Contractile Apparatus*

Muscles have been long considered as possible biological analogues of liquid crystals. The older references will be found in Lehmann (1911) and Needham (1942). A more recent work by Elliot and Rome (1969) deals with

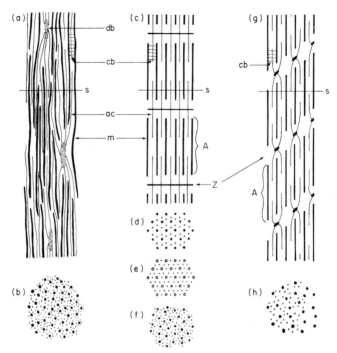

Fig. 2.   Fine structure of the contractile apparatus in muscle cells. Cross bridges are only represented in a small region of the drawing. (a) Smooth muscle in longitudinal view; (b) smooth muscle in cross section; (c) cross-striated muscle in longitudinal view; (d)–(f) cross-striated muscles in cross section [(d) vertebrates, (e) arthropods, (f) muscle cells surrounding the gut of arthropods]; (g) oblique-striated muscle in longitudinal view; (h) oblique-striated muscle in cross section. A, set of thick filaments forming A bands; ac, actin filaments; cb, cross bridges; db, dense bodies; m, myosin filaments; s, section plane corresponding to Figs. (b), (e), (f), and (h); Z, dense structure called Z line or Z element.

some of the liquid crystalline aspects of cross-striated muscles and their relevance to the contractile process.

Muscle cells contain a system of protein filaments of two sorts: thick filaments made of myosin and thin filaments made of actin (Fig. 2). These filaments lie parallel to the long axis of the cell and form separate systems that interdigitate more or less deeply according to the contraction (see Huxley, 1960, 1963). Small bridges interconnect the two sets of filaments. A very common situation is that found in the cross-striated muscles. The thin actin filaments are attached to Z bands, which are also made of proteins. The myosin filaments form wide regular A bands, which are spaced at equal intervals between the Z bands. The sets of myosin filaments forming the A bands are interpenetrated by the actin thin filaments (Fig. 2c). Contraction is the result of the relative sliding of the two sets of filaments (Figs. 3–6).

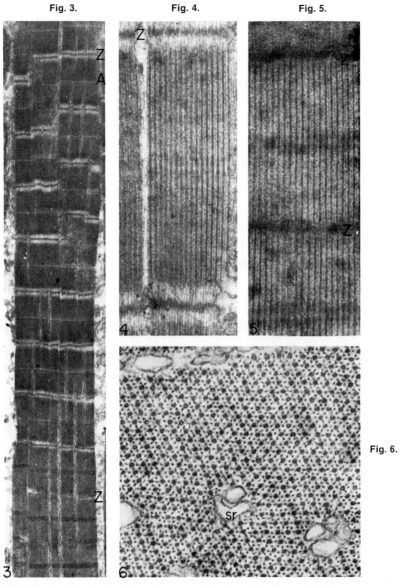

Fig. 3. Longitudinal section of a cross-striated muscle of a small fresh-water crustacea (*Acanthocyclops viridis Copepoda*). The myofibrils are more contracted in the lower part of the micrograph. A contraction wave is visualized in this picture (× 7230). (After Bouligand, 1964.)

Fig. 4. Longitudinal section of a cross-striated muscle showing the two sets of thick and thin filaments. These latter are attached to the Z bands and are not fully interdigitated between the myosin filaments (*Cyclops*, × 35,700). (After Bouligand, 1964.)

Fig. 5. Longitudinal view of a highly contracted myofibril (*Cyclops*, × 42,500). (After Bouligand, 1964.)

Fig. 6. Cross section of a cross-striated muscle with a regular hexagonal array of thick and thin filaments. We can also observe cross sections of large tubes (sarcoplasmic reticulum, sr) involved in the initiation of the contractile mechanism, namely, the calcium release (*Cyclops*, × 63,750). (After Bouligand, 1964.)

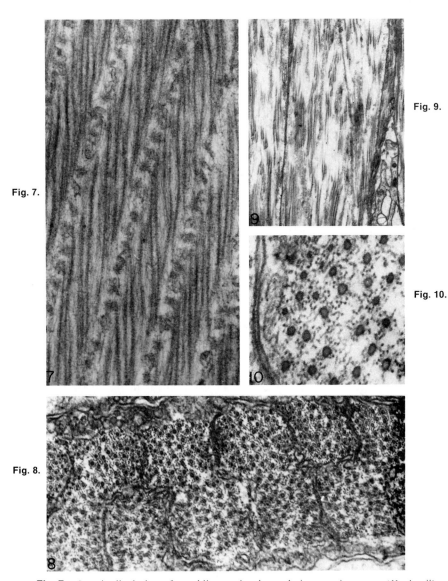

Fig. 7.   Longitudinal view of an oblique-striated muscle in a marine worm (*Haplosyllis depressa*, × 21,250). (After Bouligand, 1966.)

   Fig. 8.   Cross section of an oblique-striated muscle showing intercalated thick and thin filaments (*H. depressa*, × 34,000).

   Fig. 9.   Longitudinal section of a smooth muscle (*H. depressa*, × 8,500). (After Bouligand, 1966.)

   Fig. 10.   Cross section of a smooth muscle (*H. depressa*, × 93,500).

This motion involves the hydrolysis of an energy-rich molecule, the adenosine triphosphate (ATP), occurring at the level of the cross bridges (Fig. 2c). The chemical energy due to the oxidation of certain sugars serves to regenerate this ATP. The cross bridges are alternately broken and reformed during this process, whose mechanisms are not fully understood.

In cross section, myosin filaments often form hexagonal arrays (Fig. 2d,e). In certain animal groups namely, the arthropods, thick filaments appear to be hollow or simply clearer along their axis in the electron beam (Figs. 2d,e, 6). These patterns represented in Fig. 2c are repeated all along the cell (Fig. 3).

The myosin filaments can be selectively dissolved by certain ionic solutions (KCl, $0.6\,M$; $Na_4P_2O_7$, $0.01\,M$; $MgCl_2$, $0.001\,M$, pH, 6.5). After this extraction, the actin filaments can be dissolved in another solution (KI, $0.6\,M$). This is a process for finding out the chemical nature of the two kinds of filaments. Interestingly, these dissolutions do not result in the disruption of the myofibrils and the Z bands subsist with their regular spacing. This means that an inconspicuous system (probably proteins) of links L superimposes onto the two sets of filaments and does not prevent their mutual sliding.

In worms and mollusks, muscles are often obliquely striated as shown in Fig. 2g. In cross section, filaments have a random distribution (Fig. 2h), but in certain cases, form more or less developed hexagonal packings. Certain oblique-striated muscles are helical. The A and Z bands form a set of helices that belongs to a cylindrical myofibril. Longitudinal and cross sections of oblique-striated muscles are shown in Figs. 7 and 8. More detailed descriptions of oblique muscles will be found in Ikemoto (1963), Bouligand (1966), and Wissocq (1970).

In smooth muscles, there are thick and thin filaments lying parallel to the exerted force, but striated or banded patterns are absent. Dense bodies that are attachment structures for thin filaments (Figs. 2a,b, 9, and 10) can be observed.

B. Comparisons with Liquid Crystals

From these descriptions (Fig. 2), it appears that between smooth and striated muscles there is a structural difference resembling that found between nematics and smectics. Moreover, the difference between cross- and oblique-striated muscles corresponds closely to that found between smectics A and C. The helicity observed in certain obliquely striated muscles has its counterpart with chiral smectics C. This remark leads to the question of whether such muscles could be ferroelectric as chiral smectics C (Meyer et al., 1975).

The hexagonal arrays present in cross section in certain striated muscles are to be compared to those observed in certain variants B of smectics A and C. These structural comparisons have been discussed elsewhere with reference to the problem of muscle morphogenesis (Bouligand, 1978a).

Muscles are birefringent gels whose structures show symmetries that are those of the different groups of liquid crystals (with the exception of cholesterics). Muscles are not liquid; they are made of parallel filaments interconnected by links that are either permanent (Z bands, dense bodies, L system) or changing (cross bridges). The alternate formation and destruction of the cross bridges resemble the reversible sol–gel transition in a colloid. Muscles are accordingly permanent gels with a more or less pronounced rigidity, according to the number of cross bridges present.

The contractile apparatus of muscle cells is not a one-phase system. We have to distinguish regions that contain either myosin or actin and those where myosin and actin filaments interdigitate. Moreover, water forms a continuous phase penetrating the whole system. The nature of the water structure around proteins has been discussed by Bernal (1965) and it seems that the first 10 Å correspond to an icelike state.

Such intimate associations of several phases are called "superphases" (Zocher and Török, 1967). The striated muscle fiber is considered by Elliot and Rome (1969) as a system of two interleaving smectic superphases. These authors suggest that the interfilament distances correspond to a given balance of van der Waals attraction, electric double layer repulsion, and hydration, which can be calculated for a system of parallel cylinders. The interfilament separation depends on pH, ionic strength, and valence of the positive ions in the medium; the distances between filaments also depend on the degree of contraction. The latter might involve a perturbation of the considerable energies that equilibrate at rest.

Similar influences of pH and ionic strength have also been observed for the interdistances in the birefringent gels given by tobacco mosaic viruses (Bernal and Fankuchen, 1941).

V. Cholesteric and Nematic Analogues in Skeletal Tissues

A. *Arcs and Spirals*

1. *Arced Patterns*

Several biological materials observed in thin section present a system of stacked rows formed by parallel arcs (Figs. 11, 12). These series of bow-shaped lines are clearly visible in thin sections of the crustacean integument. We can use for instance the body wall of a crab (*Carcinus maenas*). This material is

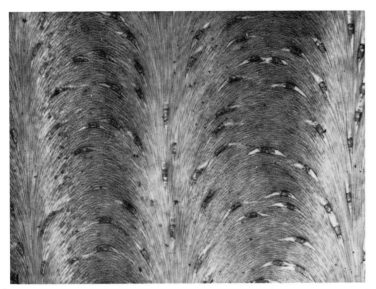

Fig. 11.   Oblique section in the organic matrix of crab cuticle. The fibrils are made of chitin and proteins. The crescent shapes correspond to the passage of cytoplasmic microvilli projecting vertically in the thickness of the cuticle (principal layer, *Carcinus maenas*, × 5950).

Fig. 12.   Oblique section in the organic matrix of the crab cuticle. In this material, the filaments form more or less regular aggregates (pigmented layer, *Carcinus maenas*, × 8950).

made of an organic matrix (mainly proteins and chitin, a linear polymer of acetylglucosamine) and a mineral (mainly calcite). The organic matrix can be studied either after decalcification (acid solution, EDTA, etc.) or before mineralization, just after one of the numerous successive molts occurring throughout the life of these animals. The bow-shaped patterns are often visible in the light microscope, but are better resolved by classical transmission electron microscopy (Bouligand, 1965).

There are many striking instances of analogous arced configurations in other biological materials both widely varied and far removed from the crustacean cuticle. Insect cuticles often present similar patterns. The organic collagenous matrix of bone tissue can show the arced figures here and there within a section. Several other cuticles, various connective tissues, and certain plant cell walls also show the arced figure (review in Bouligand, 1972a).

The problem is thus to understand this architecture, which must have a common physical origin in these widely differing materials. Numerous theoretical models had been proposed to account for the arced patterns. They all supposed that the filaments themselves followed the arcs and that, arranged in parallel order, they form distinct superimposed layers.

2. Spirals

Looking at a crab, we can hardly help noticing the many small bumps or protuberances, often called tubercles, which decorate the shell. If we look at a vertical section through a bump (Fig. 22), cut right along its vertical axis, we see that the disturbance is repeated in each subjacent layer, like a stack of Mexican hats. These deformations are reminiscent of geological structures known as anticlinal domes. Superimposed layers of sedimentary terrain form dome-shaped hills that have a common vertical axis. If the tops of these hills are eroded, one can see the oldest layers in the center and each subsequent or younger layer forming a concentric ring around the older ones, so that the most recent is on the very outside. The concentric layers actually form closed ribbons, which are in turn limited by closed curves.

Tangential sections of the carapace from regions that show bumps present very curious spirals instead of the expected closed concentric ribbons (Figs. 13–18). There is a continuous path between the oldest outermost layers and the younger innermost layers merely by following the spiral. The notion of distinct layers is compromised.

3. Orientation of Arcs

One striking aspect of the arcs is that in any two regions of the spiral on opposite sides of the center, the direction of the concavity of the arcs is always reversed. We then look for a rule for the inversion of arcs in the flat regions

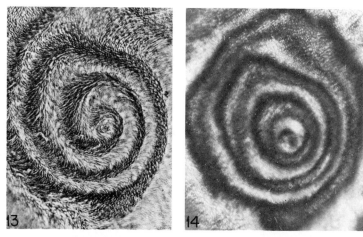

Fig. 13.   **Fig. 14.**

Fig. 13.   Spiralized figure observed in a section through a bump of the crab cuticle. The section is normal to the axis of the bump as indicated in Fig. 22. The arrangement of filaments is redrawn in Fig. 26. The series of arcs forms a double spiral. The sharply contrasted single spiral superimposes onto the double one and is due to an artifact of microtomy (*Carcinus maenas,* phase contrast microscopy, × 255). (After Bouligand, 1969, 1972a.)

Fig. 14.   The section of the bump observed in Fig. 13 is reexamined in polarizing microscopy between crossed polarizers (× 255).

of cuticle in which the layers exhibit no such protuberant deformations. The rationale for the procedure is as follows. If we look at Fig. 22, a diagram showing the plane of a section through a bump, we see that another section S normal to the axis of the deformation and parallel to the general plane of the cuticle meets many layers at an oblique angle. In the ideal case (good section orientation and revolution symmetry of the bump), the plane S cuts the layers symmetrically with respect to the axis. Clearly, the thing to do is to maneuver a similar set up in the flat regions of the cuticle and to look at oblique sections taken symmetrically about a vertical axis to determine the directions of the concavities of the arcs.

Thus a piece of decalcified cuticle is sectioned on two opposed faces as in Fig. 19, and sections are mounted between slide and coverslip so that their original orientation with respect to each other and with respect to the axis of symmetry are preserved (Fig. 20).

The invariable result is that all oblique sections, whatever their orientations, give figures of stacked rows of parallel arcs and, for sections symmetric with respect to a vertical axis V, the senses of concavities of the arcs are reversed.

Fig. 15.   **Fig. 16.**

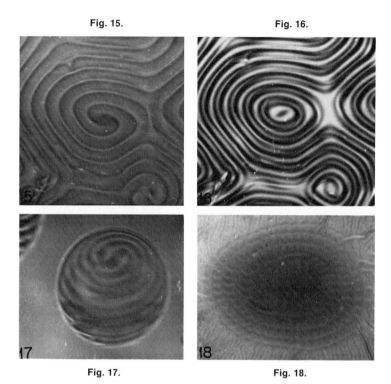

Fig. 17.   **Fig. 18.**

Fig. 15. Double spiral formed between slide and coverslip by methoxybenzylidenebutyl-aniline (MBBA) made cholesteric by addition of a small amount of cholesterol benzoate (CB). Phase contrast microscopy ($\times 280$). (After Bouligand, 1972a.)

Fig. 16. The identical region of the preparation seen in Fig. 15 but photographed between crossed polaroids. Concentric rings of extinction have replaced the double spiral ($\times 280$). (After Bouligand, 1972a.)

Fig. 17. A cholesteric droplet suspended in the isotropic phase (mixture of MBBA, CB, and toluene). A double spiralized pattern is at the opposite poles of the sphere (analyzer only, $\times 680$). (After Bouligand, 1972a.)

Fig. 18. Double spiral of commas observed in a saturated solution of BC in MBBA. These double spirals generally appear at the surface of crystals (crossed polaroids, $\times 170$).

4. Origin of the Arcs

At this point, there are enough data to allow us to figure out the structure of the arc. In flat regions of the cuticle, we define the plane of the cuticle as being horizontal. It follows that vertical means normal to the cuticle. Let us summarize the results so far:

(a) All oblique sections exhibit rows of parallel arcs that are superimposable on the original after horizontal translations in the plane of section.

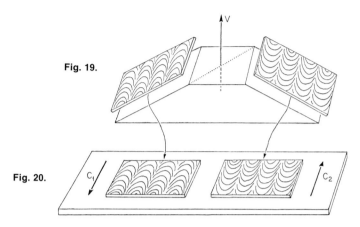

Fig. 19.

Fig. 20.

Fig. 19.   Sections sliced obliquely in a block of cuticle and symmetrical with respect to a vertical axis V.
Fig. 20.   The sections are deposited on the glass slide with their original orientation. The arrows C_1 and C_2 indicate the opposite concavities observed in the two sections.

(b)   The orientation of concavity of the arcs is the same in all rows of arcs in a given section.

(c)   In two oblique planes, symmetric about a vertical axis, the patterns obtained in section are superimposable by a rotation by 180° about the vertical axis.

From these facts come the following consequences:

(α)   The observations given in (a) allow us to conclude that in any horizontal plane the direction of the fibrils is constant. The fibrils at any two points of a single horizontal plane are parallel to each other. At this stage, we do not know if the fibril direction is horizontal or oblique, but we do know that at every point in a given horizontal plane the direction is defined and constant, since along any horizontal level of an oblique section, the projections of the fibrils onto this plane have a constant direction, and this is true for any oblique section.

(β)   Combining the conclusions of (α) and the result (c), we notice that the system is made up of fibrils of constant direction in any one horizontal plane and their set shows an axial symmetry about any given vertical axis. Now, any two straight lines that are parallel to each other and are symmetric with respect to a vertical axis must themselves be either vertical or horizontal. In the system we are studying, the fibrils cannot be vertical because oblique sections would give uniform patterns in no way resembling arcs actually observed. Thus, it follows that *the cuticle is made up of horizontal fibrils.* These are parallel to each other in a given horizontal plane. We can easily verify that horizontal sections show extensive regions where the direction of fibrils is constant.

(γ) Oblique sections show stacked rows of parallel arcs with constant orientation (b) and this is true for oblique sections very close to the horizontal, say at an angle of 10°. Such sections allow us to determine the direction of fibrils in each horizontal plane with a slight correction necessary on account of the obliqueness. This correction can be omitted in a first approximation. Considering the continuously varying directions of fibrils in the arcs, it appears that *fibrils turn in the successive horizontal levels and in only one sense.*

(δ) It follows that our system is made of fibrils that are horizontal, parallel within any given horizontal plane, and with a direction that rotates continuously as one proceeds through successive levels of the cuticle. This is exactly the geometry of cholesteric liquid crystals.

The organic matrix of the crab carapace is thus comparable to a cholesteric liquid crystal that has been hardened into a nonliquid fibrous structure by polymerization or by cross-linking. The sense of this cholesteric architecture is *left handed.* If we examine a thick horizontal section of crab carapace in the light microscope, as we change the focus, we see that the direction of the filaments rotates in regular manner. If we begin at the bottom of the preparation and focus at progressively higher levels, the direction of fibrils turns in a clockwise manner. This rotation remains obviously clockwise if we turn the preparation upside down.

B. Models

The origin of the two opposite concavities of arcs in oblique sections symmetrical with respect to a vertical axis V is shown in the model of Fig. 21.

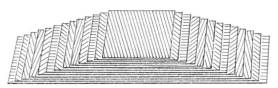

Fig. 21. Model formed by superimposed cardboards simulating oblique sections in a fibrous twisted arrangement.

The twisted fibrous materials are not arranged with the exactness of the geometrical model. In the case of the crab carapace, we have only considered the average directions of fibrils in small volumes (the biggest dimension being only one tenth of, say, the half-helical pitch). There is a certain degree of disorder defined by the deviation of the fibrils from the average direction in each volume sample. Consequently, the model does not show any fibrils interweaving between horizontal planes at different levels, but such connections could occur in the real fibrous system.

1. Continuous or Discontinous Twist

We have to note that the successive steps of our model facilitate the drawing, but the steps often do not exist. We have observed in crustacean cuticles two kinds of twisted arrangements:

(a) Systems built with superimposed layers of parallel filaments, with an almost constant angle from one layer to the following one [cuticle of small crustaceans, Copepods: *Acanthocyclops, Cletocamptus* (Bouligand, 1965; Gharagozlou-van Ginneken and Bouligand, 1973).]

(b) Systems with no recognizable distinct layers of parallel filaments. The direction of horizontal fibrils is gradually rotating as one rises in the thickness of the structure (crab cuticle).

2. Spiralized Models

When the layers are distorted by the formation of a bump in the crab cuticle (Fig. 22), a section S cut perpendicularly to the bump axis gives a spiral pattern as we have seen in Fig. 13. To represent the twisted model in that case, let us consider a series of surfaces in the form of domes revolving about a vertical axis. In a microtome section S of thickness e, our surfaces

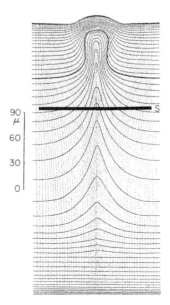

Fig. 22. Aspect of the layers in a crab cuticle in a meridian plane of a bump. The horizontal straight line corresponds to a section S normal to the axis of the bump, as that studied in Figs. 13, 14, and 26. (After Drach, 1939.)

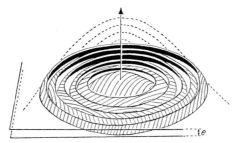

Fig. 23. Diagrammatic representation of the filaments in a horizontal section S of thickness *e* in a bump. Equidistant levels in the twisted arrangement form concentric ribbons in the thickness of the section. The orientation of filaments is indicated in the successive ribbons.

form concentric oblique ribbons (Fig. 23). One of the surfaces, the central one, is sectioned subtangentially and forms a cap shape instead of a ribbon. On these ribbons, we now draw equidistant lines (Fig. 23) that show the direction of the fibrils in each concentric ribbon. The fibril direction appears to be constant in any one ribbon in normal projection onto the horizontal plane. From one ribbon to the next, the direction of the lines rotates by a constant angle. Viewed from above, the fibrils in projection give Fig. 24. (The number of concentric ribbons is greater than in Fig. 23 in order to have a large view of the pattern.) In Fig. 25, the series of arcs is delineated by a black line and we see a double spiral.

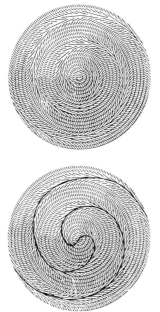

Fig. 24. Projection onto the cutting plane of the fibrous orientations in the series of concentric ribbons of Fig. 23 and resulting in double spiral of parallel arcs.

Fig. 25. The limits of the series of arcs have been delineated by a continuous line forming a double spiral.

The same double spiral actually exists in Fig. 13, which shows a section of a bump in a crab cuticle. Here, however, the double spiral is obscured by an artifactual single spiral, which arises due to the difficulty of sectioning. If we put a transparent tracing paper over Fig. 13 and trace just the directions of fibrils, we obtain the double spiral formed by the series of arcs. This is represented in Fig. 26 and the stippled zone corresponds to a microtomy artifact whose mechanism has been discussed by Bouligand (1972a). When a twisted fibrous material is sectioned, the microtome knife meets the filaments in alternate different directions. The block of cuticle and the knife have different elastic characters and are submitted to periodic tensions. Lines of compression and thickening can arise from such a situation. The complete analysis of these phenomena shows that the strongest effects occur along a single spiral.

Fig. 26. Picture drawn on a transparent tracing paper after the micrograph of Fig. 13. The stippled zone corresponds to the dark spiral due to compression and thickening arising from an artifact of microtomy. The directions of fibrils are indicated and form a double spiral of arcs.

3. Arcs and Spirals in Cholesteric Liquids

The rows of parallel arcs can be seen in certain cholesteric phases, for instance, in mixtures of methoxybenzilidenebutylaniline (MBBA) and cholesterol benzoate (CB). The MBBA is a classical nematic compound at ordinary temperatures and the addition of a small amount of CB induces a left-handed twist proportional to its concentration in the MBBA. This mixture mounted between slide and coverslip allows the observation of parallel stripes in the form of fingerprints. When the half-helical pitch varies between 10 and 15 μ and, if the thickness of the liquid crystalline slab between slide and coverslip is of the same order, we can observe with phase contrast the thermal fluctuations, and the constant movements appear to outline stacked rows of parallel arcs (Bouligand, 1972a, 1974). Series of arced decorations in cholesteric liquids have also been observed under different experimental conditions by Grandjean (1921), Friedel (1922), and Rault and Cladis (1971). These arced decorations correspond to surface distortion of the cholesteric arrangement and have been called "commas" or "virgules" (in French). We have often observed double spiralized series of commas (Fig. 18).

4. Polarizing Microscopy of Spirals

In projection, the direction of fibrils along a circular ribbon remains constant. Since the major refractive index is parallel to the fibrils, we observe alternating dark and bright concentric bands between crossed polarizers (Fig. 14). We pass from spirals clearly visible in crab cuticle with the phase contrast microscope to a set of concentric rings in polarized light. This change is also found in thin samples of cholesteric mesophases when the helical pitch is large. The previous explanations also apply to this material (Figs. 15 and 16). Cholesteric droplets floating in the isotropic phase can also exhibit these double spirals (Fig. 17).

5. Equations of Arcs

To obtain the equations of the arced curves, let us choose a system $0xyz$ (Fig. 27) such that fibrils are parallel at each point $M(x, y, z)$ to a unit vector n whose components are

$$n_x = \cos tz, \qquad n_y = \sin tz, \qquad n_z = 0$$

where t is the helical twist and is related to the helical pitch p by the identity $pt \equiv 2\pi$. We can suppose that our oblique section plane P corresponds to $z = y \tan \alpha$. The vector $n = MF$ projects normally onto P along MT which is tangent to the arced curves. Consider the s axis in P which is normal to $0x$ and corresponds to the steepest slope of this plane. The MT components

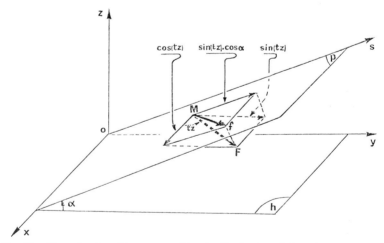

Fig. 27. The twisted model represented by the distribution of the unit vector $n = MF$ is studied in the plane P ($0xs$). The plane h ($0xy$) is supposed to be horizontal, i.e., normal to the cholesteric axis ($0z$).

along x and s are cos tz and sin tz cos α. Accordingly, $dx/ds = \cos tz/(\sin tz \cos \alpha)$ and as $z = s \sin \alpha$, the equation of the arced curves is

$$x = x_0 + \ln|\sin(t(\sin \alpha)s)|/a \cos \alpha \sin \alpha$$

Changing the units, the equation becomes $X = X_0 + \ln|\sin S|$. X_0 is an arbitrary constant indicating that X is a translation axis. If we consider the conformal mapping $Z = e^z$ in the complex plane, we can easily verify that parallel straight lines of plane Z correspond to our arced curves in the plane z.

It is clear that a vertical section of the organic matrix of crab cuticle gives no arced patterns. Filaments cut at right angle alternate periodically with filaments oblique and parallel to the cutting plane. These filaments appear to be horizontal in the section plane.

C. Optical Properties

The cholesteric liquid crystals are known to exhibit remarkable optical properties such as extremely strong rotatory power and selective reflection of circularly polarized light in a narrow band of wavelengths. These phenomena were already known for certain insect cuticles by Michelson (1911) and the relation with cholesteric phases has been emphasized by Gaubert (1924), Mathieu and Farragi (1937), and, more recently, by Robinson (1966). We have checked the existence of a twisted fibrous arrangement in the cuticle of the beetle *Cetonia aurata* at the levels that reflect circularly polarized light and show a strong rotatory power (Bouligand, 1969). On the contrary, in *Chrysocarabus splendens*, another beetle reflecting the same physical coloring, without circularly polarized light, the outer part of the cuticle is made of alternating dark and light layers with no sign of any twisted architecture (Bouligand, 1969). Another kind of periodicity has been described in certain *Cetoniinae* (g. *Heterorrhina*), which reflects iridescent colors but not circularly polarized light (Neville and Caveney, 1969).

A superb collection beetle, whose name is *Plusiotis resplendens*, reflects the right and the left circularly polarized components. This has been discovered and analyzed by Caveney (1971). The cuticle is formed by two sheets H_1 and H_2 analogous to left-handed cholesterics, separated by a layer with unidirectional orientation of fibrils, acting as a $\lambda/2$ wave plate. The incident light I can be considered as the sum of two opposite circular components. The left circular component (the extremity of the electric vector describes a right-handed helix) is reflected by H_1. The right component penetrates H_1 and the $\lambda/2$ wave plate transforms it into a left circular wave, which is in turn reflected by H_2. This left wave is retransformed into a right one by $\lambda/2$, penetrates H_1, and emerges outside. This mechanism is explained in Fig. 28. It

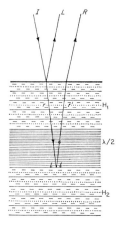

Fig. 28. Optical properties of the outer cuticle of the beetle *Plusiotis resplendens*. A half-wave plate $\lambda/2$ separates two cholesteric pseudomorphoses H_1 and H_2. The left circular component L of an incident beam I is reflected. The right circular component r penetrates H_1 and is transformed into a left circular wave by the $\lambda/2$ plate. This left wave is reflected according to l' by H_2 and retransforms into a right wave r' which emerges along the ray R.

would be interesting to know how in the course of the evolution of these insects such an amazing optical device has been selected.

D. Nematic Pseudomorphoses

The existence of a unidirectional layer forming a $\lambda/2$ wave plate in an insect cuticle is an example of a nematic analogue. The helical pitch of the arthropod cuticle can vary in a continuous manner, according to the level in the thickness of the carapace and, also according to the different regions of the body. It can happen that the twist disappears and we have a fibrous analogue of a nematic phase. This case is not rare in insects. Neville (1975) has shown that in certain locusts light and temperature can induce a preferred orientation of fibrils in the course of the cuticle deposition. The day layers are solid nematics; on the contrary, a regularly twisted cuticle forms in cold night conditions. It follows that in normal conditions, the cuticle consists of superimposed cholesteric and nematic layers. Unidirectional arrangements of fibrils also exist in the inner folds of cuticles that serve for muscle attachment. There are numerous examples of fibrous nematic analogues in skeletal and tendinous tissues in both vertebrates and invertebrates.

VI. Nematic and Cholesteric Analogues in the Cell Nucleus

A. Nuclei of Spermatozoa

Heads of spermatozoa are often birefringent and this corresponds to a good alignment of the DNA filaments. Threads drawn out from sperm of certain mollusks are often sufficient to obtain an X-ray diffraction pattern of DNA in its B form (Wilkins *et al.*, 1950). The spermatozoa have elongated heads and this makes the general alignment easier. In certain species, the nucleus of spermatozoa often form simple helices. In certain cases a marked twist can be observed in the distribution of filaments (Philips, 1976).

B. *Chromosomes of Bacteria and Dinoflagellates*

Cholesteric arrangements are found in chromosomes of certain micro-organisms, mainly dinoflagellates, a large group of protozoa, and in the nucleus of numerous bacteria.

Dinoflagellates are unicellular and their chromosomes are visible through-out the complete biological cycle. They show a fibrous structure that appears to be intermediate between that of the bacterial nucleus and the chromosomes of eukaryotes, the higher organisms (Giesbrecht, 1961). The dinoflagellate chromosomes are made of DNA filaments that can be clearly resolved in thin section. Histones or basic proteins commonly associated to DNA in the cell nucleus and in chromosomes of higher organisms seem to be absent in this material (Zubay and Watson *et al.*, 1959; Dodge, 1964; Ris, 1962). Ultrathin sections of these chromosomes reveal the presence of stacked rows of parallel arcs that are due to a cholesteric arrangement (Bouligand, 1965; Bouligand *et al.*, 1968). In longitudinal section, bands of filaments cut at a right angle alternate periodically with filaments in the cutting plane. Series of parallel arcs are obtained in oblique sections. In sections that are transversal or almost transversal, the filaments have a constant direction or draw very large arcs.

All these patterns correspond to a twisted arrangement of the DNA filaments. The helical axis (or cholesteric axis) is longitudinal and the half-helical pitch lies between 1000 and 1500 Å. The chromosomes are generally elongated, but there are exceptions. In certain cases, chromosomes are almost spherical and, in this case, the diameter normal to the cholesteric stratification will be taken to represent the longitudinal axis.

C. *Comparisons with Liquid Crystals*

These chromosomes show an interesting resemblance to the cholesteric droplets (or rodlets) given by certain mesogenic compounds when the iso-tropic phase is slowly cooled. These droplets tend to be elongated in the sense of the cholesteric axis and have exactly the same geometry as the chromosomes considered above. An important difference, however, is that for dinoflagellate chromosomes the DNA filament is a unique linear polymer, which is very long (several millimeters) and is folded back on itself a great number of times to fit in the very small volume of the chromosome. Each chromosome is made up of a unique DNA molecule and this fact corresponds to the arrangements of genes in chromosomes where they form linear groups of linkages. The classical cholesteric droplets are made of separate small molecules.

The chromosomes of dinoflagellates are suspended in the surrounding medium (nuclear sap). This situation resembles a system in equilibrium at a phase transition between the isotropic and cholesteric states.

In other types of cell nuclei and chromosomes, the fibrous network formed by the DNA is obscured by the presence of basic proteins. However, even in these cases, the condensation of chromosomes at the beginning of cell division or their dilution in the daughter cells also resembles a phase transition (Bouligand, 1972b, 1978a).

Bacterial chromosomes often show stacked series of parallel arcs whose origins are similar. DNA filaments in dinoflagellate chromosomes and in the nucleus of bacteria form either very homogenous phase, or irregular aggregates (de Haller *et al.*, 1964; Bouligand *et al.*, 1968). This probably corresponds to a sol–gel transition. We have to note that chromosomes of higher organisms also show these modifications. When studied with the help of a micromanipulator, chromosomes may appear fluid but under other circumstances they are not (Chambers and Chambers, 1961). It is worth remembering that pure DNA has been observed in a cholesteric state by several authors (Robinson, 1961; Bouligand, 1969; Lerman, 1973).

VII. Mesomorphic Self-Assembly in Living Systems

Examples of self-assembly are numerous in biology. They concern mainly the reconstitution of filaments or microtubules from their subunits, which are generally protein molecules. For instance, the protein subunits of the tobacco mosaic virus are easily dispersed in vitro or reassembled (Fraenkel-Conrat, 1956). The actin and myosin filaments of muscles can be separately dissolved and reformed (Huxley, 1963; Maruyama, 1965). In such self-assembled systems, each subunit has a given position, defined by its co-ordinates in the filament or in the lattice.

The close analogy observed between the structures of numerous biological materials and those of liquid crystals suggests that several tissues and organelles are self-assembled according to a mechanism that is very similar to the process allowing molecules to form liquid crystals. This self-assembly mechanism corresponds mainly to the spontaneous formation of a directional order. No links are necessary between subunits. These links can form but they are not responsible for the alignment. Secretions that are true mesomorphs, as mentioned in Section II, obviously follow the principle of meso-morphic self-assembly. Secretions of fibrous structures such as the organic matrix of skeletal tissues do not appear to pass through a genuine meso-morphic phase. They may however pass through a very brief one. In the case of the crab cuticle, the newly secreted layers immediately form a gel (Bouligand, 1975). The presence of links is sufficient to abolish the liquid character, but the system is very loose and this probably does not prevent the alignment from appearing and the twisting forces from being exerted. In general, the morphogenesis of mesomorphically self-assembled tissues and organelles is not always known in detail and, in particular, we do not know the nature of the first structures if they are sols or gels. We shall briefly review the great variety of materials that can be studied from this point of view.

A. Diversity of Biological Liquid Crystalline Analogues

Several skeletal tissues show a unidirectional alignment of fibrils and are analogous to typical nematic mesophases. Examples are gorgonin axis of certain octocorals, annelid bristles (Bouligand, 1972a), and numerous con-nective or tendinous structures (see Frey-Wyssling, 1953). Keratinic struc-tures of the epidermis of vertebrates such as scales, hairs, stratum corneum, also show a general alignment of filaments and are birefringent (Schmidt, 1924; Birbeck and Mercer, 1957). Most of the materials giving X-ray fiber diagrams belong to the class of nematic analogues. Interesting examples are given by association of chitin and protein in many skeletal tissues in inver-tebrates (see Rudall and Kenchington, 1973). In these materials, chitin forms microcrystals that are oriented in the same sense as polymers. We have seen in Sections II and IV that certain filaments or rods, namely, smooth muscle filaments and tobacco mosaic viruses, can show nematic alignments. In Fig. 29, the virus distribution is slightly smectic as indicated by the dotted lines. The formation of smectic layers is due to the presence of a limiting membrane.

A tentative list of the twisted fibrous arrangements in biological materials has already been published (Bouligand, 1972a) but still has to be completed. Various bodies in cytoplasm present stacked series of arcs. Beautiful cho-lesteric arrangements (Fig. 30) have been observed by Gourret in the root nodules of the legume vicia. This structure does not form spirals and corre-sponds to a cylindrical layering of the cholesteric stratification. Many twisted inclusions appear in pathological situations due either to the action of drugs or to the presence of viruses. Most of these materials show a continuous twist, but there are examples of systems with discrete steps of rotation

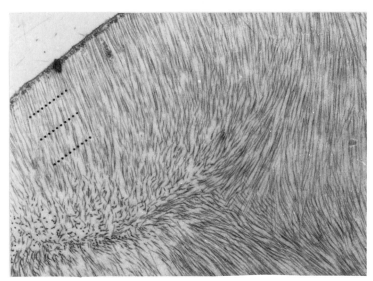

Fig. 29. Tobacco mosaic viruses accumulated in a leaf cell. Each virus is a small rod of definite length and diameter. The order is nematic. The dotted lines indicate a slight smecticity due to the limit of the viral inclusion (× 12,750). (Courtesy of Dr. J.-P. Gourret.)

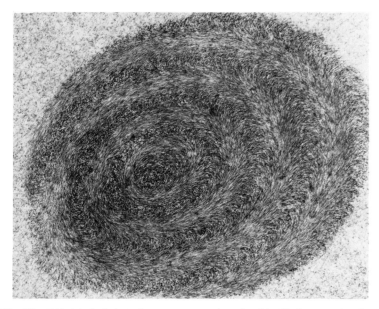

Fig. 30. Cylindrical cholesteric arrangement given by thin fibrils appearing in certain plant cells (inclusions growing in cell vacuoles of root nodules of a legume *Vicia,* × 22,950). (Courtesy of Dr. J.-P. Gourret.)

(virus of the garden pea, Rassel 1972). Alfalfa mosaic virus aggregates also give cholesteric arrangements (Hull, 1970). The narcissus mosaic virus forms very regular cholesteric liquid crystals (Wilson and Tollin, 1970).

Twisted architectures are often observed in external skeletal materials such as cuticles, or internal, such as bones (Bouligand, 1965, 1972a). The arced patterns are also found in certain connective tissues of invertebrates. Certain envelopes of eggs (chorion) often show a twisted fibrous arrangement, which has also been observed in several plant cell walls.

The biopolymers forming these twisted arrangements are extremely varied. Cellulose is mainly present in plant cell walls and in the tunica enveloping certain marine animals, the tunicates. Another polysaccharide, chitin, forms the twisted filaments of the arthropod cuticle. The other twisted arrangements are given by proteins, which are probably in the α-helix form.

Smectic analogues in living systems correspond mainly to stacked cell membranes as in myelin and in visual cells (rods and cones). Elongated supramolecular structures, such as muscle filaments, and certain viruses also form nematic and smectic phases.

B. Birefringent Colloids and Pseudomorphoses

The formation of microcrystals in an ordered polymer solution can transform a true mesophase (birefringent and fluid) into a gel (Hermans, 1941; Frey-Wyssling, 1953). As can be seen from Fig. 31, microcrystalline bundles of polymers appear when the concentration of solvent is progressively decreased. In such two-phase systems, one chain can emerge in the liquid phase and pass from one crystal to another. The whole system is then a gel with links due to microcrystallization. More recent works (Robinson et al., 1958; Samulski and Tobolsky, 1967) show that the successive steps of this process can be less simple. Certain synthetic polypeptides as poly-benzyl-L-glutamate (PBLG) form α helices and give cholesteric mesophases in solvents such as dioxane (see Chapter VI). X-ray diffraction patterns

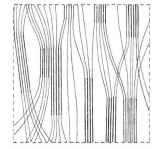

Fig. 31. Schematic representation of the microcrystallization of an anisotropic polymer solution. (After Frey-Wyssling, 1953.) An increase of the polymer concentration leads to the crystalline packing of chains, which emerge in the liquid phase and pass from one crystal to another. This process transforms an anisotropic sol (fluid) into a birefringent gel (nonfluid).

show the existence of a local hexagonal order in these liquid crystals. The distance between polymers depends on the solvent concentration. The solvent is acting as a lubricant in the hexagonal array and facilitates the mutual sliding of filaments (as in muscles). When more solvent is evaporated, the packing of polymers reaches its maximum density in certain regions where the relative positions of polymers are fixed. The solution ceases to be fluid. These transformations do not strongly affect the cholesteric twist that can subsist in the dried material (Samulski and Tobolsky, 1967). Such transformations might occur in biological mesophases and are likely in DNA and in chromosomes (see Section VI). This microcrystallization can be seen also in some cytoplasmic organelles (mitochondria, chloroplasts) where the DNA is practically pure and highly concentrated in certain zones, called DNA plasmas.

Some liquid crystals can undergo chemical hardening while retaining their birefringence and various other structural characteristics. Similarly, we can polymerize the isotropic, nematic, cholesteric and smectic phases of some mesogenic monomers. An example of this is provided by the mono- and diacrylate of a Schiff base that can polymerize and copolymerize at temperatures where it is mesomorphic (Strzelecki and Liébert, 1973; Bouligand et al., 1974). As the polymerization proceeds, the liquid character of these mesophases is rapidly abolished. Solid cholesterics are obtained by forming the terpolymer of cholesterol acrylate with the above-mentioned monomers in the twisted nematic state.

Chemical hardening and microcrystallization are frequent in biological analogues of liquid crystals and, in particular, in the insect body wall. This material is made of proteins that are hardened by a phenolic tanning, and chitin, a linear polymer of acetylglucosamine, which is microcrystalline. The chitin chains, however, form nematic and cholesteric pseudomorphoses. In this example, the two major compounds cooperate in two different ways to consolidate the material. In the case of the crustacean cuticle, whose organic matrix is a cholesteric pseudomorphose, the first deposited material is secreted in the state of a birefringent gel. This later progressively microcrystallizes and glycoprotein links appear. The phenolic tanning is generally weak.

C. Defects in Mesophases and Their Analogues

Liquid crystals show several kinds of defects, which can arrange very regularly, due to the liquid character of the medium. The elastic energies due to the presence of defects tend to equilibrate and these defects adopt a general configuration that minimizes the energy and is called texture. Liquid crystal defects have been studied in numerous papers (Friedel, 1922; Friedel and Kléman, 1970; Bouligand 1969, 1972a,b, 1973, 1974, 1975, 1978b). We recognize dislocations (edge and screw), and focal curves in smectics and cholesterics. Disinclinations are present in the three main kinds of liquid crystals. Dislocations and focal curves are relaxed in nematics. This appears clearly in smectic–nematic transitions.

The three types of defects observed in cholesteric liquid crystals have been found to occur in their biological counterpart. However, the distribution of defects in biological analogues often differs from those that are classical in cholesteric liquids. This is probably due to the fact that the first secreted deposit of a cholesteric analogue is an oriented gel, this is sufficient to prevent the defects from moving. Their positions are not due to the balance of their elastic energies; they stay in the place where they have been created.

The outermost layer of the crab cuticle is called epicuticle. This very thin film is not a cholesteric analogue. It is the part of the cuticle first secreted and shows many of the morphological details which generate the defects (Bouligand, 1975).

The experiments in the field of embryology and pattern formation often consist in an anomalous arrangement of certain grafts. For instance, in the course of development, one can change the orientation of a piece of integument and observe which structures have been affected. Defects in a developing tissue act as an experimentalist working in the best conditions. Dislocations, focal curves, and disclinations have been observed in arthropod cuticles (Bouligand, 1969, 1972a, 1975). Their careful examination allows the refinement of some of the conclusions drawn from classical experiments, and the introduction of new concepts (Bouligand, 1978a). Screw dislocations and disclinations have been observed in chromosomes (Bouligand, 1975) and probably have a functional role in chromosome cleavage (Bouligand, 1972a, 1975).

VIII. Conclusions

The subject of liquid crystals is an old one and one destined to have a rich future. The importance of liquid crystals for the life sciences was clearly indicated from the very beginning of the research in the nineteenth century. At that time and later, biologists shared the interest of physicists in liquid crystals not only for their value as model systems, but also because of their actual presence in many biological materials and their implication in certain morphogenetic processes. It would be remiss not to distinguish between liquid crystals and liquid crystal analogues in biological systems, for while the former certainly exist, there are many tissues and cellular organelles which, although they are birefringent, are by no means liquid and can even be extremely hard. They do present however supramolecular configurations which make them undeniably related to certain definite types of liquid crystals. These nonliquid analogues of the mesomorphic state can result from materials which are originally secreted while liquid and birefringent and are subsequently hardened by polymerization or by formation of chemical cross-linking. It can also happen that the material forming a liquid crystal analogue was never liquid even in its initial state, which could be a loose birefringent gel able to undergo further hardening. The first process involving a brief passage, through a genuine mesomorphic phase has not been demonstrated. In the second process, the first secreted material is a gel. The similarity between the liquid crystal and its analogue is caused by a similarity in the growth leading to the formation of both systems. These questions are similar to those which we encounter concerning colloids and polymers. There are numerous intermediate states of order between the amorphous and the crystalline and biological materials exhibit a wide selection of these.

Studies that we have made on many and diverse biological materials and the comparisons that we can draw between these materials and liquid crystals have led us to the conviction that the biocrystallography of the mesomorphic states and their polymer analogues will play an important role in the understanding of cell differentiation and organogenesis. Biochemistry and genetics provide very sure means of access to the study of these problems. Nevertheless, it is clear that the building of cellular and extracellular three-dimensional edifices calls for diverse types of self-assembly processes. One of the most fundamental of these is closely related to the mechanism that presides at the formation of a liquid crystal. Remarkable geometric and topological phenomena are exhibited by liquid crystals and by their defects. These are also found in analogous biological materials. In many cases, arrangements of mathematical interest also have a functional biological significance, particularly with respect to morphogenesis.

Acknowledgments

The author has had fruitful discussions with Dr. Charvolin and he appreciates considerable help from Dr. G. D. Mazur.

References

Bennett, S. (1956). *J. Biophys. Biochem. Cytol.* **2**, Suppl. 2, 99.
Bernal, J. D. (1965). *Symp. Soc. Exp. Biol.* **19**, 17.
Bernal, J. D., and Fankuchen, I. N. (1937). *Nature (London)* **139**, 923.
Bernal, J. D., and Fankuchen, I. N. (1941). *J. Gen. Physiol.* **25**, 111, 147.
Birbeck, M. S. C., and Mercer, E. H. (1957). *J. Biophys. Biochem. Cytol.* **3**, 203.
Bouligand, Y. (1964). *J. Microsc. (Paris)* **3**, 697.
Bouligand, Y. (1965). *C.R. Acad. Sci.* **261**, 3665, 4864.

Bouligand, Y. (1966). *J. Microsc. (Paris)* **5**, 305.
Bouligand, Y. (1969). *J. Phys. (Paris)*, *C4* **30**, 90.
Bouligand, Y. (1972a). *Tissue Cell* **4**, 189.
Bouligand, Y. (1972b). *J. Phys. (Paris)* **33**, 525, 715.
Bouligand, Y. (1973). *J. Phys. (Paris)* **34**, 603, 1011.
Bouligand, Y. (1974). *J. Phys. (Paris)* **35**, 215, 959.
Bouligand, Y. (1975). *J. Phys. (Paris)*, *C1* **36**, 173, 331.
Bouligand, Y. (1978a). *Solid State Phys.*, **34**.
Bouligand, Y. (1978b). *In* "Dislocations, Theory, a Treatise" (F. R. N. Nabarro, ed.), **5**.
Bouligand, Y., Soyer, M.-O., and Puiseux-Dao, S. (1968). *Chromosoma* **24**, 251.
Bouligand, Y., Cladis, P. E., Liébert, L., and Strzelecki, L. (1974). *Mol. Cryst. Liq. Cryst.* **25**, 233.
Caveney, S. (1971). *Proc. R. Soc., Ser. B* **178**, 205.
Chambers, R. (1930) *Ann. Physiol. Physicochim. Biol.* **6**, 233.
Chambers, R., and Chambers, E. L. (1961). "Exploration into the Nature of the Living Cell," Harvard Univ. Press, Cambridge, Massachusetts.
Chapman, D. (1965). "The Structure of Lipids," Methuen, London.
Chapman, D. (1966). *Ann. N.Y. Acad. Sci.* **137**, 745.
Chapman, D. (1970). *In* "Membranes and Ion Transport" (E. E. Bittar, ed.), Vol. 1, p. 24. Wiley, New York.
Chapman, D. (1974). *In* "Plastic Crystals and Liquid Crystals" (G. W. Gray and P. W. Winsor, eds.), Vol. 1, p. 288. Halsted Press, Ellis Horwood, Chichester, England.
Danielli, J. R., and Dawson, H. A. (1935). *J. Cell. Physiol.* **5**, 495.
de Gennes P.-G. (1974). "The Physics of Liquid Crystals," Cambridge Univ. Press, London and New York.
de Haller, G., Kellenberger, F., and Rouiller, C. (1964). *J. Microsc. (Paris)* **3**, 627.
Dodge, J. (1964). *Arch. Mikrobiol.* **48**, 66.
Döbler, J., and Bertles, J. F. (1968). *J. Exp. Med.* **127**, 711.
Drach, P. (1939). *Ann. Inst. Oceanogr. (Paris)* **19**, Part 3, 103.
Elliot, G. F., and Rome, E. M. (1969). *Mol. Cryst. Liq. Cryst.* **8**, 383.
Fraenkel-Conrat, H. (1956). *Sci. Am.* **194**(6), 42.
Frey-Wyssling, A. (1953). "Submicroscopic Morphology of Protoplasm." Elsevier, Amsterdam.
Friedel, G. (1922). *Ann. Phys. (Paris)* **18**, 273.
Friedel, J., and Kléman, M. (1970). *Natl. Bur. Stand. (U.S.), Spec. Publ.* **317**, 1.
Gaubert, P. (1924). *C.R. Acad. Sci.* **179**, 1148.
Geren, B. B. (1954). *Exp. Cell Res.* **7**, 558.
Gharagozlou-van-Ginneken, I. D., and Bouligand, Y. (1973). *Tissue Cell* **5**, 413.
Giesbrecht, P. (1961). *Zentralbl. Bakteriol., Parasitenkd., Infektionskr. Hyg., Abt. 1: Orig.* **183**, 1.
Gorter, E., and Grendel, R. (1925). *J. Exp. Med.* **41**, 439.
Gourret, J.-P. (1975). Personal communication.
Grandjean, F. (1921). *C.R. Acad. Sci.* **172**, 71.
Helfrich, W., and Deuling, H. J. (1975). *J. Phys. (Paris)*, *C1* **36**, 327.
Hermans, P. H. (1941). *Kolloid-Z.* **97**, 231.
Horne, R. W. (1971). *Symp. Soc. Exp. Biol.* **25**, 71.
Hull, R. (1970). *Adv. Virus Res.* **15**, 365.
Huxley, H. E. (1960). *In* "The Cell" (J. Brachet and A. E. Mirsky, eds.), Vol. 4, p. 365. Academic Press, New York.
Huxley, H. E. (1963). *J. Mol. Biol.* **7**, 281.
Ikemoto, N. (1963). *Biol. J. Okayama Univ.* **9**, 81.
Johansson, A., and Lindman, B. (1974). *In* "Plastic Crystals and Liquid Crystals" (G. W. Gray

and P. W. Winsor, eds.), Vol. 1, p. 192. Halsted Press, Ellis Horwood, Chichester, England.

Kenchington, W., and Flower, N. E. (1969). *J. Microsc.* **89**, 263.

Langmuir, I. (1917). *J. Am. Chem. Soc.* **39**, 1848.

Lehmann, O. (1911). "Die neue Welt der flüssigen Kristalle." Akademische Verlagsgesellschaft m.b.H., Leipzig.

Lerman, L. S. (1973). *Cold Spring Harbor Symp. Quant. Biol.* **38**, 59.

Lison, L. (1936). "Histochimie Animale." Gauthiers-Villars, Paris.

Luzzati, V. (1963). *Prog. Nucleic Acid Res.* **1**, 347.

Luzzati, V. (1968). *In* "Biological Membranes" (D. Chapman, ed.), p. 71. Academic Press, New York.

Maruyama, K. (1965). *Biochim. Biophys. Acta* **102**, 542.

Mathieu, J.-P., and Farragi, N. (1937). *C.R. Acad. Sci.* **205**, 1378.

Meyer, R. B., Liébert, L., Strzelecki, L., and Keller, P. (1975). *J. Phys. (Paris)* **36**, L-69.

Michelson, A. A. (1911). *Philos. Mag.* **21**, 554.

Milicic, D., Stefanac, Z., Juretic, N., and Wrischer, M. (1968). *Virology* **35**, 35.

Nageotte, J. (1936). *Actual. Sci. Ind.* **431/434**.

Needham, J. (1942). "Biochemistry and Morphogenesis." Cambridge Univ. Press, London and New York.

Neville, A. C. (1975). *Zoophysiol. Ecol.* **4/5**.

Neville, A. C., and Caveney, S. (1969). *Biol. Rev. Cambridge Philos. Soc.* **44**, 531.

Neville, A. C., and Luke, B. M. (1971). *J. Cell Sci.* **8**, 93.

Oncley, J. L. (1959). "Biophysical Science, a Study Program," Rev. Mod. Phys. and Wiley, New York.

Ostwald, W. (1931). *Z. Kristallogr., Kristallgeom., Kristallphys., Kristallchem.* **79**, 222.

Perrin, J. (1918). *Ann. Phys. (Paris)* **10**, 160.

Perutz, M. F., Lignori, A. M., and Eirich, S. (1951). *Nature (London)* **167**, 929.

Philips, D. M. (1976). *J. Ultrastruct. Res.* **54**, 397.

Porter, K. R., and Bonneville, M. A. (1964). "An Introduction to the Fine Structure of Cells and Tissues," Lea & Febiger, Philadelphia, Pennsylvania.

Rassel, A. (1972). *C.R. Acad. Sci., Ser. D* **274**, 2871.

Rault, J., and Cladis, P. E. (1971). *Mol. Cryst. Liq. Cryst.* **15**, 1.

Ris, H. (1962). *Symp. Int. Soc. Cell Biol.* **1**, 69.

Robertson, J. D. (1959). *Biochem. Soc. Symp.* **16**, 3.

Robertson, J. D. (1969). *In* "Handbook of Molecular Cytology" (A. Lima-de-Faria, ed.), *Res. Monogr.* No. 15, p. 1403. North-Holland Publ., Amsterdam.

Robinson, C. (1961). *Tetrahedron* **13**, 219.

Robinson, C. (1966). *Mol. Cryst. Liq. Cryst.* **1**, 467.

Robinson, C., Ward, J. C., and Beevers, R. B. (1958). *Discuss. Faraday Soc.* **25**, 29.

Rudall, K. M. (1965). *Biochem. Soc. Symp.* **25**, 83.

Rudall, K. M., and Kenchington, W. (1973). *Biol. Rev. Cambridge Philos. Soc.* **49**, 597.

Samulski, E. T., and Tobolsky, A. V. (1967). *Nature (London)* **216**, 997.

Schmidt, E. L. (1924). "Die Bausteine der Tierkörpers in polarisiertem Lichte," Cohen, Bonn.

Shikata, E., and Galvez, G. E. (1969). *Virology* **39**, 635.

Singer, S. J., and Nicholson, G. (1972). *Science* **175**, 720.

Spencer, M. (1963). *Cold Spring Harbor Symp. Quant. Biol.* **28**, 77.

Spencer, M., and Poole, F. (1965). *J. Mol. Biol.* **11**, 314.

Spencer, M., Fuller, W., Wilkins, M. H. F., and Brown, G. L. (1962). *Nature (London)* **194**, 1014.

Stewart, G. T. (1966). *Mol. Cryst. Liq. Cryst.* **1**, 563.

Stewart, G. T. (1969). *Mol. Cryst. Liq. Cryst.* **7**, 75.

Stewart, G. T. (1974). *In* "Plastic Crystals and Liquid Crystals" (G. W. Gray and P. W. Winsor, eds.), Vol. 1, p. 308. Halsted Press, Ellis Horwood, Chichester, England.

Stockem, W., and Wohlfarth-Bottermann, K. E. (1969). In "Handbook of Molecular Cytology" (A. Lima-de-Faria, ed.), Res. Monogr. No. 15, p. 1373. North-Holland Publ., Amsterdam.

Strzelecki, L., and Liébert, L. (1973). Bull. Soc. Chim. Fr. p. 597, 603, 605.

Warmke, H. E., and Christie, R. G. (1967). Virology 32, 534.

Watson, J. D., and Crick, F. H. C. (1953). Nature (London) 171, 737.

Waugh, D. F., and Schmitt, F. O. (1940). Cold Spring Harbor Symp. Quant. Biol. 8, 233.

Wilkins, M. F., Stokes, A. R., Seeds, N. E., and Oster, G. (1950). Nature (London) 166, 127.

Willison, J. H. M. (1976). J. Ultrastruct. Res. 54, 176.

Wilson, H. R., and Tollin, P. (1970). J. Ultrastruct. Res. 33, 550.

Wissocq, J.-C. (1970). J. Microsc. (Paris) 9, 355.

Zocher, H., and Török, C. (1967). Acta Crystallogr. 22, 751.

Zubay, G., and Watson, M. R. (1959). J. Biophys. Biochem. Cytol. 5, 51.

9
Mesomorphic Structure in Polyphosphazenes

N. S. Schneider

C. R. Desper

Polymer and Chemistry Division
Army Materials and Mechanics Research Center
Watertown, Massachusetts

J. J. Beres

Department of Polymer Science and Engineering
University of Massachusetts
Amherst, Massachusetts

I. Introduction

The poly(organophosphazenes) are a new class of polymers that, like the silicones, have the distinction of an inorganic backbone structure.

However, through the availability of facile substitution reactions (Allcock *et al.*, 1966; Singler *et al.*, 1974), the polyphosphazenes comprise a far larger variety of polymers. The three main groups of polymers include alkoxy (I), aryloxy (II), and amino (III) substituted homopolymers with the repeat unit structures illustrated in Fig. 1, as well as the possibility of copolymers with any variety of mixed substituent structure. The homopolymers I and II are semicrystalline materials, whereas copolymers with near equal molar proportion of the two components are generally amorphorus. No evidence of crystallinity has been found in the amino substituted homopolymers (Allcock and Kugel, 1966; White *et al.*, 1975) and, therefore, these materials will not be treated here. The unusual feature of the melting behavior of the alkoxy and aryloxy homopolymers is the fact that they display two first-order transition temperatures (Allen *et al.*, 1970; Schneider *et al.*, 1976), a lower temperature transition that will be designated $T(1)$ at which the polymer softens but some degree of order is retained and an upper temperature transition that represents passage to the isotropic liquid and, therefore, corresponds to the true melting temperature T_M. In the polyphosphazenes these two transitions $T(1)$ and T_M are separated by an unusually large temperature interval, as much as 150 to 250°C. The position of $T(1)$ exercises a decisive role in the practical application of these materials since the softening of the polymer at this temperature sets the upper limit for useful properties as a structural material.

The principal objectives of this chapter are to present information that will provide an indication of the relations between substituent composition and transition behavior, to describe observations on the phenomena concerned, and to summarize our present understanding of the nature of the transition at $T(1)$ and its relation to the mesomorphic structure observed in other polymers.

II. Structure

A. General Properties

Since the polyphosphazenes are a relatively new family of polymers, it appears useful to provide a brief summary of background information with emphasis on the factors that might affect the properties of the polyphosphazene samples and the quality of the data that is forthcoming. The reader is referred to the monograph by Allcock (1972a) for a coherent summary of work on the cyclic phosphazenes, cyclomatrix polymers, and earlier work on linear open chain polyphosphazenes, and to three recent reviews that cover the work on the synthesis, properties, and applications of the linear polyphosphazenes in more detail (Allcock, 1972b, 1975; Singler *et al.*, 1975). Most of the topics mentioned in this section are treated in the latter review in more detail, and only more recent work will be explicitly referenced.

Unlike the synthesis of most other polymers, the poly(organophosphazenes) are prepared in a two-step process. First the linear, open chain poly(dichlorophosphazene) is produced by the thermal, melt polymerization of the cyclic trimer, carried out to limited conversion to avoid the cross-linking that occurs at later stages of the reaction. Chlorine is then replaced by nucleophilic substitution, carried out in solution, to convert the hydrolytically sensitive intermediate into the desired organo-substituted polymer. The resulting highly substituted polymers are hydrolytically stable and of moderate thermal stability. However, samples will differ in molecular weight, distribution, and branching even when prepared under apparently identical conditions. This is due to lack of full knowledge or control over the polymerization conditions and to changes occurring in the process of derivatization. Significant differences in the $T(1)$ transition temperature, shape of melting endotherms at $T(1)$, or the response to annealing that cannot readily be reconciled with analytical data on the sample have been encountered (Schneider *et al.*, 1976). The presence of small amounts of residual chlorine on phosphorus or the corresponding hydrolysis products are difficult to rule out and may have an effect on the behavior under study. Care must be exercised to avoid the carryover of sodium chloride produced in the substitution reaction or of the substituted cyclics that will contribute to the X-ray diffraction pattern and to annealing behavior (Beres, 1976).

The alkoxy and aryloxy homopolymers, although semicrystalline, are generally soluble in a variety of solvents including tetrahydrofuran (THF), chloroform, and toluene at room temperature or at moderately elevated temperatures but still well below $T(1)$. Poly[bis(trifluoroethoxy)phosphazene] is soluble in THF and ketones at room temperature. Polymers with longer fluorocarbon side chains are soluble only in freon ethers but the presence of terminal CF_2H rather than CF_3 renders the polymer soluble in acetone, methanol, or THF. The polymers are high molecular weight ($M_n > 2 \times 10^5$) and extremely polydisperse ($M_w/M_n > 10$) (Hagnauer and LaLiberte, 1976a) but with no more then a small amount of long chain branching under present synthesis conditions. In TGA (thermogravametric analysis) scans the first evidence of weight loss generally occurs near 300°C depending on the nature of the substituent. However, recent evidence indicates that even at 165°C, which is in the range of the $T(1)$ transition for many polymers, chain scission can occur with a reduction in molecular weight, polydispersity, and branching (Allcock *et al.*, 1974; Hagnauer and LaLiberte, 1976b).

Data on mechanical properties are sparse and somewhat conflicting. Films cast from THF are usually permanently flexible and can be cold drawn to moderately high orientation, whereas samples compression molded above

$T(1)$ are weak materials. It should be stated that processing above $T(1)$ occurs by shear of the mesomorphic structure resulting in a crystalline sample with orientation. Melt processing is generally not possible due to the exceptionally high melting temperatures ($>300°C$) resulting in extensive degradation of the sample. The first commercial application for the polyphosphazenes has been the preparation of fuel-resistant elastomers based on fluoroalkoxide copolymers (Tate, 1975). The present interest in the polyphosphazenes stems from the exceptional fire retardant properties imparted by nitrogen and phosphorus (Quinn and Dieck, 1976). However, the relatively low $T(1)$ transition temperature and the difficulty of thermal processing may pose limitations on the use of poly(organophosphazenes) as thermoplastics.

Until very recently no method existed for the preparation of polymers with substituents linked through a phosphorus–carbon bond. However, at this time of writing Allcock *et al.* (1977) disclosed the first successful preparation of a poly[bis(phenyl)phosphazene] by reaction of phenyl lithium with poly-(difluorophosphazene). Despite the severe hydrolytic instability of the polymer intermediate, this could open the way to a new class of polyphosphazenes with properties very different from those under discussion in this chapter.

B. Chain Structure

The nature of the phosphorus–nitrogen bond in the cyclophosphazene and linear open chain polymers has been the subject of considerable controversy (Allcock, 1972a,b). Phosphorus and nitrogen each have five valence electrons. The sigma bonding at phosphorus approximates a distorted $3sp^3$ tetrahedral arrangement in which four electrons are used to bond to two nitrogen atoms and two exocyclic ligands. Nitrogen uses two electrons in bonding to two phosphorus atoms as an sp^2-type hybrid, while two lone pair electrons occupy the third sp^2 orbital. The remaining electrons, one each from nitrogen and phosphorus, probably interact in some type of $d_\pi–p_\pi$ bond by overlap of the phosphorus 3d and nitrogen $2p_z$ orbitals.

The structure is conventionally represented as shown in Fig. 1 suggesting that the molecules have aromatic character. Although theoretical and experimental arguments have been advanced both for and against delocalized electron bonding, it appears that the weight of evidence is against the occur-

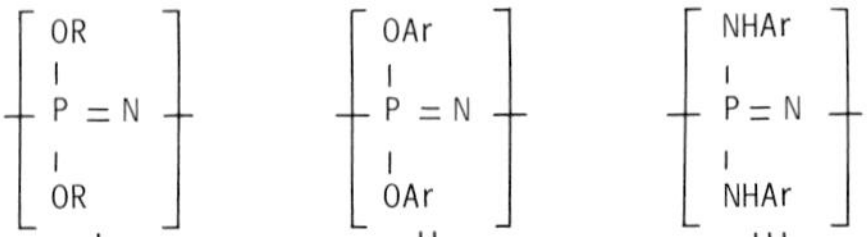

Fig. 1.   Structure of polyorganophosphazene homopolymers.

rence of a high degree of delocalized bonding. Theoretical calculations have shown that suitable hybridization of phosphorus 3d orbitals with the nitrogen p_z orbital can given rise to three center islands of π-bond character that are interrupted at phosphorus as depicted in Fig. 2 (Craig and Paddock, 1971). Among the strongest experimental evidence against delocalized bonding are the low glass transition temperatures of the poly(halophosphazenes) and many of the poly(organophosphazenes), which indicate that the main chain rotational barriers are small (see Section III.C). It has been suggested that the low rotational barriers, in addition to the absence of delocalized bonding, are due to the ease with which the large number of alternate d orbital lobes can be used to satisfy d–p_z overlap (Allcock et al., 1976). The relative magnitude of the rotational barriers for the three poly(halophosphazenes), where T_g increases with substituent size (Table III), are correctly predicted by calculations that assume that the barrier is governed solely by steric interactions between substituent or main chain atoms (Allcock et al., 1976).

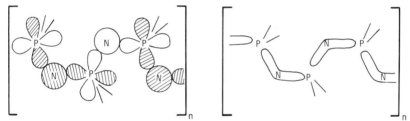

Fig. 2. Bond structure of polyphosphazenes: (a) Hybridization of suitable phosphorus 3d orbitals with nitrogen p_z orbital. Overlap is discontinuous, as indicated by alternate shaded and clear orbitals. (b) Representation of three-centered islands of π-bond character which is interrupted at phosphorus. (Singler et al., 1975.)

Bonding parameters have been gained largely from studies of the structure of cyclic species (Allcock, 1972a). The P–N bond length is about 1.58 Å, the bond angles at phosphorus are close to 120° while those at nitrogen vary in the range of 120–147°. The principal model for the polymer chain in the poly-(organophosphazenes) is a cis–trans planar conformation in the crystal (Fig. 3) identified by a repeat distance of 4.8–4.9 Å. It should be noted that in the cis–trans conformation all of the ligand bonds from phosphorus are inclined in the same direction with respect to the chain axis, thus imposing a sense of direction to the chain.

Fig. 3. Cis–trans planar conformation of the polyphosphazene chain.

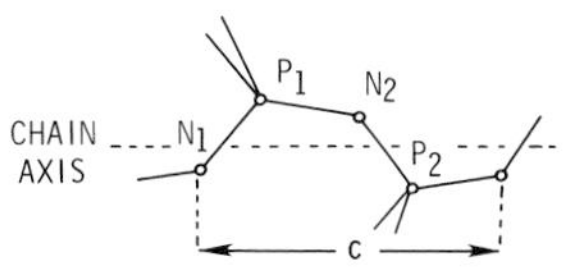

C. Crystalline Structure

Although many polyphosphazenes are known to form crystalline materials, data are not available in most instances on the unit cell or the atomic position parameters. However, fiber repeat periods in the range 4.8–4.9 Å are commonly observed where more detailed analysis is precluded by the paucity of the data. This range of c axis repeat values is taken as indicative of chain conformations of, or approximating, the cis–trans planar structure (Giglio *et al.*, 1962; Allcock *et al.*, 1972; Stroh, 1972; Bishop and Hall, 1974), the exact conformation and repeat value depending upon the substituent group. A small but significant departure from planarity for the P–N backbone is observed in poly(dichlorophosphazene) (Giglio *et al.*, 1962; Allcock *et al.*, 1972), while strict planarity is observed in the low temperature form, conformer B, of poly(difluorophosphazene) (Allcock *et al.*, 1972). For the latter polymer a high temperature conformation, designated conformer A, is also observed (Allcock *et al.*, 1972). This conformation, with a repeat of 6.49 Å, is possibly a threefold helix, but such a conformation is not observed with substituents larger than fluorine. For alkoxy and aryloxy substituted polymers, **c** values indicative of a twofold cis–trans helix with small deviations from backbone planarity are observed but the limited number of X-ray reflections often precludes the assignment of a definite structure or, in many instances, of a definite unit cell.

Bishop and Hall (1974) report a tentative solution to the structure of poly[bis(p-chlorophenoxy)phosphazene], achieved by annealing well above $T(1)$ to yield 25 usable reflections. Nonetheless, the backbone structure was assigned by inference, since the scarcity of the data and the dominating influence of side group parameters precluded a direct solution for the posi-

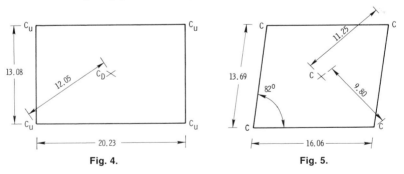

Fig. 4.   **Fig. 5.**

Fig. 4. Model for crystal structure of poly[bis(p-chlorophenoxy)phosphazene] after Bishop and Hall (1974). C_u, up chain; C_d, down chain. Dimensions in angstroms.

Fig. 5. Suggested crystal structure of poly[bis(phenoxy)phosphazene] after Stroh (1972). Dimensions in angstroms.

tions of the backbone atoms. The Bishop and Hall structure is orthorhombic ($a = 13.08\,\text{Å}$, $b = 20.23\,\text{Å}$, $c = 4.90\,\text{Å}$) with two chains per unit cell. Each chain possesses directionality (which is true for all polyphosphazenes in cis–trans conformations), and the "up" and "down" chains are placed at the corner and center of the unit cell, respectively, the sites being separated by a distance of 12.05 Å (see Fig. 4). For a similar polymer, poly [bis(phenoxy)-phosphazene], Stroh (1972) reports a monoclinic unit cell with parameters $a = 16.06\,\text{Å}$, $b = 13.69\,\text{Å}$, $c = 4.91\,\text{Å}$, and $\gamma = 82°$; the experimental density indicating two chains per unit cell. Placing the chains at the corner and center of the unit cell (see Fig. 5), the distances between chain sites are 11.25 and 9.80 Å along the longer and shorter diagonals of the **ab** parallelogram. Two conclusions are evident:

(a)   the chain sites are closer together in the phenoxy substituted polymer compared to the p-chlorophenoxy substituted polymer;

(b)   in the phenoxy substituted polymer, the chains exhibit two different distances between chain sites rather than one.

Conclusion (a) probably arises from the space required to accommodate the large chlorine atoms in the p-chlorophenoxy polymer. Conclusion (b) suggests a preferential packing of each chain in the phenoxy substituted polymer with two of its nearest neighbors, possibly involving interpenetration of the side groups, with a comparatively greater distance to the remaining two nearest neighbors. Such interpenetration is absent in the Bishop and Hall structure of the p-chlorophenoxy polymer. A solution of the structure in the unit cell of the phenoxy substituted polymer is needed, however, before firm conclusions may be drawn about the indicated differences in chain packing.

This laboratory is engaged in studies of two closely related crystalline aryloxy polyphosphazenes, containing the 3,4-dimethylphenoxy and m-chlorophenoxy substituents, respectively (Beres, 1976; Desper, 1976). Definite solution of the crystal structures are not presently available, partly because of the problem of obtaining a highly oriented crystalline sample. Both orthorhombic and monoclinic structures are under consideration at this time. In addition, the m-chlorophenoxy substituted polymer shows evidence of a crystal–crystal transformation. Another phase is also suggested by Bishop and Hall (1974) for poly[bis(p-chlorophenoxy)phosphazene] on the basis of very tenuous evidence. Crystal structures have proven to be quite elusive for the alkoxy substituted polyphosphazenes because of the small number of reflections and the large number of atoms and degrees of freedom. Stroh (1972) reports diffraction data for five alkoxy substituted polymers but offers no unit cell solutions. The available data for poly-phosphazene crystal structures are summarized in Table I.

TABLE I

Summary of X-Ray Data for Polyphosphazenes[a]

Polymer	Unit cell	Backbone conformation	Reference
$(NPCl_2)_n$	Orthorhombic, $a = 5.50$, $b = 12.72$, $c = 4.90$, $z = 4$	Nearly cis–trans planar	Giglio et al. (1962)
$(NPF_2)_n$ Conformer A	Orthorhombic or lower, $c = 6.49$	Helical, possibly threefold	Allcock et al. (1972)
$(NPF_2)_n$ Conformer B	Orthorhombic, $a = 8.69$, $b = 5.38$, $c = 4.86$, $z = 4$	Cis–trans planar	Allcock et al. (1972)
$[NP(OCH_2CF_3)_2]_n$	Orthorhombic or lower, $c = 4.86$	Probably cis–trans nearly planar	
$[NP(OCH_2C_2F_4H)_2]_n$	Orthorhombic or lower, $c = 5.19$	Possibly cis–trans nearly planar	
$[NP(OCH_2C_2F_5)_2]_n$	Debye–Scherrer pattern only	Possibly cis-trans nearly planar	
$[NP(OCH_2C_3F_7)_2]_n$	Debye–Scherrer pattern only, $c = 9.70$	Possibly cis–trans nearly planar	
$[NP(OCHC_2F_6)_2]_n$	Debye–Scherrer pattern only	Undetermined	
$[NP(OC_6H_5)_2]_n$	Monoclinic, $a = 16.06$, $b = 13.69$, $c = 4.91$, $\gamma = 82$, $z = 4$	Probably cis–trans nearly planar	
$[NP(OC_6H_4p\text{-}Cl)_2]_n$	Orthorhombic, $a = 13.08$, $b = 20.23$, $c = 4.90$, $z = 4$.	Probably cis–trans nearly planar	Bishop and Hall (1974).
$[NP(OC_6H_4m\text{-}Cl)_2]_n$	Orthorhombic or lower, $c = 4.87$	Probably cis–trans nearly planar	Desper (1976).
$[NP(OC_6H_3\text{-}3,4\text{-}[CH_3]_2)_2]_n$	Orthorhombic or lower, $c = 9.8$	Probably cis–trans nearly planar	Beres (1976).

[a] From Stroh (1972), unless otherwise noted.

III. Transition Behavior

A. Calorimetric Behavior at $T(1)$ and T_M

The results discussed in this section are taken largely from a paper by Schneider *et al.* (1976) and work not appearing in his paper will be separately referenced. Characteristic features of the behavior at the two first-order transitions are best illustrated by observations on poly[bis(trifluoroethoxy)phosphazene]. Solution cast films of this polymer form large spherulitic structures readily visible in the optical microscope. As first reported by Allen *et al.* (1970), no alteration in the gross morphology can be detected by optical microscopy on heating through $T(1)$ at 90°C and the spherulitic morphology remains intact up to the melting temperature at 240°C. However a small partially reversible increase in birefringence is noted. Recrystallization from the melt gives rise to needle-shaped crystals rather then the original spherulitic morphology. Above $T(1)$ the crystalline diffraction pattern is reduced to a single reflection with a spacing of 11 Å (Allen *et al.*, 1970), indicating the conversion to a state of partial order, as discussed in detail in Section IV. In the aryloxy samples the crystalline texture is below the resolving power of the optical microscope and the melting point is near or above the decomposition temperature so that comparable direct observations of morphology by optical microscopy are not possible. Consequently DSC (differential scanning calorimetry) was chosen to study the transformation.

The response of the transitions to thermal history has been examined by DSC on samples of a film compression molded above $T(1)$. For convenience, the term "melting" will be used to describe heating through the crystal transformation temperature but care will be taken to associate this usage with $T(1)$ so as to avoid confusion with passage to the isotropic melt at T_M. The appearance of the endotherm at $T(1)$ in a series of melting and crystallization cycles is shown in Fig. 6. The first melt produces a broad endotherm with a peak at 79.5°C, comparable to the temperature reported by Allen *et al.* (1970). In the second run the peak sharpens appreciably and the area increases by about 10%, when the sample is cooled from 385°K, and recrystallized at 20°/min. Additional small increases in area and transition temperature result from repeated scanning cycles or from annealing at temperatures between $T(1)$ and T_M. When the sample is recrystallized from the true melt after heating above T_M, there is a dramatic increase in the area and sharpness of the endotherm at $T(1)$ and the peak temperature moves to 91.5°C as indicated in Fig. 6.

Examination of the crystallization exotherm, on cooling through $T(1)$ at 10°/min, shows that the onset of crystallization is sharply defined and occurs

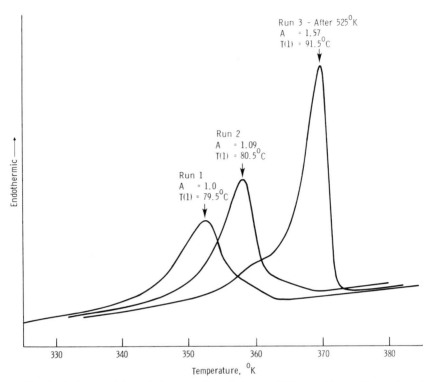

Fig. 6. Influence of thermal history on the endotherm at $T(1)$ for poly[bis(trifluoroethoxy)-phosphazene] (Schneider *et al.*, 1976). Heating rate, 10°/min; cooling rate between runs, 20°/min; A, relative area. Run 2 displaced 5°; run 3 displaced 10°.

at a supercooling of about 20°C as indicated by the temperature difference between peaks in the melting and the crystallization runs. The crystallization curve only gradually approaches the baseline with decreasing temperature indicating the presence of species that are more difficult to crystallize. The same species probably contribute to the low melting fraction of the sample that comprises the leading edge of the melting peak at $T(1)$. In Fig. 7 a comparison of the endotherms at $T(1)$ and T_M is shown. The enthalpy change at $T(1)$ is tenfold greater than at T_M (Table II) and the peak is appreciably sharper than the true melt peak. The use of a thermal mechanical analyzer in the thermal expansion mode has shown that the volume changes at $T(1)$ and T_M are well defined and have approximately the same magnitude (Table II), in contrast to the marked disparity in the heats of transition.

A similar series of determinations have been carried out on poly[bis-(p-chlorophenoxy)phosphazene] as a representative aryloxy polymer. This sample also shows the marked effect of annealing between $T(1)$ and T_M.

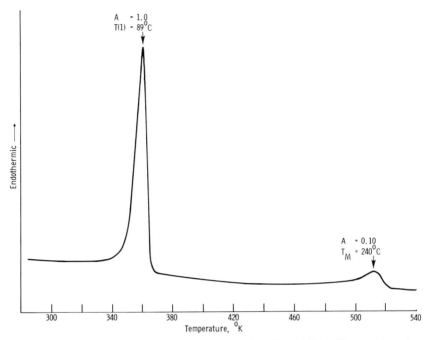

Fig. 7. Comparison of the endotherm at $T(1)$ and T_M for poly[bis(trifluoroethoxy)phosphazene] (Schneider *et al.*, 1976). Heating rate, 10°/min.

However, since the melting peak is closer to the decomposition temperature, it is not possible to crystallize the sample from above the melt. A careful search by DSC was unsuccessful in detecting T_M suggesting that the heat of fusion at T_M is close to zero. However, the volume change at T_M, in this case, is somewhat larger than at $T(1)$ as indicated in Table II. A comparison with thermodynamic data for other materials is also provided in this table and indicates some analogy to *p*-anisoxyanisole in the relative values of the heat of fusion and the volume change at the crystalline-to-meso and melting transitions (see also Section IV.C).

The foregoing results on thermal transition behavior suggest that appropriate annealing, or in the case of poly[bis(trifluoroethoxy)phosphazene], controlled crystallization from the true melt, can improve the organization of the mesomorphic state. This is manifest in the profound influence that the thermal history above $T(1)$ exerts on the characteristics of the crystalline state, as indicated by improvement in the X-ray diffraction pattern and the increase in the transition temperature, peak area, and sharpness of the endotherm at $T(1)$. Such annealing procedures are important in perfecting the crystalline organization to increase the information that can be obtained

TABLE II

Comparison of Thermal Transition Data

Sample	$T(1)$ (°C)	T_M (°C)	$\Delta H(T(1))$ (cal/g)	$\Delta H(T_M)$ (cal/g)	$\Delta V(T(1))$ (%)	$\Delta V(T_M)$ (%)	References
Poly[bis(trifluoroethoxy) phosphazene]	92	240	8.6	0.8	5	6	Schneider et al. (1976)
Poly[bis(p-chlorophenoxy) phosphazene]	169	356	6.6	0	3.5	5.7	Schneider et al. (1976)
Polyethylene	141			68		15	Brandrup and Immergut (1975)
p-Azoxyanisole	116	133	28	0.7	0.1	0.36	DuPré et al. (1971)
Poly(diethylsiloxane)	−5	20	2.4	0.36			Beatty and Karasz (1975); Beatty et al. (1975)
Poly(tetrafluoroethylene)	19–30	327	8.5	13.7	1	20	Furukawa et al. (1952); Starkweather and Boyd (1960); Mc Cane (1970)

from X-ray diffraction studies aimed at solving structure problems. However, as with other polymers the improvement in crystalline organization is brought about at the expense of embrittling the sample, whether originally solution cast or compression molded.

B. Dynamic Mechanical Behavior

In addition to the foregoing results from the thermoanalytical methods, there is a limited amount of data from the examination of polyphosphazenes using dynamic mechanical and dielectric dispersion techniques or torsional braid analysis. The upper temperature limit of the mechanical methods is set by the softening of the sample just above the crystal transformation temperature but the pronounced drop in modulus that occurs both at T_g and at $T(1)$ can be observed with poly[bis(trifluoroethoxy)phosphazene] (Allen et al., 1970). In torsional braid analysis, the sample is coated on a glass fiber braid permitting measurements to be carried through both the softening and the melting temperatures. Analysis of poly(dichlorophosphazene) (Conelly and Gillham, 1975) showed only a single loss peak with a high temperature shoulder. The two associated transition temperatures were in accord with T_g and T_M reported in Table III. Results have also been reported for several fluoroalkoxy homopolymers and their blends (Connelly and Gillham, 1976). For poly[bis(trifluoroethoxy)phosphazene], four distinct loss regions were observed including the high temperature shoulder of a small loss peak at about $-180°C$ and moderate-to-strong loss peaks near the expected T_g, $T(1)$, and T_M values listed in Table III. The dynamic mechanical spectra for the heating and cooling cycle are displayed in Fig. 8 where the hysteresis effects are associated with delayed crystallization and distinguish the two first-order transitions from the glass transition or secondary transitions. Neither $T(1)$ nor T_M could be detected in $[(HCF_2CF_2CH_2O)_2PN]_n$. It is not clear whether the apparent absence of crystallinity in this last sample is due to substitutional heterogeneity or is somehow a consequence of the proton on the terminal methyl group.

C. Relation between Composition and Transition Temperatures

Data on the thermal transitions of a large number of polyphosphazenes, mainly of the aryloxy type, are collected in Table III. In all but three of the organophosphazene samples an intermediate first-order transition is observed indicating that $T(1)$ is a characteristic feature of this major class of polyphosphazenes and is not dependent on a specific choice of substituent.

TABLE III

Summary of Transition Temperatures, Decomposition Temperatures, and Apparent Enthalpies of Transition for Various Polyphosphazenes $[N = P(R)_2]_n$

R	T_g (°C)[a]	$T(1)$ (°C)[a]	T_M (°C)[b]	T_D (°C)[c]	$\Delta H(T(1))$ (cal/g)	References
F	−96		−68[d] −40			Allcock et al. (1972)
Cl	−63		−30			Allcock et al. (1966)
Br	−15		270			Allcock et al. (1972)
CH_3O	−76					Allcock et al. (1966)
CH_3CH_2O	−84					Allcock et al. (1966)
CF_3CH_2O	−66	92	240[e]	360	8.6	Allcock et al. (1966); Allen et al. (1970); Schneider et al. (1976)
C_6H_4O	6	160	390	380	10.0	Allcock et al. (1966); Singler et al. (1974)
$o\text{-}FC_6H_4O$	−8	125			4.0	Higginbotham and Schneider (1975)
$m\text{-}FC_6H_4O$	−34	122			8.5	Higginbotham and Schneider (1975)
$p\text{-}FC_6H_4O$	−14	169	345		10.8	Allen et al. (1970); Higginbotham and Schneider (1975)
$m\text{-}ClC_6H_4O$	−24	90	370	380	5.8	Singler et al. (1974)
$p\text{-}ClC_6H_4O$	4	169	365[f]	410	6.6	Singler et al. (1974); Schneider et al. (1976)
$m\text{-}CH_3C_6H_4O$	−25	90	348	350	8.3	Singler et al. (1974); Higginbotham and Schneider (1975)
$p\text{-}CH_3C_6H_4O$	0	152	340	310	3.7	Singler et al. (1974); Higginbotham and Schneider (1975)
$m\text{-}CH_3OC_6H_4O$	−15	[g]				Higginbotham and Schneider (1975)
$p\text{-}CH_3OC_6H_4O$	15	106			2.2	Singler et al. (1974); Higginbotham and Schneider (1975)
$p\text{-}CH_3CH_2C_6H_4O$	−18	43	h	285	1.1	Singler et al. (1974); Higginbotham and Schneider (1975)
$p\text{-}C_2H_5(CH_3)CHC_6H_4O$	−16	103			0.2	Higginbotham and Schneider (1975)
$p\text{-}(CH_3)_3CC_6H_4O$	48	237	345	350		Higginbotham and Schneider (1975)
$p\text{-}C_6H_5CH_2C_6H_4O$	−3	109		320	10.4	Higginbotham and Schneider (1975)
$3,4\text{-}(CH_3)_2C_6H_3O$	−5	96	325	315	4.6	Higginbotham and Schneider (1975)
$3,5\text{-}(CH_3)_2C_6H_3O$	6	67	320	340	1.2	Beres (1976)
$3\text{-}CH_3\text{-}4\text{-}ClC_6H_3O$	6	123			5.2	Higginbotham and Schneider (1975)

[a] By DTA or DSC. [b] For poly(organophosphazenes): by TMA, penetrometer mode unless noted otherwise; open spaces sample not run. [c] For poly(organophosphazenes): by TGA, heating rate 10°/min; open spaces sample not run. [d] Crystal–crystal transformation. [e] By DSC, thermal expansion, optical microscopy. [f] By thermal expansion. [g] No transition observed. [h] Decomposes with expansion.

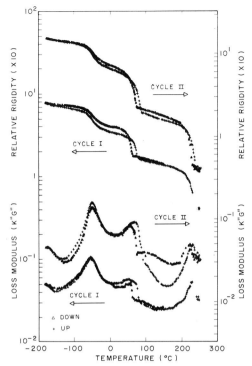

Fig. 8.   Dynamic mechanical spectra of poly[bis(trifluoroethoxy)phosphazene] by torsional braid analysis (Connelly and Gillham, 1975). He atmosphere. Prehistory: Dry, 25 → 150°C. Experiment: 150 → − 180 → 250 → − 180 → 250°C.

The absence of a $T(1)$ transition in the first two alkoxy samples may be connected with the fact that alkoxy substituents are less stable to hydrolysis than fluoroalkoxy and the resulting substituent heterogeneity would interfere with crystallinity. It will be noted that with one exception the T_g values for the aryloxy polymers are distributed over a fairly narrow range, about − 30° to 15°C. The alkoxy samples have markedly lower T_g values consistent with the more flexible and less bulky side group. The $T(1)$ transitions show a wider range of variation, from about 40 to 170°C, but there is a marked correlation between T_g and $T(1)$ transitions.

The recorded values of T_M fall in a relatively narrow range and appear to be almost insensitive to structure. It should be pointed out, however, that these T_M values are very close to and in several cases higher than the indicated temperatures of decomposition determined as the onset of weight loss by TGA. Thus, in many cases, T_M merely indicated the maximum temperature at which structure persists and supports the analyzer probe.

For poly[bis(trifluoroethoxy)phosphazene], this problem does not exist since T_M occurs at a much lower temperature than T_D. In the case of poly[bis(p-chlorophenoxy)phosphazene], T_M also undoubtedly represents true melting since T_D is higher by 45°C. After heating to the indicated T_M, the various phenoxy polymers with alkyl side group are reduced to a greasy substance but the p-chloro and p-fluoro samples become brittle. Significant decomposition has apparently occurred in these two samples even if heating has not occasioned a weight loss discernible by TGA. To the extent that the recorded values of T_M represent the melting temperature, the reduced sensitivity to composition is in keeping with the observation that almost all the enthalpy change between the crystalline and melt stage occurs at $T(1)$ (see Table II). Values of the apparent enthalpy change at $T(1)$ listed in Table III reflect both the enthalpy of the transition for the ideal crystalline lattice and variations in the degree of crystallinity, probably with the latter effect controlling. Unfortunately, no determinations of the degree of crystallinity or true heat of fusion have as yet been reported. However, the X-ray diffraction pattern for well-annealed samples of trifluoroethoxy and p-chlorophenoxy samples suggest that crystallinity is in the range of 60–80% (Desper, 1976).

We turn now to consider the relation between transition temperature and specific features of composition. The most striking observation regarding the T_g values for the poly(organophosphazenes) are the low values, in general, when compared with single-bonded carbon backbone polymers. For example, poly[bis(phenoxy)phosphazene] with two substituents on phosphorus has a T_g almost 100°C lower than that of polystyrene. These low glass transition temperatures indicate that restrictions to rotation are small, reflecting both the inherently low main chain rotational barriers and the flexibility imparted by the oxygen linkage to substituent groups.

The trends in T_g correlate to some degree with the size of the substituent; p-fluorophenoxy < p-methylphenoxy, p-chlorophenoxy < p-methoxyphenoxy < p-t-butylphenoxy. However, the fact that T_g for phenoxy is about the same as for p-chlorophenoxy suggests that some type of inductive effect as well as steric factors play a role in T_g. Substitution in the meta rather than the para position characteristically lowers T_g by 20–30°C, whereas increasing the length of the alkyl substituent from p-methylphenoxy to p-ethylphenoxy lowers T_g by 18°C. Thus, it is possible to formulate elastomers of high aromatic content by the appropriate selection of aryloxyphosphazene copolymers.

The $T(1)$ transition temperatures for the para-substituted fluoro-, chloro-, or methylphenoxy and unsubstituted phenoxy polymers differ very little. However, the strong difference between meta and para substitution observed with T_g is also evident in the $T(1)$ behavior as well as the effectiveness of the longer alkyl substituent, ethyl compared to methyl, in lowering $T(1)$. A deeper understanding of the influence of these composition variables on $T(1)$ requires a better knowledge of the crystalline organization and of the change in side group motion which occurs at $T(1)$.

IV. The Mesomorphic State

A. *Structure of the Mesomorphic State*

The first direct evidence of the structure of a mesomorphic state in a polyphosphazene is the X-ray diffraction pattern observed by Allen *et al.* (1970) for poly[bis(trifluoroethoxy)phosphazene] at 90°C. A sharp equatorial line is observed at 11 Å, indicating that order is retained above $T(1)$ in the lateral dimensions, but not in the longitudinal direction. The disorder in the longitudinal direction probably accounts for the diffuse line at 4.5–5.0 Å that appears in the same pattern. Subsequent studies in this laboratory of poly[bis(*p*-chlorophenoxy)phosphazene] and poly[bis(*m*-chlorophenoxy)-phosphazene] revealed two sharp equatorial lines and a diffuse meridional line at temperatures of 180°, 199°, and 238°C, all between $T(1)$ and T_M (Schneider *et al.*, 1976; Desper and Schneider, 1976).

The structure was referred to as pseudohexagonal (Desper and Schneider, 1976) from the fact that the two sharp lines are near the ratio 1.732 for d_{100}/d_{110} in a hexagonal structure (see Table IV). Further confirmation comes from a third sharp line that appears faintly in one instance and may be indexed as (210). Diffuse meridional scattering in the region 4.1 to 4.5 Å is attributed to (hkl) reflections broadened by rotational or longitudinal disorder. The sharp lines are remarkably insensitive to temperature and to the isomeric position of the chlorine atom. It is noteworthy that the molecular

TABLE IV

Sharp X-Ray Lines in Para and Meta
Forms of Poly[bis(chlorophenoxy)-
phosphazene] above $T(1)$[a]

Isomer	$T(°C)$	Sharp Lines (Å)	Ratio
Para	180	12.07, 7.14	1.69
Para	199	12.10, 7.04	1.72
Para	238	12.18, 7.12	1.71
Meta	180	12.13, 7.05	1.72
Estimated Error:		1.9%, 1.0%	2.9%

[a] After Desper and Schneider (1976).

orientation is maintained upon cycling from the crystalline state to well above $T(1)$ and back, indicating that fluidity of the sample above $T(1)$ is limited.

The proposed mechanism for the structural transformation at $T(1)$ from the Bishop and Hall orthorhombic crystal to the pseudohexagonal state is indicated in Fig. 9 (Desper and Schneider, 1976). The transformation involves expansion of the orthorhombic lattice by 9% in the **a** and 20% in the **b** directions, altering the **a/b** ratio from 1.55 in the crystalline state to approximately 1.73 in the mesomorphic state. At this point the diagonals of the altered orthorhombic unit cell make an angle of 120° with each other and the structure is metrically hexagonal. Static hexagonal symmetry is ruled out, since the chains do not adopt a conformation with threefold symmetry. Instead, the hexagonal nature of the lattice possibly arises from an interchain potential function that approaches cylindrical symmetry at elevated temperature. Preliminary nuclear magnetic resonance second moment data (Alexander *et al.*, 1976) indicate that the disorders causing the shift to a cylindrical form of potential function are dynamic, rather than static, in nature, and are associated with the onset of rapid rotations of the side groups and possibly of the chain backbone. Rapid backbone rotations imply a dynamic infinite-fold symmetry axis, fulfilling the requirements of hexagonal symmetry and suggesting that the qualifying prefix *pseudo* may be omitted.

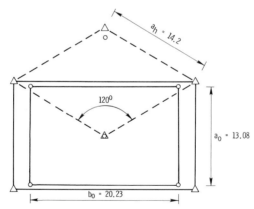

Fig. 9. Transformation of orthorhombic to pseudohexagonal structure. ○, chain sites, orthorhombic crystals; △, chain sites, pseudohexagonal crystals.

The possibility of order and disorder with respect to chain directionality is an interesting point to consider. The Bishop and Hall structure of poly-[bis(*p*-chlorophenoxy)phosphazene], as shown in Fig. 4, includes a definite pattern in the positions of the "up" and "down" chains; i.e., each down chain is surrounded by four up chains and vice versa. The diffraction data are ambiguous as to whether directional order persists in the mesomorphic state. The sharp $(hk0)$ reflections depend only upon the projection of the structure on the **ab** plane and give no evidence as to order or disorder with respect to chain directionality. The disorder preventing sharp layer reflections could arise from factors other then disorder in directionality, such as local motions with a **c**-axis component, or rotational motions about the chain backbone.

The present discussion has focused on the polyphosphazenes bearing p-chlorophenoxy and m-chlorophenoxy substituents since the two or three sharp lines observed confirm the hexagonal-type structure. Diffraction studies of three other polyphosphazenes, with trifluoroethoxy (Desper, 1976; Allen *et al* 1970), 3,4-dimethylphenoxy, and p-benzylphenoxy (Higginbotham and Schneider, 1975) side groups, have yielded only one sharp equatorial line in the mesomorphic state. This line is identified with the pseudohexagonal (100) reflection, which is by far the strongest reflection in the chlorophenoxy polymers. The lattice parameters a_h, which coincide with the distances between chain sites, are calculated from these d_{100} values. (The more accurate d_{110} value will give better a_h values but is observed in only two polymers.)

Examination of Table V shows a definite trend between increasing size of the side group and increasing a_h values. Of the side groups, trifluoroethoxy is the smallest; the two chlorophenoxy and the 3,4-dimethylphenoxy groups are larger and comparable to each other; and the p-benzylphenoxy group is even larger. The a_h values for these three catagories are approximately 12, 14, and 19 Å, respectively, indicating that side group packing is controlling the hexagonal structure parameters.

TABLE V

Pseudohexagonal Diffraction Lines for Various Polyphosphazenes above $T(1)$

Substituent	Temperature (°C)	d_{100}(Å)	a_h(Å)	Reference
CF_3CH_2O-	90	11	12.7	Allen *et al.* (1970)
	100	10.1	11.7	Desper (1976)
	120	10.2	11.8	Desper (1976)
$p\text{-}ClC_6H_4O-$	180	12.1	14.0	Desper and Schneider (1976)
	199	12.1	14.0	Desper and Schneider (1976)
	238	12.2	14.1	Desper and Schneider (1976)
$m\text{-}ClC_6H_4O-$	180	12.1	14.0	Desper and Schneider (1976)
$3,4\text{-}(CH_3)_2C_6H_3O-$	100	12.6	14.5	Higginbotham and Schneider (1975)
$p\text{-}C_6H_5CH_2C_6H_4O-$	120	16.8	19.4	Higginbotham and Schneider (1975)

B. Origin of the Mesomorphic State in Polyphosphazenes

The hexagonal-type diffraction pattern observed in the mesomorphic state could, conceivably, arise from either static or dynamic disorder. Examples of the former are phase I of poly(vinylidene fluoride) (Lando *et al.*, 1966) and poly(vinyl fluoride) (Golike, 1964), in which static pseudo-hexagonal phases are stable (or at least metastable) at room temperature. In these systems the packing of side groups of approximately spherical shape is the dominant factor controlling the lattice. Presumably the chain backbone conformation could be slightly distorted from the trans–planar position to accommodate the requirements of side group packing. The absence of extensive rotation of the chain backbone, however, is indicated by NMR line broadening, which conforms to a rigid lattice model, and by departure from true hexagonal diffraction symmetry in a rolled specimen (Lando *et al.*, 1966). A similar structure appears in irradiated polyethylene (Clough, 1970; Thomas and Sass, 1973), but in this and in the poly(vinyl fluoride) case the static nature of the structure is not confirmed.

The preferred explanation for the origin of hexagonal-type packing in polyphosphazenes is that dynamic disorder, in the form of rapid backbone rotation, causes the interchain potential energy function to approach cylindrical symmetry at elevated temperatures. Similar hexagonal-type phases are observed in the high temperature forms of poly(*p*-xylylene) (Neigisch, 1966), *trans*-1,4-polybutadiene (Iwayanagi and Miura, 1965; Suehiro and Takayanagi, 1970), and poly(tetrafluoroethylene) (Clark and Muus, 1962; Muus and Clark, 1964). In two instances, NMR data give evidence of rotational motions of molecules about their long axes in the hexagonal-type state (Iwayanagi and Miura, 1965; Muus and Clark, 1964).

The disorder introduced by the rapid rotational motions is quite likely the reason for the scarcity of diffraction lines in the mesomorphic state of polyphosphazenes. The diffraction pattern of *trans*-1,4-polybutadiene in the pseudohexagonal state is qualitatively quite similar to those of the two poly[bis(chlorophenoxy)phosphazenes]: three sharp (*hk*0) reflections plus a diffuse layer reflection. Structure factor calculations by Suehiro and Takayanagi (1970), incorporating rotation about the chain axes, predict such a diffraction pattern. Thus, it is not necessary to postulate lattice paracrystalline distortions, which are translational displacements from the ideal lattice sites, to explain the small number of sharp (*hk*0) reflections found in mesomorphic polyphosphazenes, as was earlier suggested (Desper and Schneider, 1976). The diffuse character of the layer reflection is, however, attributable to translational displacement of the molecules along the chain axes, which amounts to the same thing as paracrystalline lattice distortions. Thus, the mesomorphic state in polyphosphazenes should not be thought of as paracrystalline, but paracrystalline diffraction theory may be useful in evaluating the layer reflection.

C. *Comparison with Poly(tetrafluoroethylene) and Poly(diethylsiloxane)*

It is particularly instructive to compare the behavior of polyphosphazenes with that of two polymers, poly(tetrafluoroethylene) (PTFE) and poly(di-ethylsiloxane) (PDES), in which pseudohexagonal phases characterized by dynamic disorder have been extensively studied. In the PTFE case, the phase in question is stable in the range 30–327°C, while a second hexagonal phase showing a lesser degree of dynamic disorder is stable between 19 and 30°C. The entire disordering process, encompassing both the 19 and 30°C transitions, involves a total enthalpy change (Furukawa *et al.*, 1952) on the same order of magnitude as the heat of fusion at 327°C (Starkweather and Boyd, 1960), but a small volume change on the order of 1% (Clark, 1967) compared to the 20% volume change on fusion (Starkweather and Boyd, 1960) (see Table II). In particular, the spacing between chain sites in PTFE in the plane perpendicular to the **c** axis changes only 0.5% at the 19°C transition and 0.1% at the 30°C transition (Muus and Clark, 1964; Clark, 1967), while the corresponding distance increases 17% (14.2 versus 12.1 Å) in the $T(1)$ transition of poly[bis(p-chlorophenoxy)phosphazene] (Desper and Schneider, 1976). Thus the polyphosphazenes and PTFE are similar with regard to the wide temperature range in which the pseudohexagonal state is stable, but differ markedly in the degree of lateral expansion associated with the order–disorder process. In both instances, NMR data (Alexander *et al.*, 1976; McCall *et al.*, 1967; Hyndman and Origlio, 1960), combined with X-ray diffraction data (Desper and Schneider, 1976; Clark and Muus, 1962; Clark, 1967), indicate that rapid motion of the chain backbone occurs in the pseudo-hexagonal state. In the polyphosphazene case, however, rapid rotation of the organic side groups also occurs, while the possibility of such rotation is nonexistent in PTFE. This may explain the difference in lateral expansion and in volume change for the two systems. With the onset of both backbone and side chain rotation, the polyphosphazene chain requires considerably more volume to accommodate the increased degrees of freedom. On the other hand, the PTFE transition involves only one additional degree of freedom which apparently requires very little additional volume. Examination of the helical PTFE molecule (Clark and Muus, 1962) indicates that, even in the static state, the periphery of the molecule approximates a smooth cylinder, so the dynamic state, in which the periphery actually becomes a smooth cylinder, requires very little expansion of the lattice. Another striking difference is that for pseudohexagonal PTFE, at least six strong ($hk0$) reflections appear (Clark and Muus, 1962), while the polyphosphazenes exhibit one strong line and perhaps one or two weak ones. As Clark and Muus point out, this results from the 15_7 helical symmetry of the PTFE molecule. These equatorial reflections are sensitive only to the projection of the electron density on the **ab** plane. In the hexagonal state below 30°C, the projection of each chain has 15-fold symmetry. Above 30°C, this is transformed to infinite-fold symmetry, but the difference is slight and does not extinguish many of the lower ($hk0$) reflections. This effect does not occur with polyphosphazenes, where the static chain possesses twofold (or lower) symmetry.

In PDES there are four calorimetrically discernible events (Beatty and Karasz, 1975; Beatty et al., 1975): a glass transition temperature at -135°C; a crystal–crystal transformation, confirmed by X-ray diffraction, at -70°C; crystal melting at -5°C; and a diffuse transition, probably first order at about 20°C. Above the crystal melt, a single sharp X-ray line is observed that represents an interchain spacing of 8.7 Å (Beatty et al., 1975). Although the authors refer to this as a state of one-dimensional order, it corresponds to a state of two-dimensional order in the sense used to discuss the meso state in the polyphosphazenes. Attention has been focused on this intermediate state of order and its significance for the crystalline structure formed below -5°C (Pochan et al., 1976). On slow cooling from the melt above 20°C, weakly birefringent areas appear. Since the sample becomes qualitatively more viscous than the isotropic melt, the term viscous-crystalline has been recommended to describe this state (Beatty et al., 1975). Crystallization takes place by transformation and intensification of these weakly birefringent regions rather than by the process of spherulitic growth (Pochan et al., 1976). It has been suggested that the viscous-crystalline phase preorders the melt and facilitates crystallization, with the implication that a similar type of pre-ordering of the melt might occur in all semicrystalline polymers. The formation of the viscous-crystalline structure can be bypassed by quenching directly to the crystalline state. Under these conditions the viscous-crystalline structure obtained on melting is intensely birefringent and retains a marked resemblance to the prior crystalline texture. This is reminiscent of the behavior of a spherulitic cast film of poly[bis(trifluoroethoxy)phosphazene] on heating through $T(1)$ (Schneider et al., 1976). Low angle light scattering evidence is also presented to indicate the relation between the viscous-crystalline and crystalline structure (Pochan et al., 1976).

Neither the structure of the crystalline state nor that of the viscous-crystalline state are known in any detail. There is only limited information concerning other aspects of the viscous-crystalline state and the behavior at the isotropic melt. Some approximate heats of transition appear in Table II (Beatty and Karasz, 1975; Beatty et al., 1975). The apparent enthalpy of fusion is extremely small for the transition at -5°C, interpreted as the crystal melt. The entries for PDES in Table II emphasize a possible relation to the polyphosphazene behavior that is discussed in more detail below. Pulsed and continuous wave NMR measurements (Pochan *et al.*, 1976) above the crystal melt at $-5°C$ show that two components are present. The onset of rapid motion in the viscous-crystalline component begins at $-5°C$ and the ratio of the amorphous to the partially ordered material increases with temperature. Dielectric relaxation measurements show a linear increase in tan δ through the temperature regime of the viscous-crystalline state. Like the NMR data, this can be interpreted as evidence of a two-component structure in which the amount of the viscous-crystalline component decreases with temperature. A marked decrease in the intensity of the X-ray peak occurs in the vicinity of the transition to the isotropic melt. The position of the peak is unaltered throughout the viscous-crystalline regime and a very weak peak even appears at the same position in the melt. This behavior resembles that with cholesteryl esters as the degree of disorder increases in the transition from the smectic to cholesteric states and finally to the isotropic melt (McMillian, 1972). However, in this case, the X-ray reflection is a measure of the molecular length rather than an interchain spacing.

There appear to be strong qualitative similarities between the behavior observed in PDES and the polyphosphazenes, if one identifies T_M at $-5°C$ with $T(1)$ and the diffuse transition at 20°C with T_M as used in the discussion of the polyphosphazenes. Then, in both cases $T(1)$ marks the transition to a state of partial order, and T_M the transition to the isotropic melt. There are, however, strong differences of a quantitative nature which indicate that the viscous-crystalline state is less ordered and more mobile than the meso state in polyphosphazenes. This is evident in the greater temperature range of stability, more nearly solidlike properties and, in certain cases, the larger number of diffraction peaks observed for the polyphosphazenes. These differences probably can be rationalized in terms of a higher degree of chain mobility occurring in the crystalline as well as the viscous-crystalline regions of PDES which, in turn, may be related to the smaller and more mobile ethyl pendant group and the weaker interchain forces. It is also worth noting that there is a strong similarity between the siloxane and phosphazene backbone structure. It will be of interest to learn whether evidence of a partially ordered state can be detected in poly(dimethylsiloxane) or in silicones with pendant groups that are longer than ethyl and whether further evidence supports the unique role suggested for the viscous-crystalline state as a precursor of the crystalline structure.

D. The Problem of Classification

The term mesomorphic has been used here to suggest an intermediate level of organization between the crystalline and the glassy or liquid states in polyphosphazenes. Beyond that, the term is not specific. The term liquid crystal holds a broader range of implications: the various liquid crystal models incorporate explanations of calorimetric, diffraction, optical, and flow properties. The corresponding properties of polyphosphazenes are reminiscent of liquid crystal properties, but fitting the present substances into the liquid crystal classification schemes (deVries, 1975; Brown and Doan, 1973) is difficult. The cholesteric type is quite unlikely, so one must choose between nematic and smectic types. Sharp diffraction peaks of the type observed are characteristic of the smectic type, but not the nematic class (deVries, 1975). Pseudohexagonal packing occurs in certain smectic structures (Brown, 1973; Brown and Doan, 1973), and the term smectic has been used to describe pseudohexagonal phases in polymers (Neigisch, 1966; Blumstein et al., 1975). However, the term is more properly reserved to structures showing layering in the third dimension as well as ordered packing in two dimensions, as occurs in certain acrylic polymers with long side chains (Blumstein et al., 1975), rather than the more common pseudohexagonal structures (Neigisch, 1966).

The term plastic crystal, used for the pseudohexagonal phase of trans-1,4-polybutadiene (Iwayanagi and Miura, 1965), is also borrowed from nomenclature used in connection with low molecular weight compounds. Westrum and McCullough (1963) describe four properties commonly found in plastic crystals of low molecular weight compounds:

(a) low entropy of fusion;
(b) high triple point temperature and pressure;
(c) crystals usually of cubic or hexagonal symmetry; and
(d) one or more quite energetic transitions in the solid state.

These properties are observed in polyphosphazenes with the exception of (b), which is not relevant to polymers in general. The significant difference between liquid crystals and plastic crystals is that in liquid crystals loss of positional order occurs well before the loss of rotational order, while in plastic crystals, the reverse is true. In particular, Smith (1975) prefers to reserve the term plastic crystal for a phase in which complete orientational disorder occurs, rather than rotation about a single axis. Thus he applies the term to substances possessing globular molecules, such as adamantane, but not to polymers.

The problem with both of the preceding major classifications liquid crystal and plastic crystal is that they do not take into account the macromolecular nature of mesomorphic polymers such as the polyphosphazenes. It might be noted, in particular, that these polymers are incapable of displaying the layered structure characteristic of smectic or cholesteric liquid crystals and deformation does not occur by slip of a layered structure but rather by shear along the molecular axis. This limitation in the polyphosphazenes may be a factor contributing to the brittle nature of samples that are thermally processed between $T(1)$ and T_M, unavoidably producing an oriented morphology on crystallization. The term viscous-crystalline is a suggestive term for polymers with partial order and meltlike flow behavior (Beatty *et al.*, 1975). Offhand, this designation does not appear appropriate to the polyphosphazenes or the wider group of polymers considered here with quite different properties in the mesomorphic state. Beatty *et al.*, however, recognize the need for a term that applies to polymers with liquid crystal order but chain-folded morphology, and in this sense the designation viscous-crystalline has a more general connotation. Although it would seem preferable to agree upon a term that highlights the chain motion, which is the common characteristic of this group of polymers, the term viscous-crystalline in its more general sense may prove acceptable.

V. Conclusions

It is clear that the appearance of mesomorphic structure in the polyphosphazene homopolymers, and in several of the other polymers with which comparisons have been made, is not due to a rigid-chain structure, such as the examples considered by Preston, or to geometric factors, such as the polymers with mesomorphic side groups treated by Blumstein in other chapters of this book. These are polymers in which the backbone structure is flexible, although there is considerable difference in the rotational barriers for the polyphosphazenes compared to Teflon. The characteristic behavior involves a transition from a crystalline state to a mesomorphic state marked by a large enthalpy change while passage to the isotropic state may be less conspicuous calorimetrically. The mesomorphic state displays at best a small number of X-ray reflections representing interchain spacings that correspond to hexagonal packing. The transition to the mesomorphic state involves the onset of some degree of backbone motion, which averages out intermolecular forces into a cylindrical potential. Although the behavior of some individual polymers has been discussed in similar terms, this common basis for a class of polymers displaying mesomorphic structure has not, to our knowledge, been pointed out previously. Very likely an examination of the literature will reveal other polymers that can usefully be identified as displaying viscous-crystalline behavior, for want of a better term.

There are a number of vexing problems if one accepts the classification as suggested. The polymers that are included display a striking range of behavior and properties associated with the mesomorphic state, from near solidlike to near fluid, and wide differences in the temperature range of stability. How does one explain these differences? An equally fundamental question is, how does one account for the stability of a phase that is characterized by extensive side chain and backbone motion? One should not lose sight of the fact, in the more general discussion, that many of the conclusions for the polyphosphazenes are based on preliminary data and much remains to be done experimentally in defining the structure, the extent of motion, and the thermodynamics of the transition. In addition, there are opportunities for studying other aspects of polyphosphazene mesomorphic behavior including the kinetics of crystallization from the mesomorphic state, the influence of a diluent on the transition temperature and transition behavior, and the examination of unusual features of the transition behavior which have been observed in certain samples.

Acknowledgments

The authors wish to express their appreciation to Robert E. Singler who has provided imaginative leadership to the polyphosphazene program at the Army Materials and Mechanics Research Center, has supplied the samples on which many of the measurements reported here were made, and has assisted with some details and a careful review of the manuscript. Our thanks also to Carolyn Shimansky who has carried out the several cycles of correction and typing with patient good humor.

References

Alexander, M. N., Desper, C. R., Sagalyn, P. L., and Schneider, N. S., (1977). *Macromolecules* **10**, 721–723.
Allcock, H. R. (1972a). "Phosphorus Nitrogen Compounds." Academic Press, New York.
Allcock, H. R. (1972b). *Chem. Rev.* **72**, 315–355.
Allcock, H. R. (1975). *Chem. Tech.* **5**, 552–560.
Allcock, H. R., and Kugel, R. L. (1966). *Inorg. Chem.* **5**, 1716–1718.
Allcock, H. R., Kugel, R. L., and Valan, K. J. (1966). *Inorg. Chem.* **5**, 1709–1715.
Allcock, H. R., Kugel, R. L., and Stroh, E. G. (1972). *Inorg. Chem.* **11**, 1120–1123.
Allcock, H. R., Moore, G. Y., and Cook, W. J. (1974). *Macromolecules* **7**, 571–575.
Allcock, H. R., Allen, R. W., and Meister, J. J. (1976). *Macromolecules* **9**, 950–955.
Allcock, H. R., Patterson, D. B., and Evans, T. L. (1977). *J. Am. Chem. Soc.* **99**, 6095–6096.
Allen, G., Lewis, C. J., and Todd, S. M. (1970). *Polymer* **11**, 44–60.
Beatty, C. L., and Karasz, F. E. (1975). *J. Polym. Sci., Polym. Phys. Ed.* **13**, 971–975.
Beatty, C. L., Pochan, J. M., Froix, M. F., and Hinman, D. D. (1975). *Macromolecules* **8**, 547–551.
Beres, J. J. (1976). Unpublished work.
Bishop, S. M., and Hall, I. H. (1974). *Br. Polym. J.* **6**, 193–204.
Blumstein, A., Blumstein, R. B., Clough, S. B., and Hsu, E. C. (1975). *Macromolecules* **8**, 73–76.
Brandrup, J., and Immergut, E. H., eds. (1975). *In* "Polymer Handbook," p. III–7. Wiley (Interscience), New York.
Brown, G. H. (1973). *J. Electron. Mater.* **2**, 403–429.
Brown, G. H., and Doan, J. W. (1973). *Appl. Phys.* **4**, 1–15.
Clark, E. S. (1967). *J. Macromol. Sci., Phys.* **1**, 795–800.

Clark, E. S., and Muus, L. T. (1962). *Z. Kristallogr., Kristallgeom., Kristallphys., Kristallchem.* **117**, 119–127.

Clough, S. B. (1970). *J. Polym. Sci., Polym. Lett. Ed.* **8**, 519–523.

Connelly, T. M., Jr., and Gillham, J. K. (1975). *J. Appl. Polym. Sci.* **19**, 2641–2644.

Connelly, T. M., Jr., and Gillham, J. K. (1976). *J. Appl. Polym. Sci.* **20**, 473–488.

Craig, D. P., and Paddock, N. C. (1971). *In* "Nonbenzenoid Aromatics" (J. P. Snyder, ed.), II pp. 273–357. Academic Press, New York.

Desper, C. R. (1976). Unpublished work.

Desper, C. R., and Schneider, N. S. (1976). *Macromolecules* **9**, 424–428.

deVries, A. (1975). *Proc. Int. Conf. Liq. Cryst., Bangalore, December, 1973* (S. Chandrasekhar, ed.), pp. 93–113.

DuPré, D. B., Samulski, E. T., and Tobolsky, A. V. (1971). *In* "Polymer Science and Materials" (A. V. Tobolsky and H. Mark, eds.), pp. 123–160 Wiley (Interscience), New York.

Furukawa, G. T., McCoskey, R. E., and King, G. J. (1952). *J. Res. Natl. Bur. Stand., Sect. A* **49**, 273–278.

Giglio, E., Pompa, F., and Ripamonti, A. (1962). *J. Polym. Sci.* **59**, 293–300.

Golike, R. C. (1964). *J. Polym. Sci.* **42**, 583–584.

Hagnauer, G. L., and LaLiberte, B. R. (1976a). *J. Polym. Sci., Polym. Phys. Ed.* **14**, 367–371.

Hagnauer, G. L., and LaLiberte, B. R. (1976b). *J. Appl. Polym. Sci.* **20**, 3073–3086.

Higginbotham, E., and Schneider, N. S. (1975). Unpublished work.

Hyndman, D., and Origlio, G. F. (1960). *J. Appl. Phys.* **31**, 1849–1852.

Iwayanagi, S., and Miura, I. (1965). *Rep Prog. Polym. Phys. Jpn.* **8**, 303–304.

Lando, J. B., Olf, H. G., and Peterlin, A. (1966). *J. Polym. Sci., Polym. Phys. Ed.* **4**, 941–951.

McCall, D. W., Douglass, D. C., and Falcone, D. R. (1967). *J. Phys. Chem.* **71**, 998–1004.

Mc Cane, D. I. (1970). *In* "Encyclopedia of Polymer Science and Technology" (H. F. Mark, N. G. Gaylord, and N. M Bikales, eds.), Vol. 13, pp. 628–631. Wiley (Interscience), New York.

McMillian, W. L. (1972). *Phys. Rev. A* **6**, 936–947.

Muus, L. T., and Clark, E. S. (1964). *Polym. Prepr., Am. Chem. Soc., Div. Polym. Chem.* **5**, 17–21.

Neigisch, W. D. (1966). *J. Appl. Phys.* **37**, 4041–4046.

Pochan, J. M., Hinman, D. F., and Froix, M. F. (1976). *Macromolecules* **9**, 611–616.

Quinn, E. J., and Dieck, R. L. (1976). *J. Fire Flammability* **7**, 5–18.

Schneider, N. S., Desper, C. R., and Singler, R. E. (1976). *J. Appl. Polym Sci.* **20**, 3087–3103.

Singler, R. E., Hagnauer, G. L., Schneider, N. S., LaLiberte, B. R., Sacher, R. E., and Matton, R. W. (1974). *J. Polym. Sci., Polym. Phys. Ed.* **12**, 433–444.

Singler, R. E., Schneider, N. S., and Hagnauer, G. L. (1975). *Polym. Eng. Sci.* **15**, 321–330.

Smith, G. W. (1975). *Adv. Liq. Cryst.* **1**, 189–266.

Starkweather, J. W., and Boyd, R. H. (1960). *J. Phys. Chem.* **64**, 410–414.

Stroh, E. G., Jr. (1972). Ph.D. Thesis, Pennsylvania State Univ., University Park.

Suehiro, K., and Takayanagi, M. (1970). *J. Macromol. Sci., Phys.* **4**, 39–46.

Tate, D. P. (1975). *Rubber World* **172**, 41–43.

Thomas, E. L., and Sass, S. L. (1973). *Makromol. Chem.* **164**, 333–341.

Westrum, E. F., and McCullough, J. P. (1963). *In* "Physics and Chemistry of the Organic Solid State" (D. Fox, M. N. Labes, and A. Weissberger, eds.), Vol. 1, pp. 3–178, Wiley (Interscience), New York.

White, J. E., Singler, R. E., and Leone, S. A. (1975). *J. Polym. Sci., Polym. Chem. Ed.* **13**, 2531–2543.

Index